Recent Developments in Mass Spectrometry in Biochemistry and Medicine

——————————Volume 2——————————

Recent Developments in Mass Spectrometry in Biochemistry and Medicine

―――――――――― Volume 2 ――――――――――

Edited by
Alberto Frigerio
"Mario Negri" Institute
Milan, Italy

Plenum Press · New York and London

Library of Congress Cataloging in Publication Data

International Symposium on Mass Spectrometry in Biochemistry and Medicine, 5th, Rimini, 1978.
Recent developments in mass spectrometry in biochemistry and medicine.

Organized by the Italian Group for Mass Spectrometry in Biochemistry and Medicine.
Includes index.
1. Mass spectrometry—Congresses. 2. Biological chemistry—Technique—Congresses. 3. Drugs—Analysis—Congresses. I. Frigerio, Alberto. II. Italian Group for Mass Spectrometry in Biochemistry and Medicine. III. Title. [DNLM: 1. Spectrum analysis, Mass—Congresses. W1 RE106AM]

QP519.9M3157 1978   612'.01585   79-19982

ISBN-13: 978-1-4613-3020-2   e-ISBN-13: 978-1-4613-3018-9
DOI: 10.1007/978-1-4613-3018-9

Proceedings of the 5th International Symposium on Mass Spectrometry in
Biochemistry and Medicine, held at Rimini, Italy in June 1978

© 1979 Plenum Press, New York
Softcover reprint of the hardcover 1st edition 1979
A Division of Plenum Publishing Corporation
227 West 17th Street, New York, N.Y. 10011

Preface

The papers collected in this volume were presented at the 5th International Symposium on Mass Spectrometry in Biochemistry and Medicine held at Rimini, Italy, in June 1978.

This meeting was organized by the Italian Group for Mass Spectrometry in Biochemistry and Medicine, which was founded in Milan, Italy, in 1975 by researchers working in different fields (chemistry, pharmacology, medicine, biology).

I wish to thank Mrs. Anna Bernard, Mrs. Donata Castoldi and Miss Vanna Pistotti for their invaluable technical assistance and countless efforts in the preparation of this volume.

Alberto Frigerio

Milan,
March, 1979

Contents

THE PLACE OF MASS SPECTROMETRY IN DRUG METABOLISM STUDIES

A. Benakis

Laboratory of Drug Metabolism, Department of Pharmacology

University, Geneva, Switzerland

A vast fund of published information on the use of mass spectrometry and associated techniques for the study of drug metabolism has accumulated in recent years. The annual "Specialist Periodical Reports" on mass spectrometry of the London Chemical Society is particularly worthy of mention. This publication reviews all the literature in the field. In the issue covering the two year period from 1972 to 1974 (1), there are more than one hundred and forty reviews of papers devoted to mass spectrometry applications for drug metabolism. From 1974 to 1976 (2), there were a hundred and twenty.

The papers presented at the annual international symposium on mass spectrometry are published by Dr. Frigerio and greatly appreciated (3-8). Other valuable publications in this field are "Advances in Mass Spectrometry" put out by the London Institute of Petroleum, and "Quantitative Mass Spectrometry in Life Sciences", a report of the proceedings of the Ghent International Symposium in 1976 (9).

Interesting mass spectrometric data can also be found in numerous pharmacological journals, particularly in those specializing in drug metabolism studies (10, 11, 12). The journal "Biomedical mass spectrometry", in which papers on drug metabolism are well represented, can also be mentioned (13).

Finally, new developments in mass spectrometry applied to drug metabolism studies have been widely reviewed by other authors (14-20).

The comments I shall be making in this paper will be those of

one who habitually uses mass spectrometry techniques in drug
metabolism studies and pharmacokinetics and will concern these
fields rather than the technical and technological progress in mass
spectrometry.

During the years from 1958 to 1968, the study of drug metabolism
was relatively limited; it was conducted mainly by university
research teams in the context of fundamental studies and was
essentially concerned with the detoxication mechanism.

The first period could be called the "period of scientific
curiosity", during which some precise results were obtained and
certain techniques were devised and refined.

The period between 1968 and 1973, which I consider the second
period, was one of growing awareness and realization of the
importance of the problem.
Metabolic and pharmacokinetic studies became more generalized and
were performed not only in university laboratories but also and
especially by the pharmaceutical industry. During the seventies,
most of the drug metabolism laboratories of the pharmaceutical
industry, especially in the United States, centered their attention
on mass spectrometry and used the instruments of their company's
analytical laboratories. In just a few years, however, both industrial and university laboratories had acquired mass spectrometers,
and frequently, GLC/MS as well.

Although this growing interest was certainly due in part to
the natural desire for a better understanding of the mode of action
of certain drugs by studying their biotransformation mechanisms,
it cannot be denied that the determining role in the development
of this field was played by the governmental authorities of an
ever-increasing number of countries who began to insist upon this
type of study.

It is my feeling that there is a need at present to better
define the objectives of the metabolic and pharmacokinetic studies
required by the authorities in many countries. I believe that this
problem is of interest to everyone performing studies of this kind.
Some controversial aspects are the ever-increasing demands of the
authorities with regard to results, and the consequent increasing
complexity and cost of the studies. This situation could be
disastrous in the long run since it is technically and economically
impossible to keep increasing, ad infinitum, the number of studies
required to introduce a new drug on the market.

The period from 1973 to the present has been very rich, both
in experimental results and perspectives.

By 1976, metabolic studies represented 18% of all the pharmaco-

logical studies published, as compared with only 6% in 1971. While
the number of classical pharmacological studies doubled during the
period from 1971 to 1976, the number of metabolic and pharmacokinetic
studies quintupled. Furthermore, there has been an equally impressive
development in the quality of these studies, which, thanks to many
new techniques available, have become increasingly precise and
complete. Among these new techniques, that of labelling drugs with
radioactive isotopes is undeniably a most important one, and has
considerably enhanced the possibility of detecting drugs and their
metabolites, particularly in in vivo experiments. The use of
labelled drugs, however, has not provided a satisfactory answer to
the important problem of the determination of metabolite structure.

Although labelled molecules provide a very sure method for
detecting the presence of labelled drugs (unchanged product and
metabolites) particularly in plasma and tissue, it is practically
impossible, without separation procedures, to know whether the
molecules correspond to unchanged product or metabolites. Another
problem is that results can also be affected by the labelling
position. Considering the difficulty of labelled synthesis,
particularly with radioactive isotopes, the specialized laboratories
and technology, as well as the high cost of the molecules, it is
understandable that new techniques have been developed to eliminate
some of these problems. Furthermore, it must be mentioned that the
use of molecules labelled with radioactive tracers is being
regulated with increasing severity in most countries, especially in
the United States, and has become quite difficult due to the severe
restriction of μCi that may be administered to man. As to the use
of molecules labelled with stable isotopes, which is very desirable
for mass spectrometric studies, the problem of the isotopic effect
on drug metabolism must be borne in mind, particularly when
molecules labelled with deuterium are used in in vivo experiments.
The labelling position must be chosen with care in order to avoid
excessive metabolism which would lead to non significant results.
It must be verified that the elimination kinetics are not affected
in any significant way by the substitution of hydrogen by deuterium.
When it is not possible to insert the deuterium label at a
metabolically inert position, the rare and more expensive ^{15}N, ^{13}C
and ^{18}O isotopes can be used. Some of these problems have been
recently discussed at the Second International Conference of Stable
Isotopes in the United States (21).

The determination of metabolic structure is a most important
and exciting area of study for a number of reasons. It continues
to fascinate scientists, not only because of a desire to understand
a basic biochemical phenomenon but also because the findings could
be important in the detection of metabolites that might be
pharmacologically active and thus lead to the discovery of new and
useful medicinal drugs.

The enormous difficulties involved in finding new drugs is
well known as is the great number of synthesized products necessary
to produce just one with undeniable pharmacological value. Recently,
there has been considerable interest in and vogue for studies
related to drug design, which, nevertheless, have not yet borne
fruit and it is difficult to imagine, with the present fund of
knowledge, that new drugs could be created in this way. This hope
persists, however, and it is with the idea that metabolites could
be as active or even more active than the original drug, that a
great deal of effort has been made, especially in industrial laboratories, to determine their structure. It must be remembered,
however, that in many cases, like those of phenylbutazone, imipramine,
and the benzodiazepines, the structures of the most important
metabolites had been predicted by organic chemists before they were
determined in _in vivo_ experiments. At present, many metabolite
structures are determined but the results are not published, either
because the structure had not been predicted and therefore was not
yet protected by patents or because the pharmacological studies
have not yet been performed.

Moreover, publication of metabolite structure determined by
in vivo studies is likely to provoke the authorities to ask
additional questions concerning toxicology, teratogenicity and
mutogenicity of these metabolites found _in vivo_, which is, in some
cases,unjustified. In any case, it is evident from a review of the
literature, that there is considerable interest in the determination
of structures of well known products like barbiturates, hydantoines,
imipramine, phenothiazines, benzodiazepines etc.

It is obvious that without mass spectrometry, structure
determination would be extremely difficult, if not impossible.
Although determinations of numerous metabolites has revealed that
some of them are pharmacologically active, it has not yet been
determined whether or not the metabolic processes affect the part
of the molecule which gives it its pharmacodynamic properties.

Another problem related to structure determination of metabolites
concerns N-oxidized products whose structure is difficult to detect
due to the great lability of the N-O bond. However, this may yet
become possible with the aid of certain ionization techniques.

In the past, it was sometimes not suspected that the labelled
position, especially for ^{14}C labelled compounds, was not the one
that had been chosen. Due to the relatively low specific activity
of the synthesized compounds, it was difficult to verify the
labelled position by mass spectrometry. However, the progressive
increase in specific activity of molecules labelled with ^{14}C
resulting from the necessity of working with therapeutic doses has
led to the synthesis of molecules with high specific activity that
are well adapted to mass spectrometry, making it possible to detect

twice $^{12}C/^{14}C$ and thus confirm the correctness of the labelling position.

Radioactive ^{14}C labelled synthesis can sometimes have unexpected results. In one case that we know of, the labelled position obtained was not the desired one. For reasons that we have not been able to determine, the neighbouring carbon was labelled instead. This incorrect labelled-position was confirmed by peak matching using high resolution mass spectrometry. Fortunately, this error had no effect on the results since neither carbon was affected by the metabolic process. However, it is not impossible that this has happened in other cases and it is disturbing to realize that metabolic studies may have been conducted with molecules whose labelled position had not been accurately determined.

Although structure determination has been the most important application of mass spectrometry, quantification has become an increasingly important one in recent years. This application has become possible thanks to GLC/MS, selected ion monitoring (SIM) and modern computerized techniques.

The development of these highly sensitive techniques was stimulated by the growing need for precise measurement of the blood level of drugs and metabolites. Pharmacokinetic studies, first conducted only in animals, even though in many different species, soon began to be made on human subjects, due especially to the growth of clinical pharmacology. The present tendency is to rely increasingly on experimental data for decisions involving the use of drugs.

Of greatest importance among these data are the blood level values as related to therapeutic effect. For certain types of drugs, such as anti-convulsants, this relationship is already established but the information is unfortunately not yet generally applied. For other types of drugs, however, there has been considerable controversy regarding the relationship between therapeutic effect and blood level values, possibly because the latter have not always been determined with sufficient precision. Precise measurement, however, is an absolute necessity, not only to satisfy government requirements, but also to determine a possible relationship between blood level values and the pharmacodynamic effect of drugs.

One of the reasons that it has been difficult, in the past, to measure blood levels precisely is that, in some cases, they are extremely low; sometimes because the product has been extensively metabolized, and sometimes, especially in the case of new drugs, because a very low dose has been given.

Clonidine, and its hypotensive derivatives, for example, are

administered in doses of only 100 µg. It is understandable,
therefore, that it is extremely difficult, at this low dosage, to
quantify blood levels as a function of time. It would, in fact,
be practically impossible without the use of fragmentometric techniques.   These techniques are also indispensable for the quantification
of rapidly metabolized drugs such as Atarax, which we have recently
studied (22). In this case, the amount of unchanged product present
in the plasma was so low that it was extremely difficult to quantify
in human subjects, even when the radioactive labelled product was
used at authorized radioactivity and in therapeutic doses.
However, with SIM, we were able to quantify the blood level of
Atarax, which, incidentally, we found to be related to the
pharmacodynamic activity of the product (23). Although the value
of fragmentometry in metabolic and pharmacokinetic studies has
been demonstrated, I should like to mention an important reservation
that I have regarding fragmentometric determinations made on samples
obtained by extraction of biological materials.

Since many drugs are known to bind to plasma proteins, it is
important to take this binding into consideration in interpreting
pharmacokinetic data.

Although certain extraction techniques allow for some
displacement and extraction of unchanged product and metabolites
bound to circulating proteins, quantitative displacement is, in
most cases, highly improbable. The possibility of quantitative
displacement depends on both the chemical structure of the drug and
the nature of the protein binding.

At present, far too many of the results obtained by fragmentometry, pertain only to the extractable part of the product contained
in the biological sample. It is difficult to believe, however,
that valid data for protein binding levels of drugs and metabolites
can be obtained without the use of radioactive labelled molecules,
unless new spectrometric techniques were to be developed in the
future which would take into account the bound part. Until such
time, however, it is essential that in reports and publications,
the extraction conditions be precisely described, and that it be
clearly mentioned that quantification by MS techniques concerns
only the unbound part.

Finally, it should be remembered that drugs bound to plasma
proteins are not necessarily inactivated but may simply be placed
in reserve for subsequent use by the organism. Naturally, these
comments concerning the binding of plasma proteins are also
applicable to the binding of drug and metabolites to target organs.

Another area in which mass spectrometry can be very valuable
is in confirming the pharmacokinetic results obtained by immunological

methods such as RIA and EMIT techniques which are very rapid and highly sensitive. However, the very important problem of specificity must now be raised and it is very likely that both the possibilities and limitations of immunological techniques can be established by means of mass spectrometry.

Mass spectrometry, and GLC in particular, have also proved very valuable in perinatal pharmacology because of the small size of the samples available.

Interesting results in this area, particularly concerning the facile transfer of drugs from mother to foetus, have recently been reported by Horning and co-workers (24, 25).

Another type of pharmacological study in which drugs labelled with ^{13}C and ^{15}N could be particularly useful is in the evaluation of bioavailability by comparing the serum levels of the labelled and unlabelled drug after one form has been administered intravenously and the other orally, as has been done by Dutcher and co-workers (26).

Other interesting related problems are the use of stable isotopes, the different types of ionization used in mass spectrometry, the association of GLC/MS-HPLC/MS techniques, clinical applications for the monitoring of anesthetic gases during surgery, and the highly important role of mass spectrometry in toxicology.

Also of great interest is the sequencing of endogeneous peptide hormones which is particularly applicable to peptides produced in the body through the action of certain drugs.

Finally, I should like to insist upon the fact that the quality of the results obtained in the field of mass spectrometry, just as in so many other areas, depends less on sophisticated equipment than on human competence and adequate conceptualization of the problems.

REFERENCES

1) R.A.W. Johnstone, A Specialist Periodical Report. Mass Spectrometry, Vol. 3, A review of the Literature Published between July 1972 and June 1974, The Chemical Society, Burlington House, London.
2) R.A.W. Johnstone, A Specialist Periodical Report. Mass Spectrometry, Vol. 4, A Review of the Literature Published between July 1974 and June 1976, The Chemical Society, Burlington House, London.
3) A. Frigerio, Ed., Proceedings of the International Symposium

on Gas Chromatography-Mass Spectrometry, Tamburini Publisher, Milan, 1972.

4) A. Frigerio and N. Castagnoli, Eds., "Mass Spectrometry in Biochemistry and Medicine", Raven Press, New York, 1974.

5) A. Frigerio and N. Castagnoli, Eds., "Advances in Mass Spectrometry in Biochemistry and Medicine", Vol. 1, Spectrum Publications, New York, 1975.

6) A. Frigerio, Ed., "Advances in Mass Spectrometry in Biochemistry and Medicine", Vol. II, Spectrum Publications, New York, 1977.

7) A. Frigerio and E.L. Ghisalberti, Eds., "Mass Spectrometry in Drug Metabolism", Plenum Press, New York, 1977.

8) A. Frigerio, Ed., "Recent Developments in Mass Spectrometry in Biochemistry and Medicine", Vol. I, Plenum Press, New York, 1978.

9) A.P. de Leenheer and R.R. Roncucci, Eds., "Quantitative Mass Spectrometry in Life Sciences", Amsterdam, 1977.

10) Drug Metabolism Reviews, Dekker, New York.

11) European Journal of Drug Metabolism and Pharmacokinetics, Medicine et Hygiène, Genève.

12) Xenobiotica, Taylor and Francis, London.

13) Biomedical Mass Spectrometry, Heyden, London.

14) J.T. Watson, F.C. Falkner and B.J. Sweetman, Biomed. Mass Spectrom., 1974, 1, 156.

15) B.J. Millard, in "Progress in Drug Metabolism", Vol. 1, J.W. Bridges and L.F. Chasseaud, Eds., Wiley Interscience, New York, 1976, p. 1.

16) G. Horvath, in "Progress in Drug Research", I.E. Jucker, Ed., Birkhauser Verlag, Basle, 1974, p. 339.

17) R. Roncucci, M.J. Simon, G. Jacques and G. Lambelin, Eur. J. Drug Metab. Pharmacokin., 1976, 1, 9.

18) T. Walle, Princ. Tech. Hum. Res. Ther., 1974, 3, 83.

19) G. Spiteller and G. Remberg, Naturwiss., 1974, 61, 491.

20) R. Carrington and A. Frigerio, Drug. Metab. Rev., 1977, 6, 243.

21) E.R. Klein, P.D. Klein, Eds., in "The Proceedings of the Second International Conference on Stable Isotopes", Oak Brook, III, 20-23 Oct. 1975, National Technical Information Service, US Department of Commerce, Springfield, VA, US Energy Research and Development Administration, CONF.-751027, 1975.

22) Métabolisme de l'hydroxyzine chez l'homme. To be published.

23) M. Colon and J. Polderman, Effect of Hydroxyzine on Human Vigilance, Eur. J. Clin. Pharmacol., submitted for publication.

24) M.G. Horning, J. Nowlin, C.M. Butler, K. Lertratanangkoon, K. Sommer and R.M. Hill, Clin. Chem., 1975, 21, 1282.

25) M.G. Horning, J. Nowlin, M. Stafford, K. Lertratanangkoon, K. Sommer, R.M. Hill and R.N. Stillwell, J. Chromatogr., 1975, 112, 60S.

26) J.S. Dutcher, J.M. Strong, W.K. Lee and A.J. Atkinson, in "Proc. 2nd Int. Conf. Stable Isotopes", Oak Brook, 1975, E.R. Klein and P.D. Klein, Eds., USERDA, Rep. Conf-751027, 1975, p. 186.

IN VITRO AND IN VIVO STUDIES ON THE METABOLISM OF DIBENZO $\lceil$c,f$\rfloor$-
-$\lceil$1,2$\rfloor$ DIAZEPINE

A. Frigerio, P. Negrini, D. Rotilio, A. De Pascale and

E. Rossi, Istituto di Ricerche Farmacologiche "Mario Negri"

Via Eritrea 62, 20157 Milan, Italy

Aromatic and olefinic compounds can be metabolized by mammalian microsomal mono-oxigenases to arene oxides and epoxides respectively.

These oxygenated species represent electrophilic intermediates in the formation of phenols and dihydrodiols, which are considered the real detoxification products of these compounds.

In order to explain the toxic, carcinogenic and mutagenic effects of these "bioactivated intermediates" many factors have to be taken into consideration, as their rate of formation and the high reactivity towards nucleophilic agents, such as DNA, RNA and proteins, together with the sufficient stability to reach target sites in the organism. It therefore becomes important to identify the presence of these intermediates and to study their effects in the different sites, where their toxic action may occur.

The concern that clinically useful drugs containing aromatic and/or olefinic groups may be transformed to bioactivated intermediates, during the last years led us to examine the metabolism of several tricyclic compounds. The substances that have been chosen for these studies belong to two different classes of compounds: the derivatives of 5H-dibenzo (b,f) azepine and the derivatives of 5H-dibenzo (a,d) cycloheptene. Biotransformation studies of these compounds were carried out both in vitro, by incubation with rat liver microsomes, and in vivo, in rat and, for clinically used drugs, also in man (1-6). The original hypothesis of the involvement of an epoxide-diol metabolic pathway common to all the listed compounds has been confirmed by means of chromatography and mass spectrometry techniques.

Recently we have been studying the metabolism of 5H-dibenzo
/ b,f_/azepine and 5H-dibenzo/ a,d_/cycloheptene, the parent
compounds of the two classes of substances mentioned above (5,6).

As part of a systematic study on the biotransformation of
tricyclic compounds and drugs, this paper deals with the metabolism
in vitro and in vivo of a new compound with tricyclic structure,
dibenzo/ c,f_/-/ 1,2_/diazepine (I), in order to study the influence
exerted by the hetero atom on the metabolic pathway. Its chemical
structure is reported in Fig. 1.

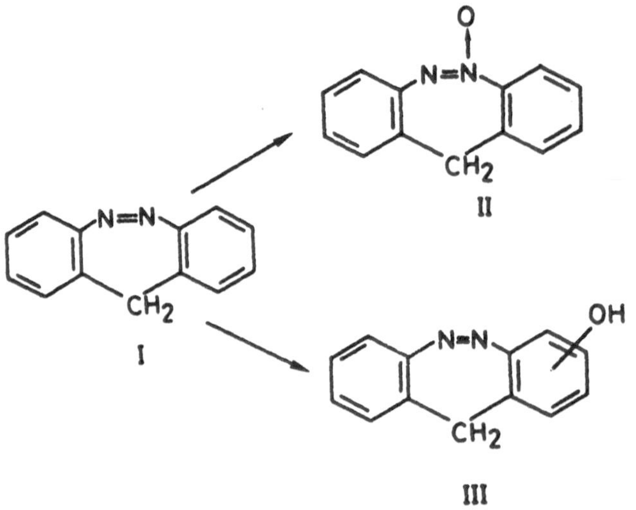

FIG. 1. In vitro metabolic pathway of dibenzo
/ c,f_/-/ 1,2_/diazepine (I).

The metabolic pathway of I in vitro, as it is presently known,
is shown in Fig. 1.

The metabolic transformation of the compound probably occurs,

in_a first step, by oxidation of the nitrogen to give dibenzo/⁻c,f_7-
-/⁻1,2_7diazepine-1-N-oxide (II), while another oxidative step gives
rise to a monohydroxyderivative in the aromatic ring (III).

In the in vitro study the extract of the rat liver microsomes
incubated with I was analyzed by thin-layer chromatography (TLC).
The Rf values of I and of its in vitro metabolites are reported in
Table 1.

TABLE 1
Rf values of dibenzo/⁻c,f_7-/⁻1,2_7diazepine (I) and
its in vitro and in vivo metabolites.

Rf	compound	in vitro	in vivo	in vivo after incubation with β-glucuronidase
0.42	I	+	+	−
0.36	IV	−	+	−
0.33	VI	−	+	+
0.30	II	+	+	−
0.24	VII	−	+	+
0.13	III	+	+	+
0.10	V	−	+	−

Silica gel F_{245} glass plates, 0.25 mm layer thickness
Solvent system: Ethyl acetate/Carbon tetrachloride (16:84 V/V)

The ethyl acetate extract of rat liver microsomes incubated with
I, was analyzed also by gas chromatography (GLC): in Fig. 2 the gas
chromatogram of the treated sample shows three peaks not present in
the control sample. Further analyses were carried out by utilizing
a gas chromatography coupled with mass spectrometry (GLC-MS) for a
more precise identification of the peaks.

In this way, it was possible to establish that the first GLC
peak corresponded to unchanged I, the second to the compound II,
and the third to the compound III (Fig. 1).

The GLC analyses were performed using a glass column 2 meters
long, packed with OV-17 (3%) on Gas Chrom Q (100-120 mesh).

The oven temperature was 220°C and the injection port temperature
290°C. A mass spectrometric control was performed also on the
eluates of the spots from TLC. The spot at Rf 0.42 eluted with
methanol and introduced by the D.I.S. into the mass spectrometer,
and the first GLC peak, gave a mass spectrum identical to that

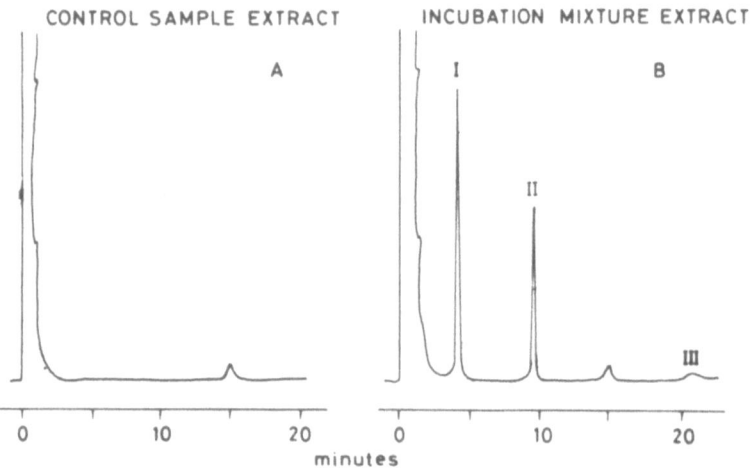

FIG. 2. Gas chromatograms of (A) control microsomes and (B)microsomes
         incubated with dibenzo[c,f]-[1,2]diazepine. Peaks: I,
         dibenzo[c,f]-[1,2]diazepine; II: dibenzo[c,f]-[1,2]-diazepine-
         1-N-oxide; III: monohydroxyderivative of I in the aromatic
         ring.

obtained with authentic dibenzo/¯c,f¯/-/¯1,2¯/diazepine.

The mass spectrum of I (Fig. 3) shows a molecular ion at m/e
194 and a peak at m/e 166 (base peak), characteristic of the
tricyclic structure.

The mass spectrum obtained from the eluate of the spot at Rf
0.30 (Fig. 4) resulted identical to the mass spectrum of the second
GLC peak: the molecular ion at m/e 210, compared with that of I,
suggested the gain of one oxygen atom in the molecule, that is 16
mass units, as a result of the in vitro biotransformation. The loss
of 30 mass units from the molecular ion to give a peak at m/e 180,
with a metastable ion for this transition, suggests for this
compound the loss of nitrogen oxide. High resolution mass spectrometric measurements confirmed this hypothesis, and suggested for
this compound a structure corresponding to II (Fig. 1).

This structure was confirmed by comparison of TLC Rf, GLC
retention time and MS behaviour, which resulted to be identical to
that of a synthetic sample of II.

The spot at Rf 0.13 eluted with methanol and introduced by
the D.I.S. into the mass spectrometer gave a mass spectrum (Fig. 5)
with a molecular ion of 210 which, by comparison with I, suggests
the gain of one oxygen atom in the molecule, that is 16
mass units, as a result of the in vitro biotransformation.

Moreover, the molecular ion of this metabolite at m/e 210 loses
29 mass units (that is HCO radical) to give a peak at m/e 181,
with a metastable ion for this transition. This suggests for this
molecule the introduction of an hydroxy group in the aromatic ring
(III); this was also confirmed by the formation of a methylether
of the compound after an-column reaction with trimethylanilinium
hydroxide (7).

In order to ascertain that the formation of the in vitro
metabolites was not due to a chemical reaction occurring within
the incubation mixture, but to an enzymatic process, I was incubated
under various experimental conditions as described in Table 2.

The metabolic transformation of this compound was studied also
in vivo, after administration of I to rats. In Fig. 6 is reported
the metabolic pathway of I in vivo (in rat).

The overall general experimental procedure, used in extracting
these metabolites, was the following: three male Sprague Dawley rats,
each weighing about 200 g., were injected i.p. with 40 mg/Kg
of I, dissolved in peanut oil, and the urine collected from
these animals at various times over a 48 hour period. Urine

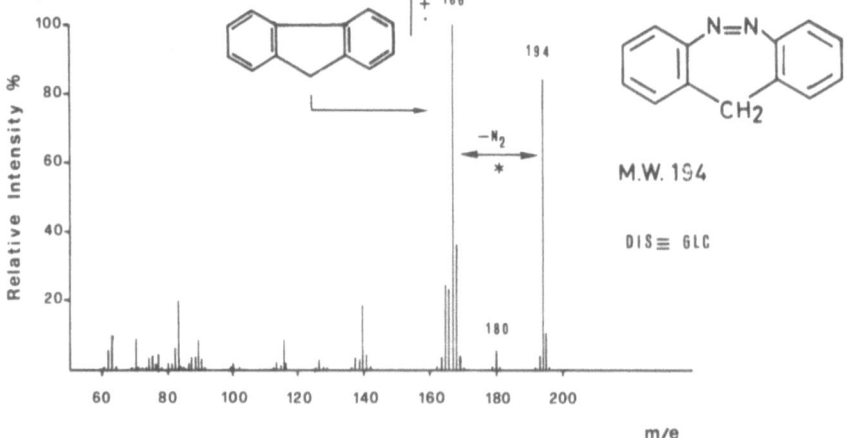

FIG. 3. Mass spectrum of dibenzo$\underline{/}^-$c,f$\underline{/}^-$-$\underline{/}^-$1,2$\underline{/}$diazepine (I).

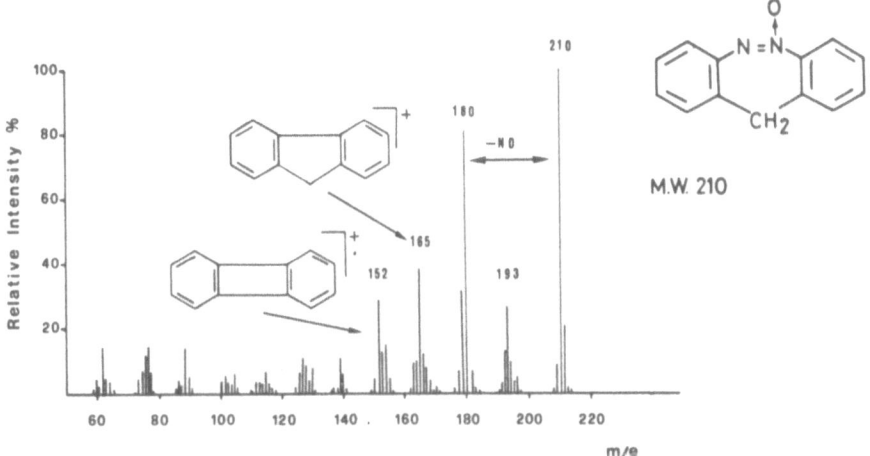

FIG. 4. Mass spectrum of dibenzo/⁻c,f⁻/-/⁻1,2⁻/-1-N-oxide (II).

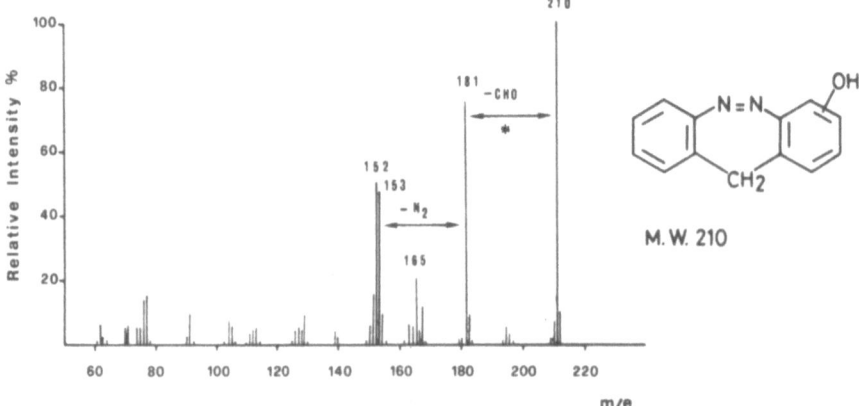

FIG. 5. Mass spectrum of monohydroxyderivative of I in the aromatic
ring (III).

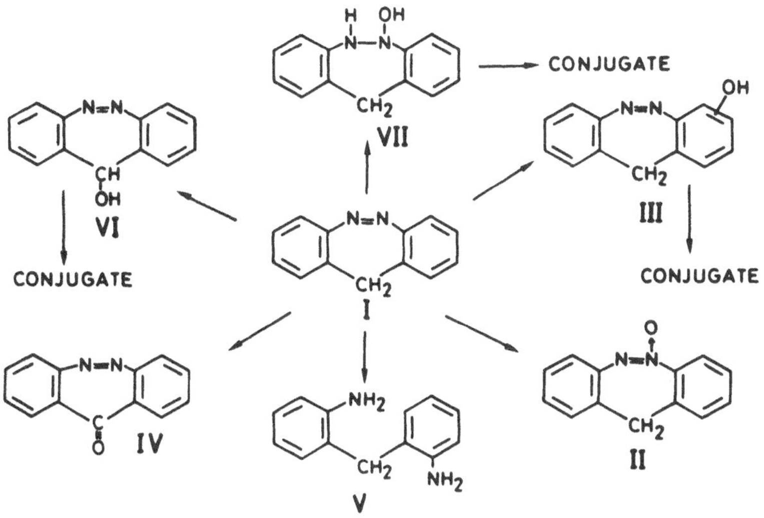

FIG. 6.   In vivo metabolic pathway of dibenzo/ ̄c,f_7-
/ ̄1,2_7diazepine (I).

TABLE 2.
Experimental conditions for in vitro metabolite
formation.

Experimental conditions	II	III
Microsomes + cofactors	+	+
Boiled microsomes^ + cofactors	−	−
Microsomes + cofactors−NADP	−	−
Cofactors only	−	−

^ Boiled microsomes were obtained by heating microsomes,
resuspended in phosphate buffer (0.2 M), at 100°C for
10 min.

was kept frozen until the analyses were performed. The urine collected
from rats was adjusted to pH 9 with sodium hydroxide, and extracted
twice with ethyl acetate.

The organic phase was taken to dryness and the material obtained,
redissolved in methanol, was then used for TLC, GLC and MS analyses.

After extraction of the free metabolites, the urine samples
were incubated with β−glucuronidase at pH 4.5, at 37°C for 12 hours.

The incubate was extracted twice with ethyl acetate, taken to
dryness and analysed by the techniques described above, that is TLC,
GLC and MS.

The urine extract of rats treated with I, analysed by TLC,
contained, in contrast with the extract of control urine, seven
spots, when examined under U.V. light at 254 and 365 nm. The
corresponding Rf are reported in Table 1. The GLC analyses with a
flame ionization detector showed that the ethyl acetate extract of
rat urine treated with I, in contrast with controls, gives rise to
four GLC peaks (Fig. 7).

Further analyses were carried out by utilizing a GLC−MS
procedure for a more precise identification of the peaks.

The spot at Rf 0.42 eluted with methanol and introduced by the
D.I.S. into the mass spectrometer, and the first GLC peak, gave a
mass spectrum identical to that obtained with authentic I (Fig. 3).
The mass spectrum of the eluate of the spot at Rf 0.30, and the
mass spectrum of the third GLC peak gave a mass spectrum identical
to that obtained with authentic II (Fig. 4). The mass spectrum of
the eluate of the spot at Rf 0.36, and the mass spectrum of the

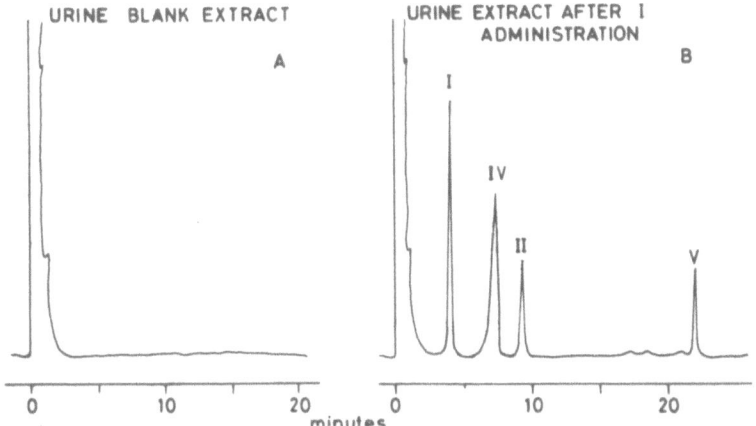

FIG. 7. Gas chromatograms of (A) control urine and (B) urine of rats
treated with dibenzo[c,f]-[1,2]-diazepine(40 mg/Kg i.p.).
Peaks: I, dibenzo[c,f]-[1,2]-diazepine; IV, dibenzo-
[c,f]-[1,2]diazepine-11-one; II, dibenzo[c,f]-[1,2]-diazepine-
1-N-oxide; V, 2,2'-diaminodiphenylmethane.

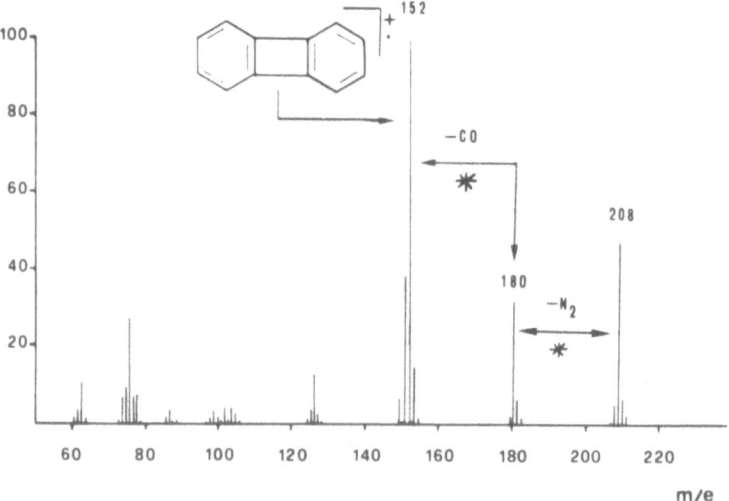

FIG. 8. Mass spectrum of dibenzo[c,f]-[1,2]diazepine-11-one (IV).

second GLC peak (Fig. 8), exhibited a molecular ion at m/e 208, followed by the loss of 28 mass units, to give a peak at m/e 180, with a metastable ion for this transition.

The ion 180 loses then 28 mass units, to give the base peak 152, for which there is a metastable ion.

This fragmentation pathway suggests for this compound the structure of dibenzo/ c,f /-/ 1,2 /diazepine-11-one (IV).

The TLC, GLC and MS behaviour were found to be identical to that of a synthetic sample of IV.

The mass spectrum of the eluate of the spot at Rf 0.10, and the mass spectrum of the fourth GLC peak (Fig. 9) exhibited a molecular ion at m/e 198.

The presence of the ions at m/e 152 and 165 suggests that this is a metabolite of the parent compound. Moreover, the presence of the aniline ion 93 and the presence of the ion 106 (198-92) suggests for this metabolite the structure of 2,2'-diaminodiphenylmethane (V).

In addition, the spot isolated from TLC reacts with fluorescamine, a specific reagent for primary and secondary amines (8, 9).

The mass spectrum of the eluate of the spot at Rf 0.33 (Fig. 10) shows a molecular ion at m/e 210. The molecular ion of this metabolite loses 2 mass units to give a peak at m/e 208, with a metastable ion for this transition. The ion 208 loses 28 mass units to give a peak at m/e 180, with a metastable ion for this transi- tion. The ion 180 loses 28 mass units to give the base peak 152, for which there is also a metastable ion.

This is in accordance with the introduction of an hydroxy group with the formation of the 11-hydroxy-derivative of I (VI).

The mass spectrum of the eluate of the spot at Rf 0.24 (Fig. 11) shows a molecular ion at m/e 212.

In this spectrum there is an increase of 18 mass units compared with the parent compound, together with the presence of characteristic ions at m/e 152 and 165, and the mass spectrum is similar to that of II. Moreover, this compound reacts with trimethylanilinium hydroxide. In addition, it is partially conjugated, probably with glucuronic acid. From these data, it is reasonable to assume the structure of a dibenzo/ c,f /-/ 1,2 /diazepine-1-N- -hydroxy derivative (VII). The formation of methyl ethers of VI, VII and III, obtained after on column-reaction with trimethylanilinium

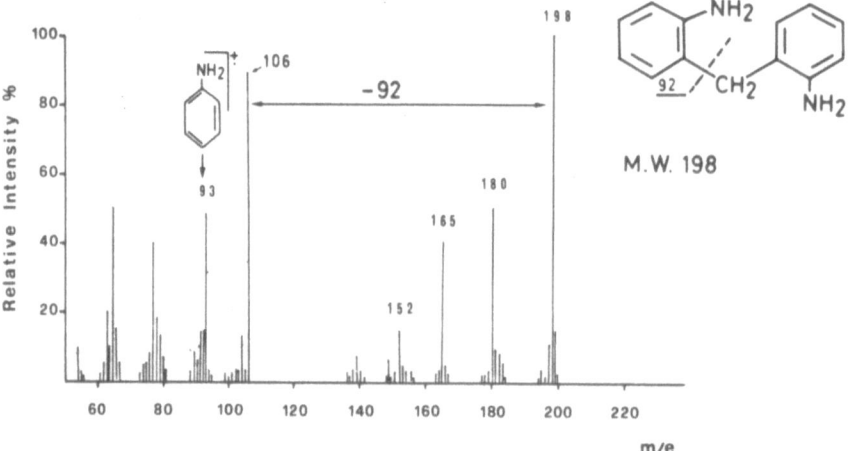

FIG. 9. Mass spectrum of 2,2'-diaminodiphenylmethane (V).

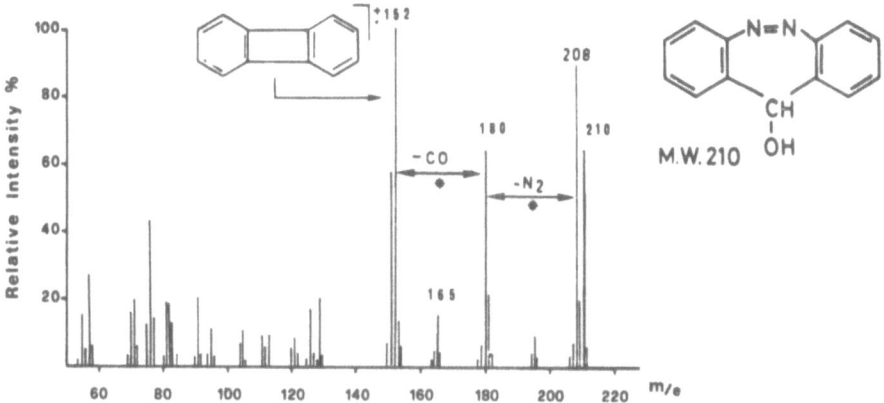

FIG. 10. Mass spectrum of 11-hydroxy-derivative of I (VI).

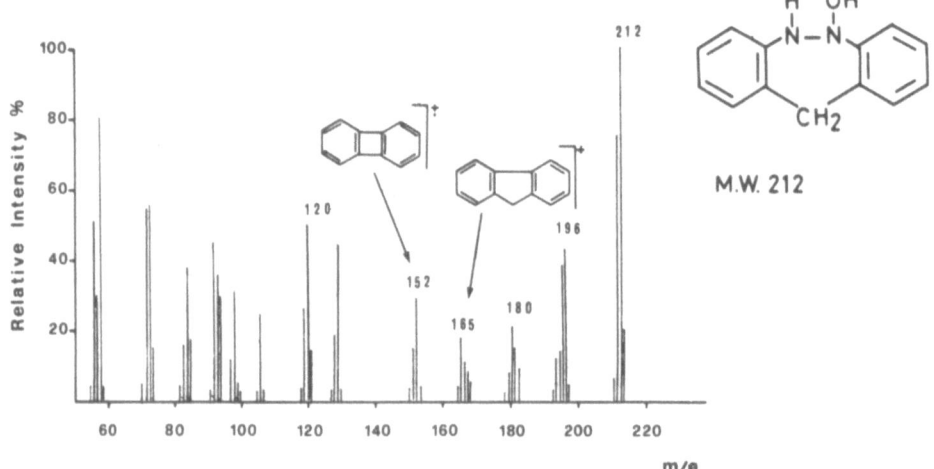

FIG. 11. Mass spectrum of dibenzo[c,f]-[1,2]diazepine-1-N-hydroxy
        derivative (VII).

hydroxide was a further confirmation of the hypothesized structures.

Further studies are in progress for synthesizing an authentic sample of VII to compare its spectroscopic properties with those of the metabolite.

In conclusion, it is interesting to note that while I can be transformed in vivo into VI and into IV as shown in Fig. 6, so far we have not been able to identify in vivo these metabolites in 5H-dibenzo(a,d)cycloheptene (5).

ACKNOWLEDGEMENTS

The authors wish to thank Drs. R. Roncucci and C. Gillet for synthesis of the compounds.The technical assistance of Dr. M.C. Tagliabue and Mr. L. Cobelli is also gratefully acknowledged.

REFERENCES

1) A. Frigerio, N. Sossi, G. Belvedere, C. Pantarotto and S. Garattini, J. Pharm. Sci., 1974, 63, 1536.
2) G. Belvedere, V. Rovei, C. Pantarotto and A. Frigerio, Xenobiotica, 1975, 5, 765.
3) P.L. Morselli and A. Frigerio, Drug Metab. Rev., 1975, 4, 97.
4) A. Frigerio, M. Cavo-Briones and G. Belvedere, Drug. Metab. Rev., 1976, 5, 197.
5) C. Pantarotto, L. Cappellini, P. Negrini and A. Frigerio, J. Chromat., 1977, 131, 430.
6) C. Pantarotto, L. Cappellini, A. De Pascale and A. Frigerio, J. Chromat., 1977, 134, 307.
7) E. Brochmann-Hanssen and T.O. Oke, J. Pharm. Sci., 1969, 58, 370.
8) S. Udenfriend, S. Stein, P. Böhlen, W. Dairman, W. Leimgruber and M. Weigele, Science, 1972, 178, 871.
9) P. Böhlen, S. Stein, J. Slone and S. Udenfriend, Analyt. Biochem., 1975, 67, 438.

GAS CHROMATOGRAPHIC-MASS SPECTROMETRIC IDENTIFICATION OF URINARY

METABOLITES OF PROPILDAZINE IN RAT

L. Simonotti, R. Colombo and G. Pifferi

I.S.F. - Italseber Research Laboratories

Trezzano s/N, Milano, Italy

INTRODUCTION

Propildazine (ISF 2123) is a new antihypertensive drug selected among a series of 6-substituted 3-hydrazinopyridazines synthesized in our laboratories (1-3) (Fig. 1).

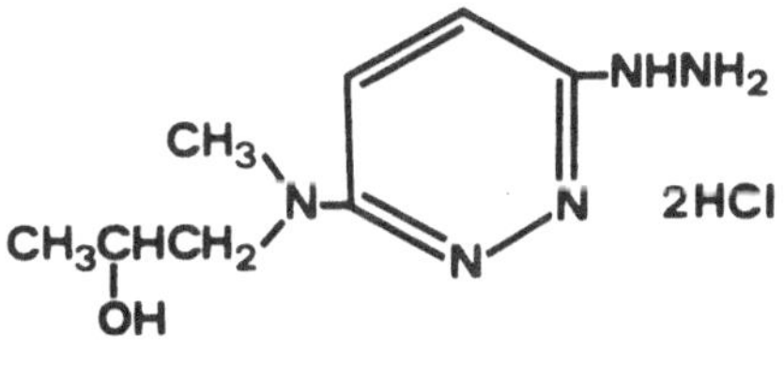

PROPILDAZINE (I S F 2123)

3-Hydrazino-6-[(2-hydroxypropyl) methylamino] pyridazine dihydrochloride

FIG. 1. Propildazine.

Haemodynamic and antihypertensive studies in various animal species indicated that ISF 2123 is a potent and long-lasting peripheral vasodilator (4, 5). Preliminary clinical trials confirmed its activity at dosage levels at least ten times lower than those required for hydralazine (6).

This paper deals with the identification by gas chromatography--mass spectrometry of some metabolites of propildazine in rat urine. The structural indications obtained by GC-MS were used to perform the unequivocal synthesis of standards for comparison and conclusive identification of the potential metabolites. A preliminary metabolic pathway is also proposed and discussed.

EXPERIMENTAL

Reagents. Silylating reagent (BSTFA) was obtained from Pierce (Rockford, Ill., USA). Other analytical grade reagents were purchased from Carlo Erba (Milan, Italy); solvents from Baker (Deventer, Holland) and Merck (Darmstad, Germany), Amberlite XAD-2 resin from BDH Italia (Milan, Italy).

Animal experiments. Five male Sprague Dawley rats (200 g body weight) were overdosed orally with 80 mg/kg of propildazine and a urine sample of 60 ml was collected into hydrochloric acid (pH below 2) over a 24h period.

Treatment of urine samples. The pooled urine was adjusted with NaOH to pH 9 at low temperature and rapidly adsorbed on an Amberlite XAD-2 resin column. After washing with water, metabolites were eluted with methanol (Scheme 1).

Aliquots of dried residues were derivatized following two different procedures: - by silylation with BSTFA in acetonitrile; - by acylation with a mixture of propionic anhydride/triethylamine (2/1).

Standards. 3-/⁻(2-hydroxypropyl)methylamino_/pyridazine was synthesized from the corresponding 3-chloro derivative by catalytic dehalogenation at room temperature. The same starting 3-chloropyridazine was refluxed with anhydrous potassium acetate in glacial acetic acid and subsequently hydrolized to yield 6-/⁻(2-hydroxypropyl)methylamino_/pyridazin-3-one (7). Both these synthetic samples were silylated for GC-MS analysis (Scheme 2).

Synthesis of the 6-/⁻(2-hydroxypropyl)methylamino_/triazolo /⁻4,3-b_/pyridazines was performed starting from propildazine which was selectively acylated on the N^2-hydrazino position by means of acetoformic anhydride, acetyl chloride and propionyl chloride, respectively (Scheme 3).

The obtained 3-hydrazidopyridazines (A) where then thermally

POOLED **RAT URINE**

(80 mg **Kg po**)

↓

aq. NaOH to pH 9

↓

adsorption on Amberlite X AD-2 resin

↓

washing (H$_2$O) and elution (MeOH)

↓

dried residues

silylation   acylation

BSTFA/CH$_3$CN   (EtCO)$_2$O/TEA

GC-MS

SCHEME 1. Rat urine treatment.

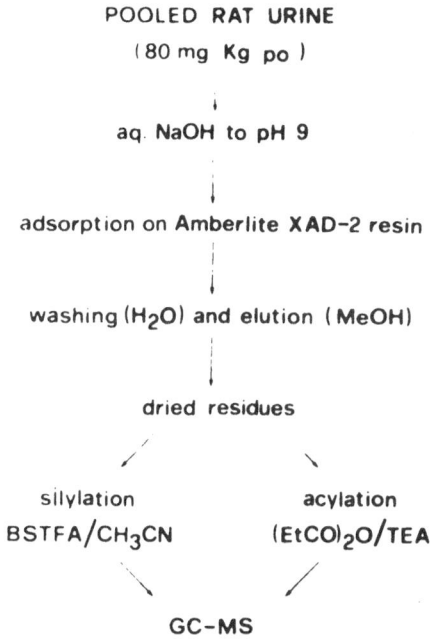

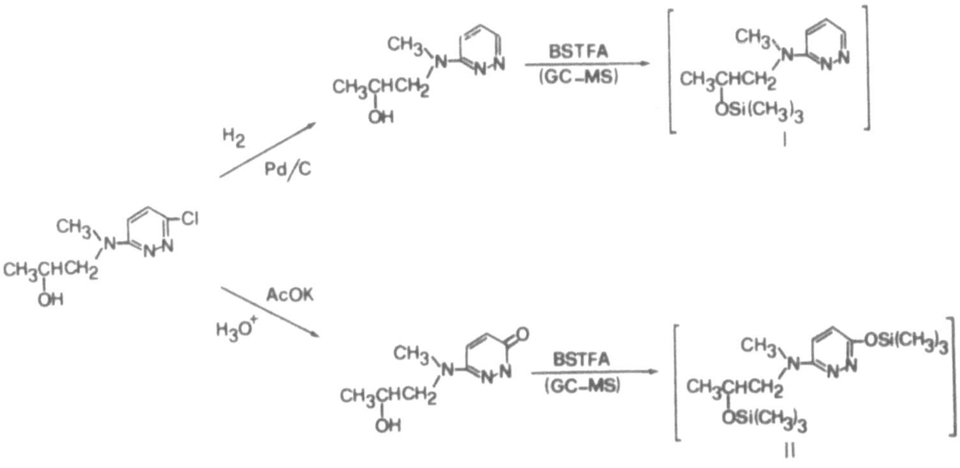

SCHEME 2. Synthesis of pyridazine standards.

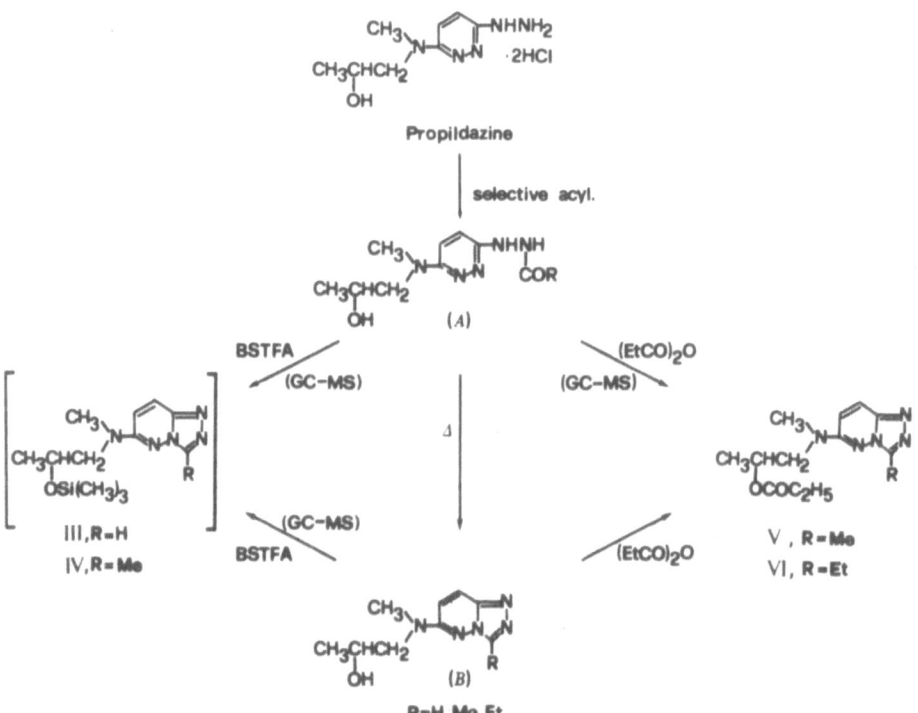

SCHEME 3. Synthesis of triazolo $\lfloor$¯4,3-b_$\rfloor$pyridazine standards.

converted to the desired triazolo / 4,3-b /pyridazines (B) (8). These
standards were derivatized with BSTFA to give III and IV, as
evidenced by GC-MS analysis.

Alternatively, compounds III and IV have been obtained directly
by submitting silylated samples of hydrazidopyridazines (A) to
GC-MS; a flash cyclization took place during injection.

Propionyl derivatives V and VI were unequivocally prepared by
acylating the corresponding triazolo / 4,3-b /pyridazines (B) with
propionic anhydride at room temperature. The direct propionylation
at room temperature of pyridazine (A) in which R=CH$_3$ followed by
GC-MS analysis,yielded a mixture of V and VI.

Instrumentation. A Varian MAT 112 mass spectrometer interfaced
with an Aerograph 1400 gas-chromatograph was used. Gas chromatography
was carried out on a 180 cm x 3 mm ID Pyrex column packed with 3%
OV-17 on Chromosorb W-HP, AW-DMCS 80-100 mesh. Oven temperature was
programmed from 200° to 260° at 6° c/min for silylated extracts
and kept at 240°C for acylated extracts analysis; helium flow rate
was 30 ml/min. Mass spectra were recorded at 70 eV, emission
current 1.5 mA, source temperature 250°C.

RESULTS AND DISCUSSION

The mass spectrum of propildazine (Fig. 2) shows that non
volatile dihydrochloride undergoes thermal decompostion in the mass
spectrometer source, giving rise to the more volatile free base
with molecular ion at m/e 197. The mass spectrum of the base does not
show appreciable differences thus excluding further thermal
contributions to the fragmentation of propildazine.

Most of the total ion current derives from an intense
fragmentation related to the amino-alcohol moiety as a consequence
of an easier extraction of the nitrogen lone-pair electrons (9).
The ion corresponding to the base peak at m/e 152 directly derives
from an α-cleavage at the carbon-carbon bond adjacent to the
nitrogen atom. Ions at m/e 153 and 139 are rearrangement structures
already found in a number of analogous 6-/ (2-hydroxypropyl)alkyl-
amino /pyridazines and 1,2,4-triazolo/ 4,3-b /pyridazines (10).

Because of the high water solubility of the drug as a dihydrochloride, direct extraction from aqueous solutions is difficult. The
titration curve, moreover, shows that the dihydrochloride is
converted, at pH 7.4 in water, into an equilibrium mixture of the
corresponding monochloride and free base.

The U.V. absorbance curves in simulated physiological conditions
of pH and concentration are indicative of a rapid degradation of
the drug which is evident even after two hours (Fig. 3).

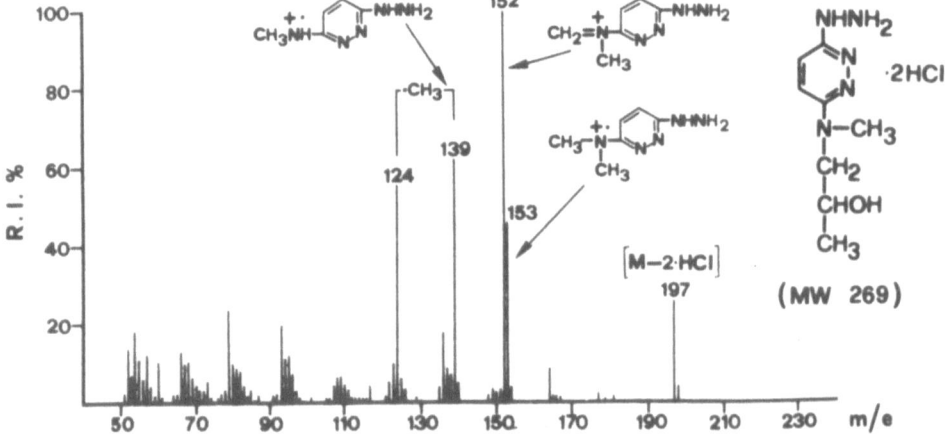

FIG. 2. Propildazine mass fragmentation.

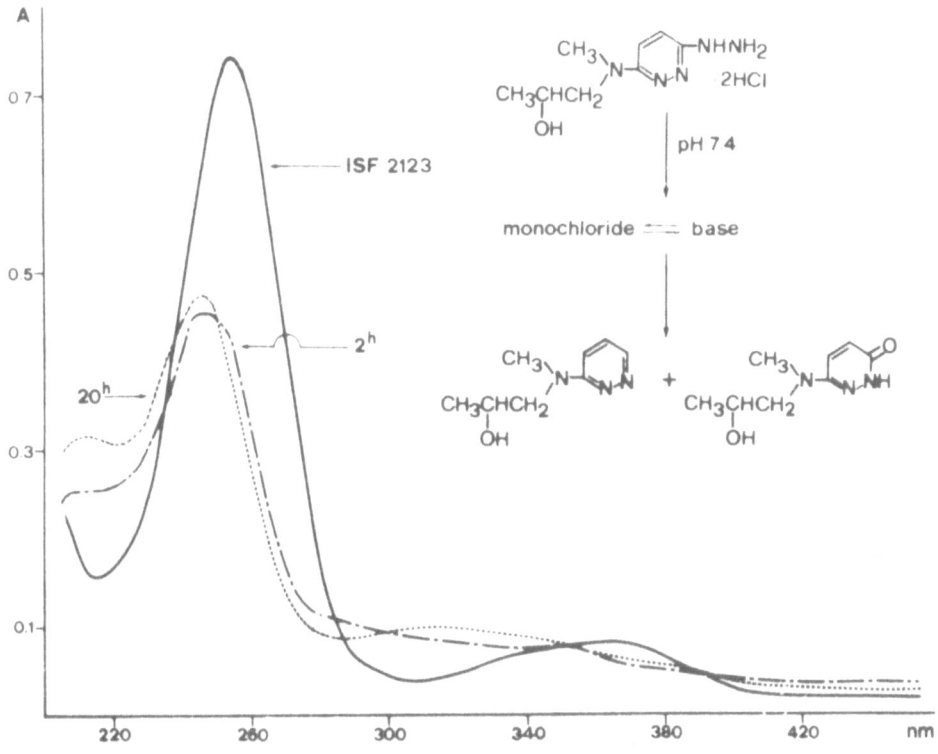

FIG. 3. Propildazine stability (10^{-5} g/ml water, pH 7.4).

An "in vitro" decomposition study has been carried out and, in the complex mixture of several decompostion products, two main structures were identified: the 3-substituted pyridazine and the corresponding pyridazin-3-one. Splitting of the hydrazine moiety by hydrolytic and oxidative enzymatic reactions has already been observed in the metabolism of other drugs (11, 12). For these reasons special attention has been paid to the acid storage of urine samples to reduce decomposition artifacts. Subsequent basic urine extraction at low temperature permitted a rapid and direct isolation of some metabolites of propildazine. In Fig. 4 the gas chromatographic profile of silylated and acylated urine extracts are depicted in comparison with the corresponding blanks.

In the left chromatogram, three peaks, not present in the

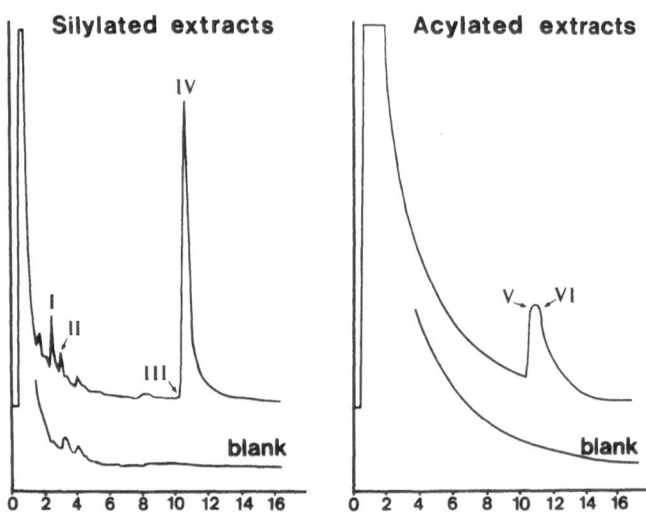

FIG. 4. GC profiles of urine extracts.

control, are evident. Mass analysis of the main peak during elution
clearly revealed the presence of two compounds, III and IV, with
similar retention times.

Overlap of non-resolved peaks was also noted in the chromatogram
of acylated urine extracts on the right side. Mass monitoring of the
broad peak showed it to be the result of the contribution of two
chromatographic peaks attributed to compounds V and VI.
The mass spectrum of derivative I (Fig. 5) displays an almost
abundant molecular ion at m/e 239. Mass fragmentation parallels that
of propildazine, with a shift of 30 a.m.u. due to the absence of the
hydrazino group. Part of the ion current is in fact related to ions
triggered by charge retention on the side chain nitrogen as indicated
by peaks at m/e 109, 122 and 123. Contemporaneously, ions due to the
presence of a trimethylsilyl group, at m/e 73 and 224 (loss of a
methyl radical) contribute to the fragmentation.

A structure of 3-hydroxy-6-substituted pyridazine silylated on
both oxygen atoms has been attributed to derivative II (Fig. 5).
The fragmentation is very similar to that of derivative I, even if
the peaks are shifted at 112 units, corresponding to the silanol
group.

The alternative structure of a pyridazin-3-one, silylated on
the nitrogen atom in position 2, was ruled out based on evidence
in the literature (13) and by means of spectroscopic data. NMR and

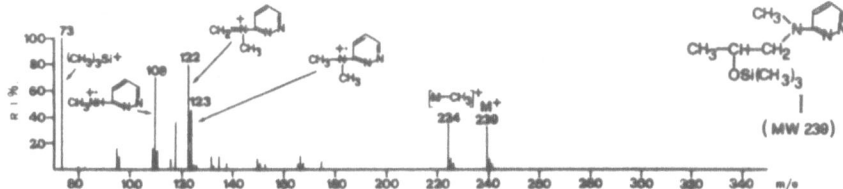

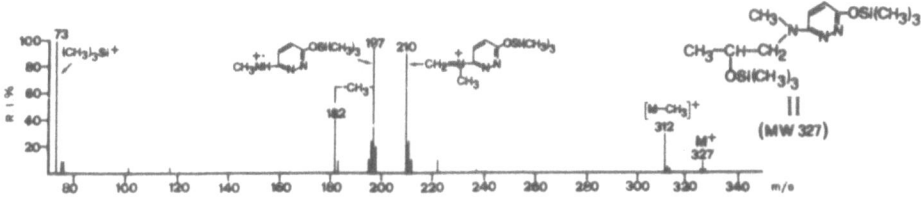

FIG. 5. Mass spectra of silyl derivatives I and II.

UV spectra indicate an increase of heteroaromaticity after silylation of an authentic sample of pyridazin-3-one consequent to a displacement of the tautomeric equilibrium to 3-hydroxypyridazine.

Mass spectrometric data of silyl derivatives III and IV are in agreement with the silylated triazolopyridazine structures (Fig. 6). A common feature for all mass spectra of these heterocycles is, in fact, their great stability especially if compared to the other heterocycles, such as 1,2,4-triazoles, and tetrazolopyridazines (14). Therefore, the character and the position of the substituents generally play an important role in the fragmentation pattern.

Concerning the present mass spectra, the abundance of the molecular ions at m/e 279 and 293, in addition to ions at m/e 121 and 135 respectively, is indicative of this stability.

The structure corresponding to protonated triazolopyridazines has been supported by high resolution measurements of the peaks at m/e 135 ($C_6H_7N_4$; calc. 135.0671, found 135.0688 ± 0.002).

Here too, the dominant fragmentation pattern originates from the substituent in position 6 with a high contribution of the silyl group; namely, base peak at m/e 73, loss of a methyl radical, α--cleavage of amino alcohol moiety. This operates with charge retention on both fragments, that is, ions at m/e 162, 176 and the complementary ion at 117.

Finally, an enhancement in the intensity ratio has been observed for ions deriving from methyl rearrangement (that is, at m/e 163 and 177) as a consequence of the trimethylsilyl group present in the molecule.

The spectra of propionyl derivatives V and VI are simpler, due to the absence of silyl groups (Fig. 7). Most of the total ion current is in fact carried by ions at m/e 176 and 190 respectively, deriving from the routinely seen amino α-cleavage with total charge retention on the amino moiety. On the other hand, ions due to competitive rearrangements appear to an appreciably lower extent (i/e at m/e 135, 177 and respectively, the homologs at m/e 149 and 191). An alternative fragmentation pathway of the molecular ion is triggered by the propionyl ester undergoing a McLafferty rearrangement with charge retention on the unsaturated fragment, as can be seen by peaks at m/e 203 and 217. These odd electron ions undergo a further fragmentation with loss of a methyl radical.

Once again the perfect correspondance of the two spectra with a shift of a methylene group suggests that the fragmentation does not involve the heterocycle moiety.

The structural indications obtained by GC-MS were used to perform

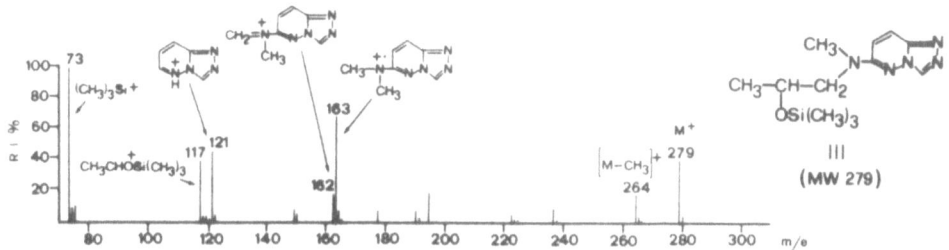

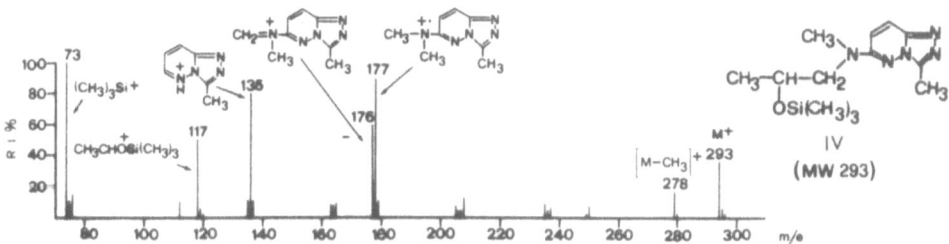

FIG. 6. Mass spectra of silyl derivatives III and IV.

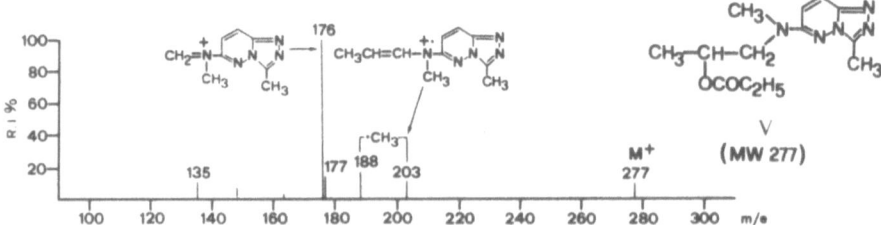

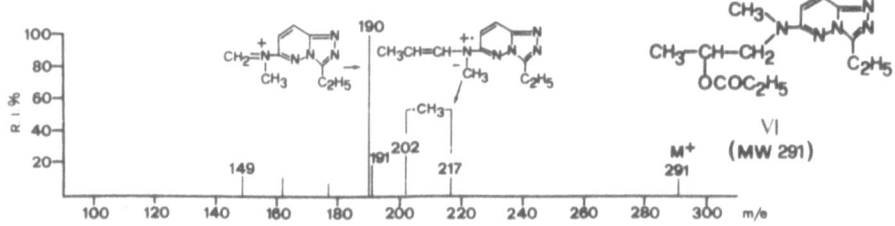

FIG. 7. Mass spectra of acyl derivative V and VI.

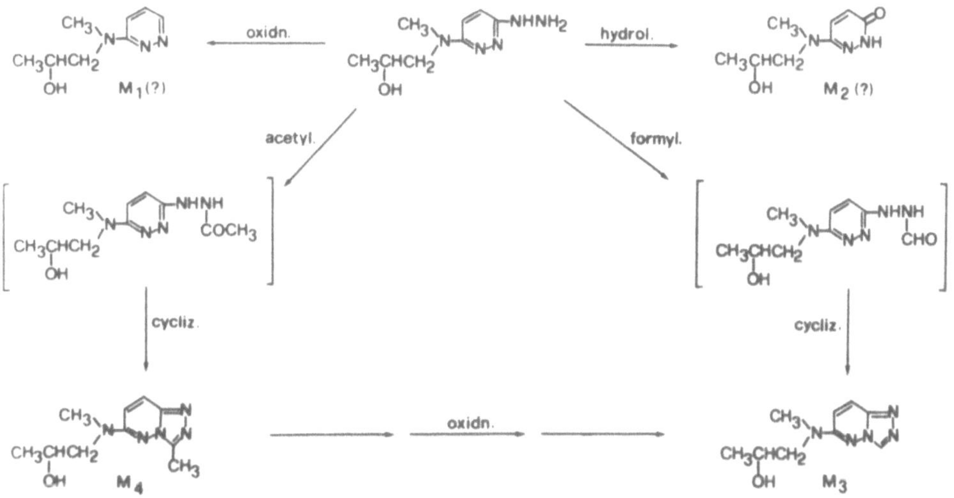

FIG. 8. Proposed metabolic pathway of propildazine in rat.

the unequivocal synthesis of standards (see Scheme 2 and 3) for
comparison and conclusive identification of the potential metabolites.

CONCLUSIONS

On the ground of analytical results performed on the basic
extracts and chemical behavior of the authentic samples, it is
possible to draw preliminary considerations on the biotransformation of propildazine (Fig. 8), omitting, at present, implications
due to overdose.

M1 and M2 may be formed by oxidative and hydrolitic processes
and are here referred to only as "possible metabolites" since we
cannot exclude, at least in part, the formation of artifacts during
urine storage and analytical procedures as verified by "in vitro"
experiments.

Isolation of the M4 metabolite clearly indicates a metabolic
acetylation of propildazine in the rat. On the other hand,
determination of the compound VI (3-ethyl-triazolo/ 4,3-b /pyridazine)
makes it possible to recognize the presence of structures referred
to the unchanged drug and/or its acetyl derivative not yet cyclized.

Concerning the metabolite M3 two alternative hypothesis of formation are here proposed. First, multistep oxidation of metabolite M4
followed by final decarboxilation could lead to M3. This pathway is
supported by other authors' experiments dealing with isolation of
the hydroxymethyltriazolophtalazine (15) as an oxidation intermediate
and with ^{14}C labelled methyltriazolophtalazine (16). With regard
to our experiments, we have not yet identified the analogous
hydroxymethyl intermediate.

Alternatively, metabolite M3 can also derive from a direct
"in vivo" formylation of propildazine with subsequent cyclization
to triazolopyridazine. This would also be in agreement with the
results of Japanese authors supporting direct formylation of the
hydrazino moiety in the budralazine metabolism (12).

In addition to the acylating pathways, it is worthwhile
mentioning that studies with ^{14}C labelled propildazine indicate the
presence of a further metabolic route via glucuronic acid
conjugation (17).

Identified metabolites have been pharmacologically tested for
their antihypertensive activity; the pyridazine derivatives M1
and M2 and the triazolopyridazines M3 and M4 did not show any
appreciable activity.

On the contrary, hydrazidoacyl derivatives were as active as
propildazine itself (18).

REFERENCES

1) G. Pifferi, F. Parravicini, C. Carpi and L. Dorigotti, J. Med. Chem., 1976, 18, 741.
2) F. Parravicini, G. Scarpitta, L. Dorigotti and G. Pifferi, Farmaco, ed. Sci., 1978, 33, 99.
3) Drugs of the Future, 1976, 1, 290.
4) L. Dorigotti, R. Rolandi and C. Carpi, Pharmacol. Res. Commun., 1976, 8, 295.
5) L. Dorigotti, C. Semeraro and C. Carpi, Riassunti 7° Congresso S.S.F.A., Varese, 11-12 ottobre 1977.
6) R. Pellegrini and G. Abbondati, Farmaco, ed. Sci., 1977, 32, 19.
7) L. Simonotti et al., to be published.
8) F. Parravicini, G. Scarpitta, L. Dorigotti and G. Pifferi, Farmaco, ed. Sci., in press.
9) H. Budzikiewicz, C. Djerassi and D.H. Williams, in "Mass Spectrometry of Organic Compounds", Holden Day Inc., 1967, p. 297.
10) L. Simonotti et al., to be published.
11) K.D. Haegele, H.B. Skrdlant, N.W. Robie, D. Lalka and J.L. McNay Jr., J. Chromatogr., 1976, 126, 517.
12) R. Moroi, K. Ono, T. Saito, T. Akimoto and M. Sano, Chem. Pharm. Bull., 1977, 25, 830.
13) A.E. Pierce, in "Silylation of Organic Compounds", Pierce Chemical Company, Rockford, Ill., 1968, p. 63.
14) V. Pirc, B. Stanovnik, T. Tisler, J. Marsel and W.W. Paudler, J. Het. Chem., 1970, 7, 639.
15) H. Zimmer, R. Glaser, J. Kokosa, D. Garteiz, E.V. Hess and A. Litwin, J. Med. Chem., 1975, 18, 1031.
16) Z.H. Israili and P.G. Dayton, Drug Metab. Rev., 1977, 6, 283.
17) Life Science Research, internal report, 1976.
18) G. Pifferi, F. Parravicini, G. Scarpitta and C. Semeraro, VIth Int. Symposium on Medicinal Chemistry, Brighton, 4-7 Sept., 1978.

MASS SPECTROMETRIC CHARACTERIZATION OF SYDNOCARB ® AND ITS MAJOR

METABOLITES IN RAT

J. Tamás^, M. Polgár¨, G. Czira^ and L. Vereczkey¨

^ Hungarian Academy of Sciences, Budapest;

¨ Gedeon Richter Ltd.,Budapest, Hungary

INTRODUCTION

Sydnocarb, N-phenylcarbamoyl-3-(β-phenylisopropyl)-sydnonimine (1) is a new psychostimulator. From earlier examinations it was established (1) that the effect of sydnocarb on the central nervous system, e.g. as compared to that of Amphetamine, develops more gradually, lasts longer and the therapeutic doses do not cause euphoria or locomotor excitation. Studies concerning the pharmacokinetics of sydnocarb in rat have been carried out using sydnocarb labelled with ^{14}C at various positions (2, 3), but no report on metabolic fate of this drug has appeared in literature.

In our laboratories the metabolic pathways of sydnocarb have been investigated in rat by ^{14}C sydnocarb labelled in α-position of the isopropyl part of the molecule.

The metabolites present in urine and bile were isolated by organic solvent extraction or by an ion-exchange procedure and were purified by thin layer chromatography using several solvent systems.

The isolated conjugates of metabolites were hydrolysed by a treatment with β-glucuronidase – aryl sulfatase enzymes.

For structural characterization and identification of metabolites, mass spectrometry together with IR and NMR spectroscopy were successfully applied.

The results showed that the main metabolic route of sydnocarb in rat is the conversion of the drug into hydroxylated derivatives 2 and 3, followed by formation of conjugates, i.e. majority of metabolites contained in the intact sydnocarb skeleton.

Herewith, as a part of the study, some details and results of
mass spectrometric investigation of sydnocarb (1) and its major
metabolites (2, 3), as well as some of close related model compounds
(4-7) are presented also showing the difficulties we faced in
characterising compounds of sydnocarb skeleton by mass spectrometry.

SCHEME

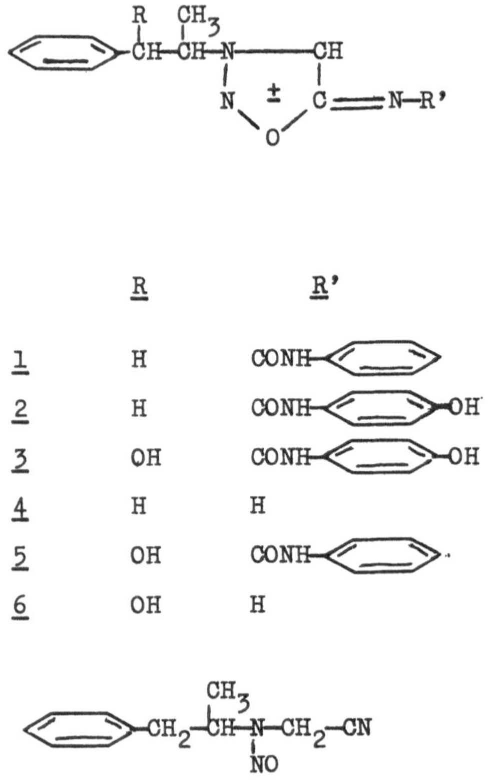

	R	R'
1	H	CONH–⟨ ⟩
2	H	CONH–⟨ ⟩–OH
3	OH	CONH–⟨ ⟩–OH
4	H	H
5	OH	CONH–⟨ ⟩.
6	OH	H

7

Other detailed results of the examination and certain conclusions concerning pharmacology of the drug will be published elsewhere (4).

EXPERIMENTAL

Sydnocarb (1) used was a product of Gedeon Richter Ltd., Budapest. Compounds 2 and 3 were isolated as metabolites. For details of isolation see(4)Compounds 4 to 7 were synthetised and kindly given by the Chemical Pharmacological Research Institute, Moscow.

The electron-impact (EI) mass spectra of 1 to 7 were taken and the exact mass measurements were carried out using an AEI MS-902 instrument. The samples were evaporated using no direct heating of the sample probe. The ion source temperature was kept at 160°C.

The isobutane chemical ionization (CI) mass spectra were run on a VG Micromass 12F1A mass spectrometer equipped with a dual EI/CI source. In this case a directly heated probe was used at 140°C. The ion source temperature was usually kept at 160°C.

The low resolution EI spectra of these compounds taken on both instruments showed no significant difference.

RESULTS AND DISCUSSION

The EI mass spectrum of sydnocarb is shown in Fig. 1. It can be seen that abundant ions appear only in the low mass region,

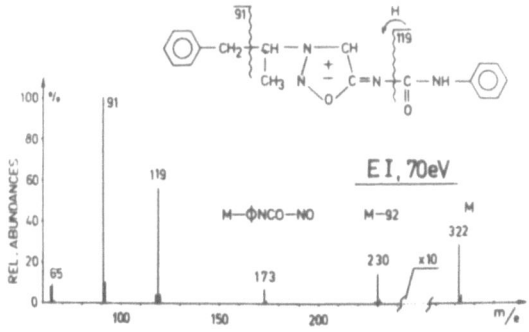

FIG. 1. The EI mass spectrum of sydnocarb
(compound 1).

namely at m/e 91 and 119, corresponding to the two terminal parts
of the molecule as depicted in the figure.

The spectrum exhibits a molecular peak of low abundance and
the fragments containing sydnonimine moiety are detectable as weak
peaks too. Some compounds containing a sydnonimine ring were studied
by electron-impact mass spectrometry to a slight degree (5). These
are much more simple molecules as compared with sydnocarb, and it
was found that the main fragmentation pathways of sydnonimine ring
substituted in N-3 position is the loss of an NO molecule followed
by subsequent elimination of two HCN molecules. The ion at m/e 230
is caused by elimination of a toluene molecule or an anilino-group
from the molecular ion, and the peak at m/e 173 is interpreted as
a product of subsequent eliminations of phenylisocyanate and NO
molecules.

In order to get further and more convincing information by
mass spectrometry for identification of sydnocarb from isolated
samples, as a complementary technique, chemical ionization (CI)
mode was chosen. In this case, using isobutane as reactant gas, a
surprisingly large temperature effect on the mass spectrum was
observed.

Fig. 2 presents CI mass spectra of sydnocarb taken at three
different ion source temperatures. As can be seen from comparison
of the observed mass spectra, sydnocarb underwent thermal decomposition during the experiments and the extent of decomposition strongly
depends on the source temperature. Since the direct probe temperature
was kept at the same value, 140°C, in all the three experiments,
we concluded that thermal decomposition occurred in the gas phase,
due to the presence of isobutane molecules as a heat transfer
medium. Keeping the temperature of the ion source at about 100°C,
sydnocarb exhibited a well detectable M+1 peak allowing us to
identify sydnocarb in this way.

The two products of the thermal decomposition denoted by M and
M_2 are most probably phenylisocyanate and the corresponding sydnonimine, unsubstituted in the exocyclic N atom.

The low abundance of the M_2+1 peak and the easy loss of an H
atom from it is attributable to the low proton affinity of phenylisocyanate.

The appearance of an ion at m/e 148, i.e. the elimination of
NOCN from m/e 204, confirms that the compound M_1 has an intact
sydnonimine ring in the gas phase. This is of interest in relation
to earlier findings (6) that due to heating, sydnonimines having
no substinent at the exocyclic N atom can isomerise into the

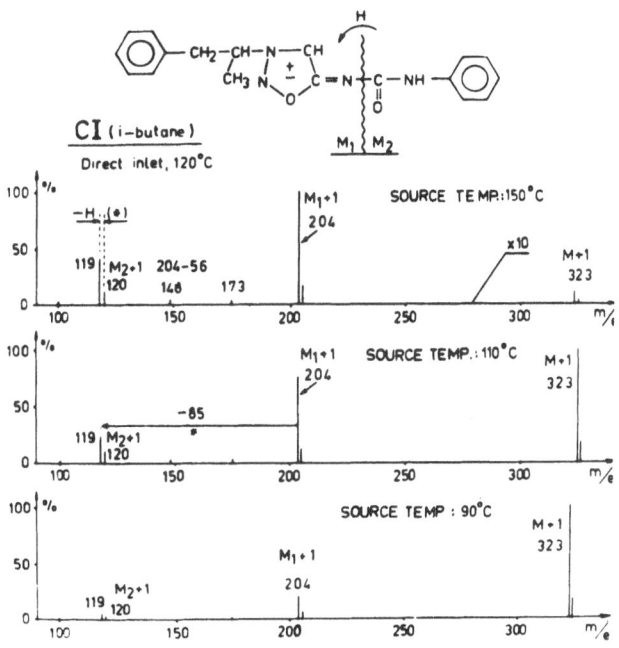

FIG. 2. The CI mass spectrum of sydnocarb (compound 1).

corresponding acetonitrile derivatives by opening the ring. For further investigations the mass spectral behaviour of model compounds 4 and 7, i.e. a sydnonimine derivative corresponding to M_1 and its substituted acetonitrile isomer, were compared.

In Fig. 3 the electron-impact mass spectra of the two isomers (4 and 7) are presented. Although the two mass spectra are very similar to each other, they have distinct features, too. For example: the cleavage a in the case of sydnonimine leads to ions at m/e 119, while for its isomer, this process involves an H migration giving rise to a peak at m/e 118. The differences observed indicate that this isomerisation is not a significant process during the mass spectral investigation.

In fig. 4 the electron-impact mass spectrum of 2 is shown. Similarly to sydnocarb, in this case again two abundant peaks are observable. One of them appears at the same mass units in both spectra and the other is shifted here by 16 mass units to higher masses.

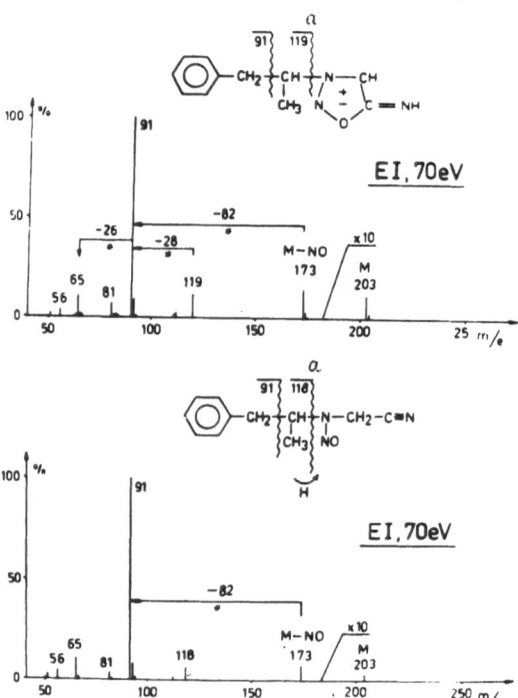

FIG. 3. The EI mass spectra of compounds 4 and 7.

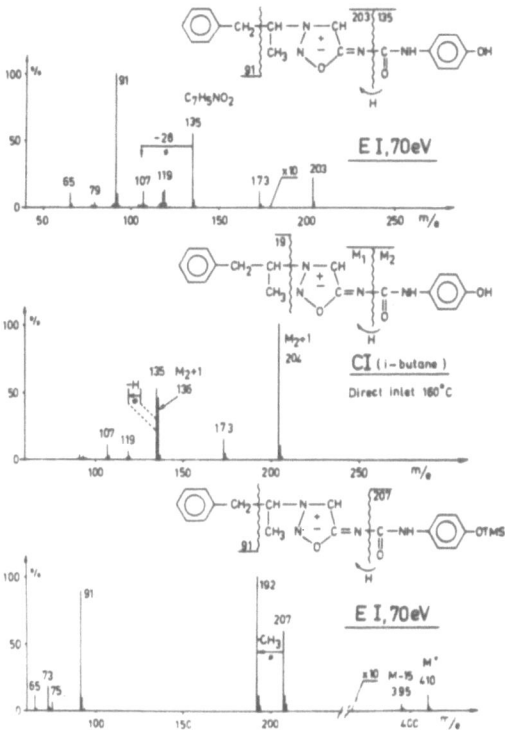

FIG. 4. The EI and CI mass spectra of compounds
2 and EI mass spectrum of its silylated
derivative.

From exact mass measurements this ion corresponds to the molecular
ion of a hydroxylated phenylisocyanate fragment. Ions at m/e 91
and 203 are due to the complementary part of the molecule, but no
molecular ion of the original molecule appeared in the spectrum.

The chemical ionization mass spectrum (Fig. 4) showed the
presence of two compounds, M_1 and M_2, in the gas phase. It is in
accordance with the EI spectrum, and is due to thermal decomposition.
The para-substitution of phenylisocyanate by an OH group, the
position of which having been established from its IR spectrum, is
reflected in an enhanced proton affinity and hence in a higher
abundance of M_1+1 peak.

Silylation of this metabolite (Fig. 4) caused a shift of the
peak from m/e 135 to 207 m.u.. This ion is able to lose a methyl
group, supporting its radical ion character. The other terminal

part of the molecule appears as fragment ion at unchanged m.u.,
i.e. at m/e 91. Furthermore, due to the higher volatility, molecular
ion and M-15 ions are also observable in the spectrum, and give
more direct information on the molecular weight of this metabolite.

The analogous mass spectra of the other metabolite (compound
3) are presented in Fig. 5.

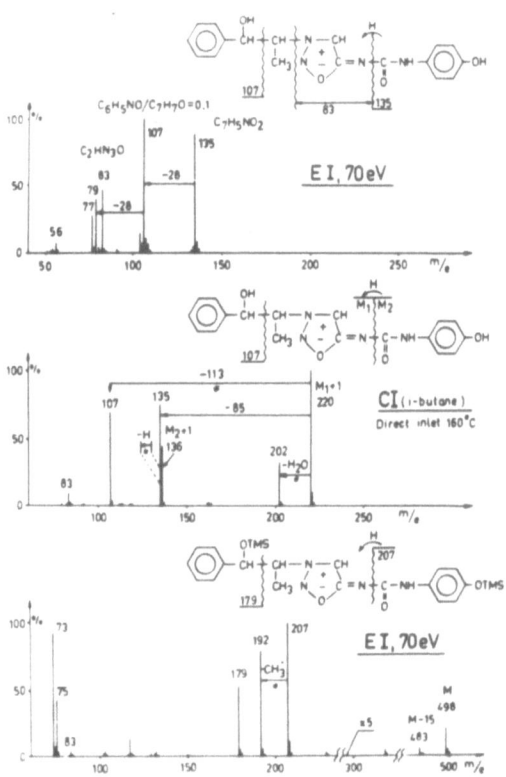

FIG. 5. The EI and CI mass spectra of compound 3
and EI mass spectrum of its silylated
derivative.

In the EI mass spectrum (Fig. 5) the two most abundant peaks again
originate from the two terminal parts of the molecule, both of them
shifted by 16 mass units, as compared with those of sydnocarb.
Their chemical compostion was established by high resolution mass
measurements. In this case the abundant ions appearing at m/e 83
is caused by the sydnonimine part of the molecule.

The CI mass spectrum of this compound (Fig. 5) compared with that of 2 shows that M_1+1 ion is here shifted by 16 mass units from m/e 204 to 220. This ion can easily lose a water molecule indicating that the OH group is bonded to an aliphatic carbon atom, and hence it follows that from m/e 107 this OH group must be in α position to the phenyl group.

The mass spectrum of the silylated derivative shows that both terminal fragments are shifted due to silylation, i.e. each has an OTMS group. Furthermore, peaks corresponding to M and M-15 ions were well detectable, too, giving again direct information about the molecular weight of the metabolite.

The appearance of the sydnonimine ring as an abundant fragment peak in the EI spectrum at m/e 83 is due to the presence of the OH group in α position to the phenyl group, since this ion is completely absent in the spectra of 1 and 2.

This interesting substituent effect has also been observed in the case of model compounds 5 and 6 (Fig. 6).

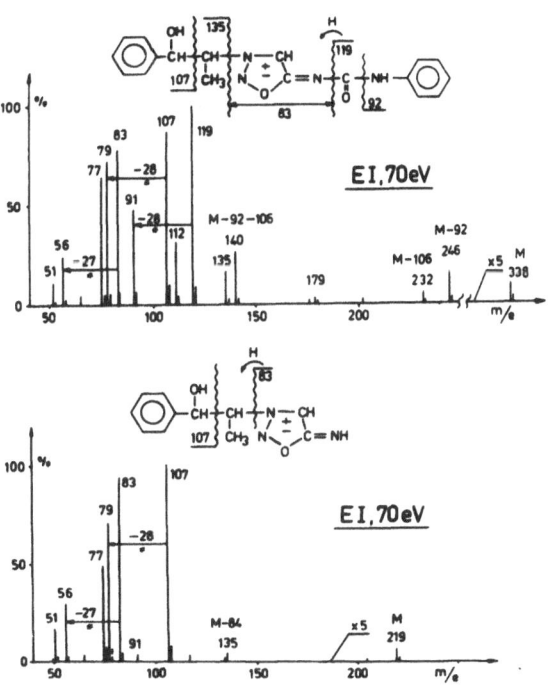

FIG. 6. The EI mass spectra of compounds 5 and 6.

In this case compound 5, having no phenolic OH group in the
phenylcarbamoyl part, was evaporable without significant thermal
decomposition. It can be seen that both spectra exhibit abundant
peaks at m/e 83, the further fragmentation of which is connected
with the elimination of an HCN molecule leading to ion at m/e 56.

CONCLUSION

From the results concerning mass spectral investigations of
sydnocarb and its related compounds it can be concluded that
application of EI and CI methods of mass spectrometry together with
silylation could give detailed and convincing information for
characterisation and structure analyses of these compounds from
isolated samples in spite of their thermal instability.

REFERENCES

1) R.A. Altshuler, M.D. Mashkovskii and L.F. Roshchine, Farmakol.
   Toksikol. (Moscow), 1973, 36, 18.
2) R.A. Altshuler, L.E. Holodov, D.V. Meirena, S.B. Seredenin, I.A.
   Sultanov, N.V. Stelletskaya and Yu.A. Belednov, Khim. Farm.
   Zh., 1977, 11, 16.
3) R.A. Altshuler, L.E. Holodov, A.P. Gulev, D.V. Meirena and A.G.
   Odinets, Khim. Farm. Zh., 1977, 11, 12.
4) M. Polgár, L. Vereczkey, L. Sporny, G. Czira and J. Tamás, to be
   published.
5) Shima Takashi, Onchida Akira and Asaki Yutaka, Shitsuryo Bunseki,
   1969, 17, 661.
6) H.U. Daeniker and J. Druey, Helv. Chimica Acta, 1962, 45, 2426.

METABOLISM OF AN ARYLOXY BETA BLOCKING DRUG

G.L. Passetti, E. Grassi and A. Trebbi

Laboratori di Ricerca Zambeletti, Milano, Italy

INTRODUCTION

Zami 1305: 1-(2-nitro-3-methyl phenoxy)3-terbutylamino-propan--2-ol (Fig. 1) is a potent new β-adrenergic blocker recently synthesized in our laboratories (1).

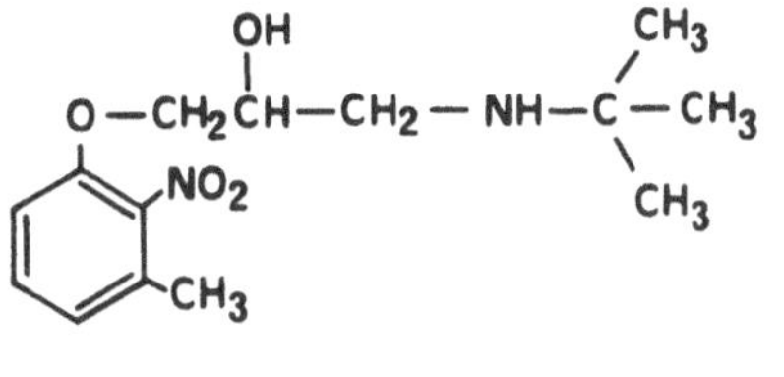

1- (2-nitro,3-methyl-phenoxy)3 -terbutylamino -propan- 2-ol

FIG. 1. Formula of ZAMI 1305.

The results of previous studies on the chronic toxicity of this compound showed differences between male and female rats in the toxic effects of very high doses.

These findings prompted us to investigate whether divergent toxicological effects can be attributed to different biological fates of this substance in animals of different species and sexes.

The aim of this study was to identify the metabolites in
biological fluids of dogs and rats of both sexes and also to check
whether some of the metabolites had pharmacological activity, which
is the case for many metabolites of other β-blocking drugs.

The drug and its metabolites were identified by thin layer
chromatography (TLC) coupled with mass-spectrometry (MS), while the
serum levels and urinary recovery were quantified by gas-chromato-
graphy (GLC) of a fluorated derivative after the peak had been
identified by gas-chromatography-mass-spectrometry (GLC/MS).

MATERIALS AND METHODS

Reagents and standards. All the chemicals used were analytical
reagent grade and were tested for purity in blank runs.
Animals. Beagles (10-12 Kg) and Wistar rats (200-250 g) bred
in our animal house quarters were used. Zami 1305 was chronically
administered orally to dogs (60 mg/kg:day) and rats (100 mg/kg/day)
for twelve months.
Isolation of metabolites. Urine and blood sera were collected
separately from male and female rats and dogs (Fig. 2), hydrolyzed
at pH 2 and pH 4.5 at 60°C. After hydrolysis, the samples were
extracted and then the aqueous layer was adjusted to pH 11 and
extracted again with ethyl acetate.

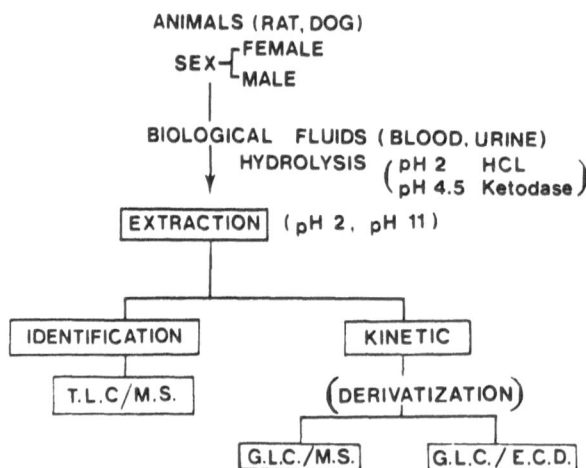

FIG. 2. Scheme of the method used for the
        isolation of the metabolites of
        ZAMI 1305.

The organic extracts (drug and metabolites) were chromatographed by TLC with reference standards of probable metabolites
and appropriate blanks from biological fluids of drug-free animals.
(The plates used were Silica Gel Merck F 254 2 mm thickness). The
plates were developed with methanol:conc. ammonium hydroxide
(100:1.5). The spots with Rf corresponding to those of the reference
standards were eluted with methanol and the solvent was then
evaporated off under nitrogen at room temperature. The residue was
then analyzed by mass spectrometry using a direct inlet system
(DIS). The mass spectra from biological samples were compared with
those of the authentic standards.

Mass-spectrometry (DIS). An LKB Model 900 mass spectrometer was
used. The operative conditions were: probe temperature 70°C; gain
7; slits 1.09-0.09; filters 60 c.p.s.; electron energy 70 eV;
accelerating voltage 3.5 kV; ion source temperature 290°C.

Gas-liquid chromatography (GLC). Analyses were carried out
with a Perkin Elmer Model F 30 gas-chromatograph equipped with a
^{63}Ni electron capture detector.

The column was a glass tube (90 cm x 6 mm I.D.) packed with
80-100 mesh gas-chrom Q (Applied Science Lab., State College, Penn-
sylvania, USA) coated with 3% XE GE 60.

The column was conditioned for 1 h at 250°C with nitrogen, flow
rate of 60 ml/min; for 4 hr. at 200°C without flow; 48 hr. at 250°C
with a flow rate of 60 ml/min. The following operating conditions
were used for analyses of the trimethyl-silyl-heptafluorobutyryl
derivatives of the drug; column temperature 200°C; injection port
250°C; electron capture detector 300°C; nitrogen flow rate 50 ml/min.

The derivative was obtained as follows: the substance was at
first silylated with N-O-bis-(trimethylsilyl)-trifluoroacetamide
(BSTFA), from Pierce), which reacts with both the hydroxyl and the
amino groups. The compound obtained is further reacted with heptafluorobutyric anhydride, which replaces the trimethylsilyl group
on the amino function with the heptafluorobutyril group.

Gas-chromatography-mass spectrometry (GC/MS). A Varian model
MAT 112 mass spectrometer equipped with gas-chromatograph and with
Spectrum System 101 MS was used to confirm the structure of the
derivative described above. A glass tube (2 m x 6 mm I.D.) packed
with 80-100 mesh Chromasorb G.H.P. coated with 3% OV 25 was used.
The operating conditions were column temperature 250°C; injection
port 250°C; molecular separator temperature 270°C; ionization beam
energy 70 eV; flow rate: 30 ml/min.

RESULTS AND DISCUSSION

The metabolic fates of ZAMI 1305 after chronic oral administration (12 months) to the rat and the dog is shown in Fig. 3.

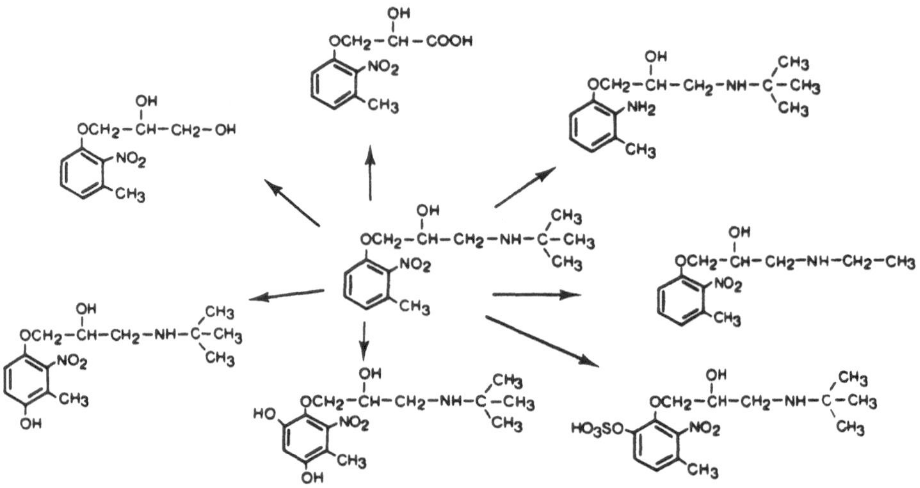

FIG. 3. Schematic rapresentation of ZAMI 1305 metabolism
         in the rat and the dog.

The pathways are different for different sexes and species.

The metabolites extracted from the biological fluids of the
male rat at two different pHs, pH 2 and pH 11, are shown in Fig. 4.

Zami 1305 was the only metabolite found in serum, while in urine,
in addition to unchanged drug we found an hydroxylated derivative:
2-nitro-3-methyl-4-hydroxy-phenoxy-ter butylamino-propan-2-ol and
the glycol metabolite: 2-nitro-3-methyl-phenoxy-propan-1,2-diol.

In Fig. 5 it is shown that in serum and urine of female rats
only the unchanged drug was present.

Zami 1305 is more extensively metabolized in the dog than in the
rat. In fact, in male dog serum (Fig. 6) we found the unmetabolized
drug plus a product resulting from reduction of nitro groups: 2-
-amino-3-methyl-phenoxy-ter butylamino-propan-2-ol, and another
metabolite resulting from partial dealkylation, 2-nitro-3-methyl
phenoxy-ethylamino-propan-2-ol.

In urine, besides these three metabolites there were also the
product obtained by the oxidation of the side chains 2-nitro-3-
-methyl-phenoxy-acetic acid, and a sulfuric derivative, 2-nitro-3-
-methyl-6-hydroxy-ter butylamino-propan-2-ol-6-sulfuric ester.

In Fig. 7 are shown the metabolites found in fluids from the
female dog.

METABOLISM OF ZAMI 1305 IN MALE RAT

BIOLOGICAL FLUID	pH EXTRACTION	T.L.C. Rf	NAME	FORMULA
SERUM	pH 2			
	pH 11	0.40°	ZAMI 1305	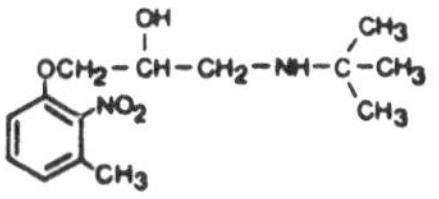
URINE	pH 2			
		0.40°	ZAMI 1305	
	pH 11	0.31	2-NITRO-3-METHYL-4-HYDROXY-PHENOXY-ter-BUTYL·AMINO·PROPAN-2ol	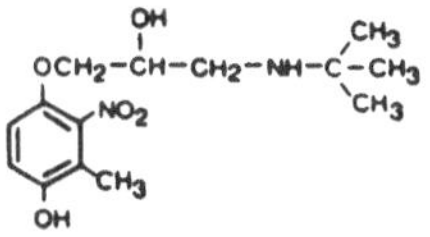
		0.20	2-NITRO-3-METHYL-PHENOXY-PROPAN-1,2-DIOL	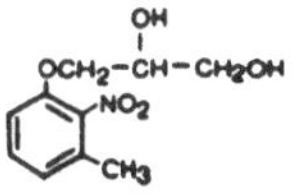

FIG. 4. Metabolites in male rat.

METABOLISM OF ZAMI 1305 IN FEMALE RAT

BIOLOGICAL FLUID	pH EXTRACTION	T.L.C. Rf	NAME	FORMULA
SERUM	pH 2			
	pH 11	0.40°	ZAMI 1305	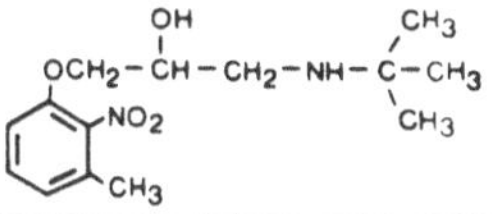
URINE	pH 2			
	pH 11	0.40°	ZAMI 1305	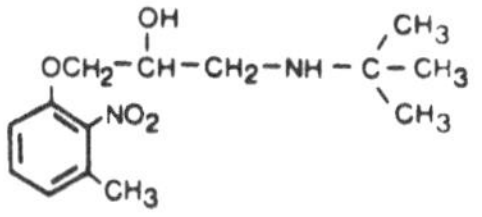

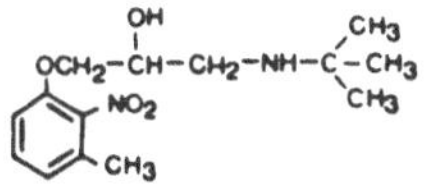

FIG. 5. Metabolites in female rat.

METABOLISM OF ZAMI 1305 IN MALE DOG

BIOLOGICAL FLUID	pH EXTRACTION	TLC Rf	NAME	FORMULA
	pH 2			
SERUM	pH 11	0 29°	2·AMINO·3·METHYL·PHENOXY· ter·BUTYL·AMINO·PROPAN·2ol	
		0 4	ZAMI 1305	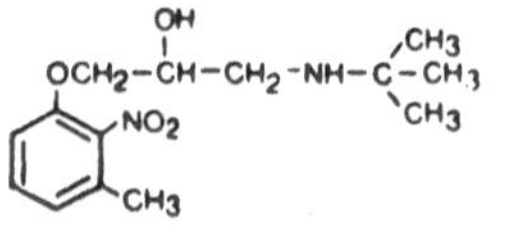
URINE	pH 2	0.53	2·NITRO·3·METHYL·6·HYDROXY· ter·BUTYL·AMINO·PROPAN·2 ol 6·SULPHURIC ESTER	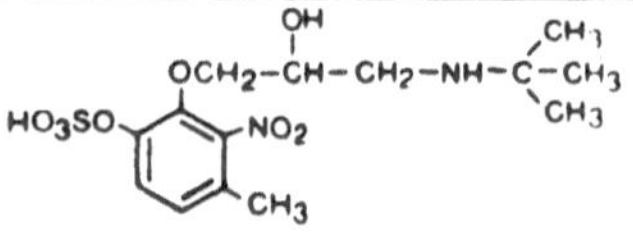
		0.80	2·NITRO·3·METHYL·PHENOXY ACETIC ACID	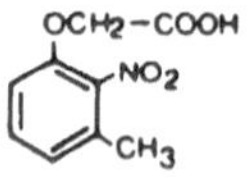
	pH 11	0.29°	2·AMINO·3·METHYL·PHENOXY· ter BUTYL·AMINO·PROPAN·2ol	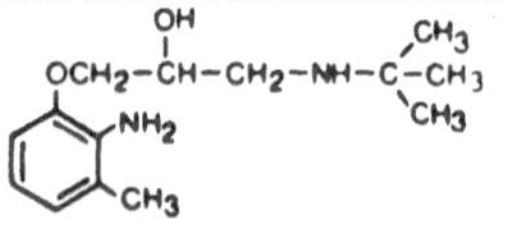
		0 4	ZAMI 1305	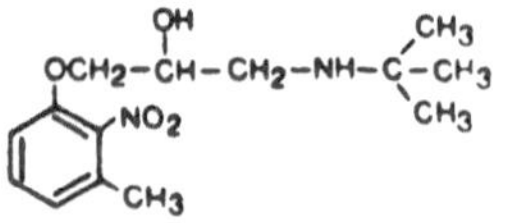
		048	2·NITRO·3·METHYL·PHENOXY· ETHYL·AMINO·PROPAN·2 ol	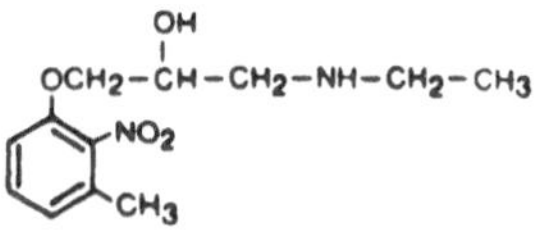

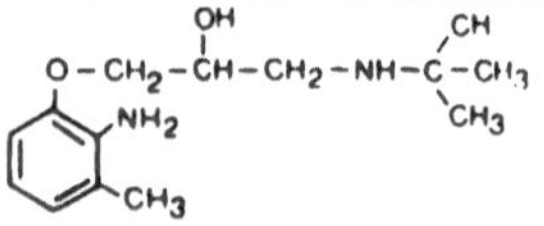

FIG. 6. Metabolites in male dog.

METABOLISM OF ZAMI 1305 IN FEMALE DOG

BIOLOGICAL FLUID	pH EXTRACTION	T.L.C. Rf	NAME	FORMULA
SERUM	pH 2			
	pH 11	0.4	ZAMI 1305	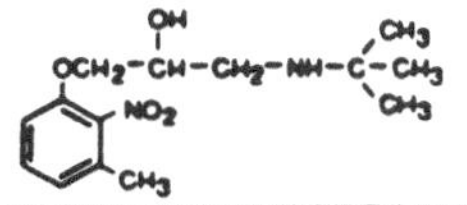
URINE	pH 2	0.37	2-NITRO-3-METHYL-4,6-DIHYDROXY PHENOXY-ter-BUTYL-AMINO-PROPAN-2-ol	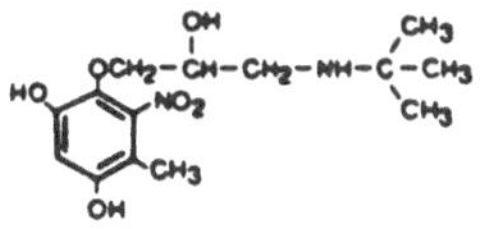
		0.8	2-NITRO-3-METHYL-PHENOXY ACETIC ACID	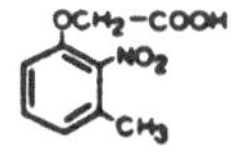
	pH 11	0.60	ZAMI 1305	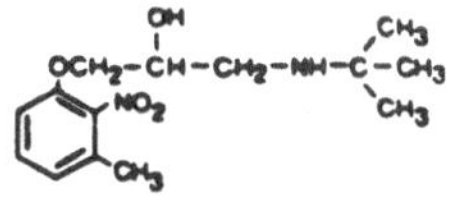
		0.48	2-NITRO-3-METHYL-PHENOXY-ETHYL-AMINO-PROPAN-2-ol	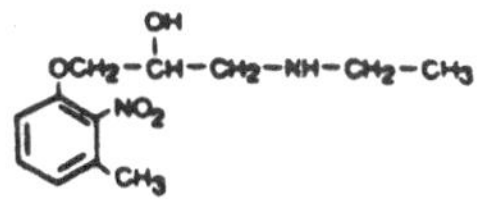

FIG. 7. Metabolites in female dog.

These were ZAMI 1305 and 2-amino-3-methyl-phenoxy-ethylamino-
-propan-2-ol in serum, while in urine, in addition to these two
metabolites we found 2-nitro-3-methyl-phenoxy-acetic acid and 2-
-nitro-3-methyl-4,6-dihydroxy-ter buthylamino-propan-2-ol.

From the data obtained, it is clear that ZAMI 1305 is extensively metabolized with ring and side chain reactions, aromatic
hydroxylations, oxidative deamination of the side chain, and even
reduction of the nitro group. The biotransformation of the drug
follows different patterns, which depend on the sex and the animal
species.

As stated above the identities of the metabolites were confirmed
by comparison of the Rf and the Mass Spectra with those of
separately synthesized standards.

The mass spectrum of ZAMI 1305 is shown in Fig. 8 , where the
peak at 282 a.m.u. is the molecular ion, and ion at 238 a.m.u.
results from the loss of 44 a.m.u., that is, of a vinyl radical,
from the molecular ion. It is interesting to note that a fragment
with 44 a.m.u. less than the molecular ion is a characteristic

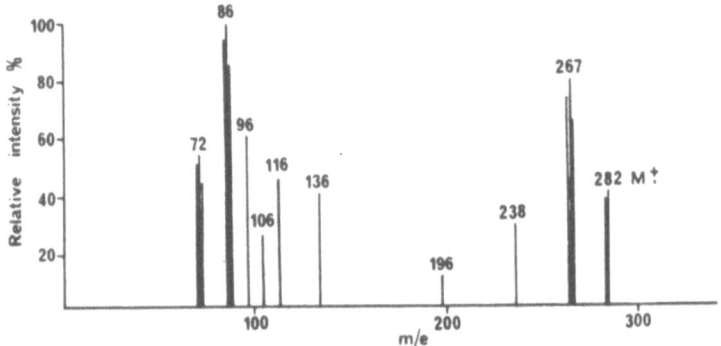

FIG. 8. Mass spectrum of the ZAMI 1305.

finding in the mass spectra of many β-blocker drugs, which have an
aryloxy-amino-propanol side chain. In fact, in this case, during
demolition of the molecule, a fragment of the central part of the
side chain is eliminated, while the two terminal parts bond again
(2). The other fragmentations are shown in the figure.

The first aromatic hydroxylation (or ring oxidation) product
is 2-nitro-3-methyl-4-hydroxy-phenoxy-ter butylamino-propan-2-ol,
(Fig. 9) for which besides the molecular ion at 298 a.m.u., there
are two characteristic peaks present at 280 a.m.u., due to the loss
of the hydroxyl group, and at 240 a.m.u. (loss of the terbutyl
group).

The second hydroxylated metabolite is the dihydroxy derivative
of the drug, 2-nitro-3-methyl-4,6-dihydroxy-phenoxy-terbutylamino-
-propan-2-ol.

In its mass spectrum (Fig. 10), we can note a very strong
molecular ion at 314 a.m.u. and three peaks, at 280 a.m.u. (loss
of the two hydroxyl radicals), at 240 a.m.u. and at 165 a.m.u., as
in the spectrum of the monohydroxylated metabolite.

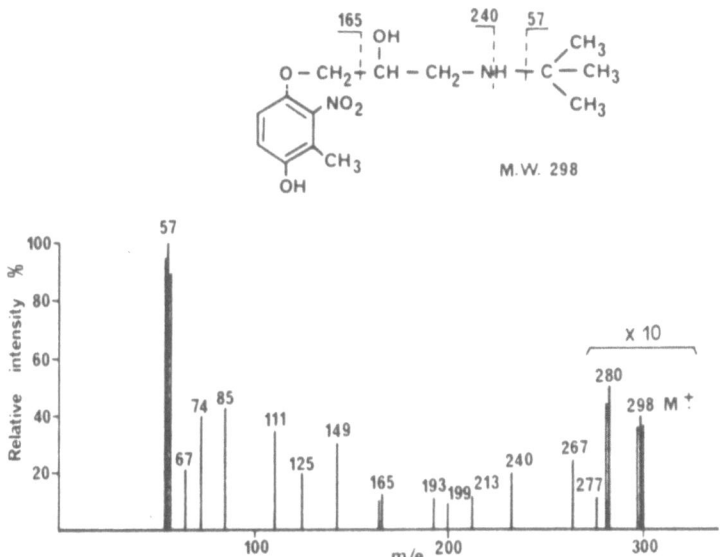

FIG. 9. Mass spectrum of 2-nitro-3-methyl-4-hydroxy-
-phenoxy-ter butyl amino-propan-2-ol.

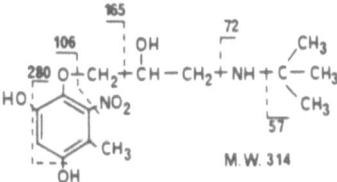

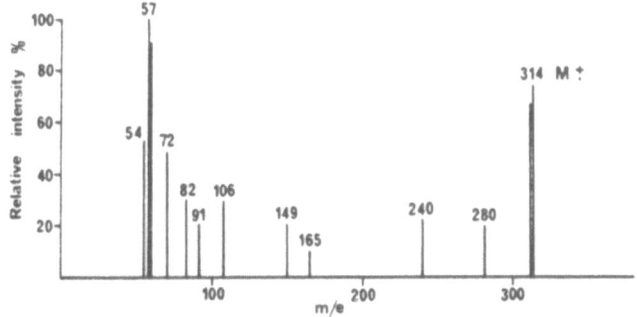

FIG. 10. Mass spectrum of 2-nitro-3-methyl-4,6
dihydroxy-phenoxy-terbutylamino-propan-
-2-ol.

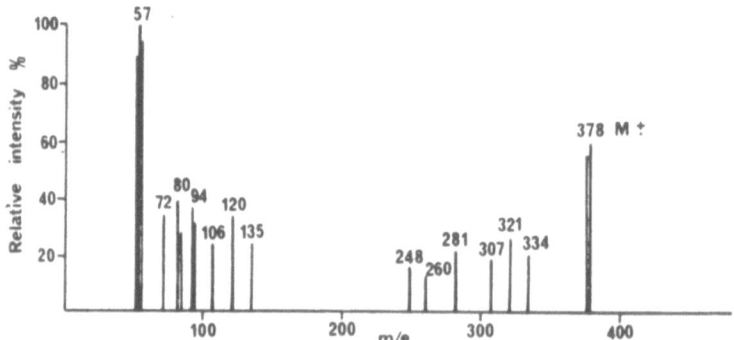

FIG. 11. Mass spectrum of 2-nitro-3-methyl-6-hydroxy-
-ter butyl-amino propan-2-ol-6-sulfuric ester.

Fig. 11 shows the formula and the mass spectrum of the
sulfuric derivative. The substitution is on position 6 of the aromatic ring. Besides the molecular ion at 378 a.m.u., there is a characteristic peak at 281 a.m.u., corresponding to the loss of the
sulfuric radical, and a peak at 334 a.m.u. arising from the loss
of 44 a.m.u., that is of the vinyl group, as we said before.

We must say here that the position of the phenolic and sulfuric
substitutions in the aromatic ring of these three metabolite were
not found by the study of mass spectra, but by comparing the Rf of
the metabolites extracted from biological fluids with the Rfs of
reference standards.

Two more metabolites arise from the oxidative deamination of
the side-chain.

The first is the glycol derivative whose mass spectrum is
presented in Fig. 12. The molecular ion is at 227 a.m.u.. The ion
at 153 a.m.u. is the base peak, which corresponds to a fragment
formed by a cleavage of the carbon-oxygen bond of the ether.

The second derivative of the oxidation of the side chain is
the 2-nitro-3-methyl-phenoxy-acetic acid (Fig. 13). Its mass spectrum shows a very strong molecular ion at 211 a.m.u.

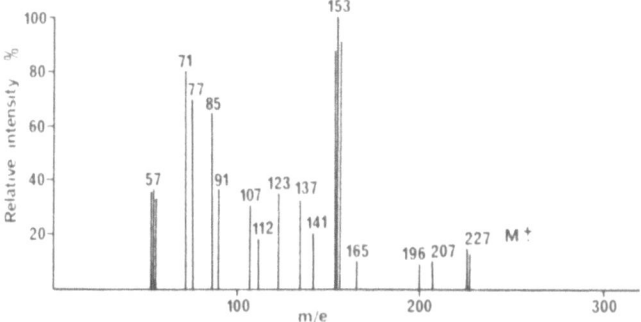

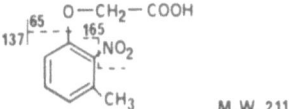

FIG. 12. Mass spectrum of 2-nitro-3-methyl-phenoxy-
-propan diol.

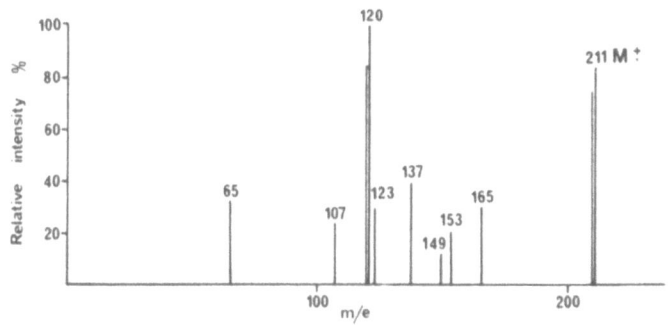

FIG. 13. Mass spectrum of 2-nitro-3-methyl-phenoxy
acetic acid.

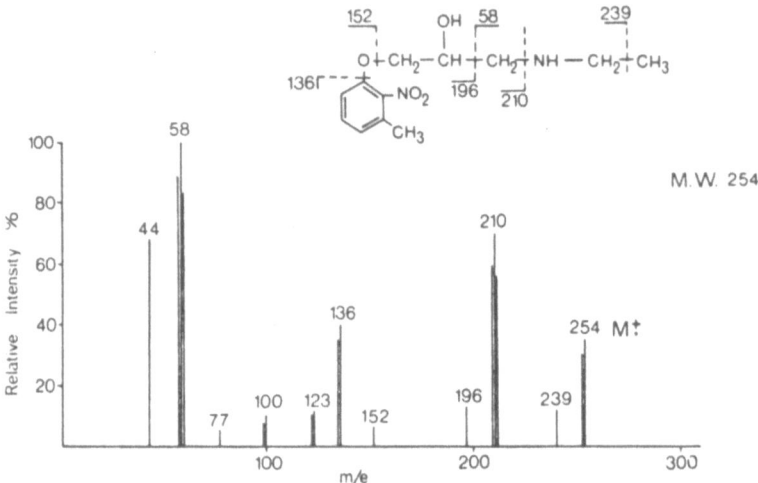

FIG. 14. Mass spectrum of 2-nitro-3-methyl-phenoxy
       ethyl-amino-propan-2-ol.

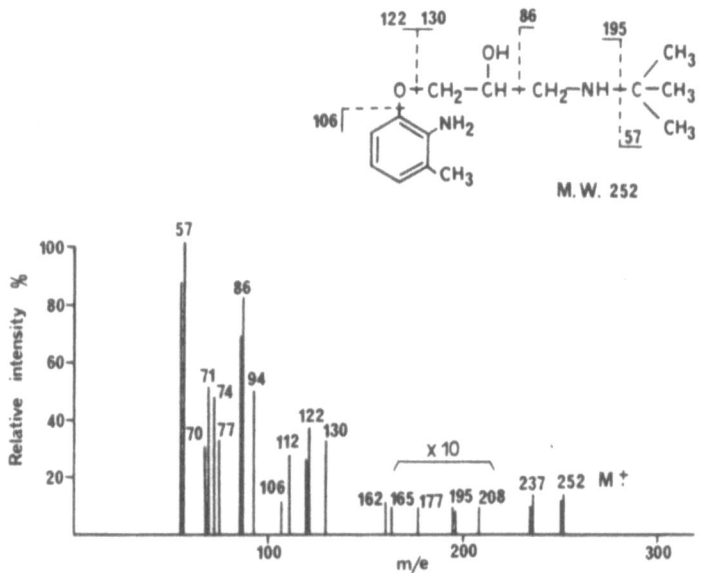

FIG. 15. Mass spectrum of 2-amino-3-methyl-phenoxy-
       -terbutyl amino propan-2-ol.

Another metabolite resulting from a partial dealkylation of
the side chain is 2-nitro-3-methyl-phenoxy ethylamino-propan-2-ol
(Fig. 14).

The last metabolite found was 2-amino-3-methyl-phenoxy-ter
butylamino-propan-2-ol, arising from reduction of the nitro group
in ZAMI 1305 to an amino group (Fig. 15). Its characteristic peaks
in the mass spectrum are the molecular ion at 252 a.m.u. and the peak
at 208 a.m.u., arising from a loss of 44 a.m.u..

PHARMACOKINETICS

As we wrote before the pharamcokinetics of ZAMI 1305 in the
rat and the dog, as determined from serum levels and the urinary
recovery, were studied separately.

This work was carried out with derivatized extracts, using a
gas-chromatographic method after the relative peak had been identified by gas-chromatography-mass spectrometry.

The drug was derivatized not only to improve its volatility,
but also to enhance the sensitivity of the method, because β-blocker
drugs are generally administered in small doses.

The minimum detectable amount of the trimethylsilyl-heptafluorobutyryl derivative is 0.1 ng., as can be seen from the
calibration curve for this gas-chromatographic method shown in
Fig. 16.

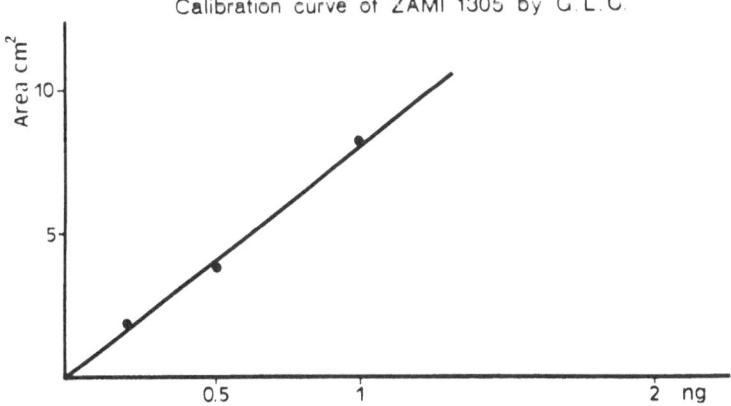

FIG. 16. Calibration curve of ZAMI 1305.

Moreover, the gas-chromatographic peak of this derivative does not interfere with the peaks of other metabolites or endogenous substances. Fig. 17 shows the formula and the mass spectrum for the trimethylsilyl-heptafluorobutyryl derivative of ZAMI 1305.

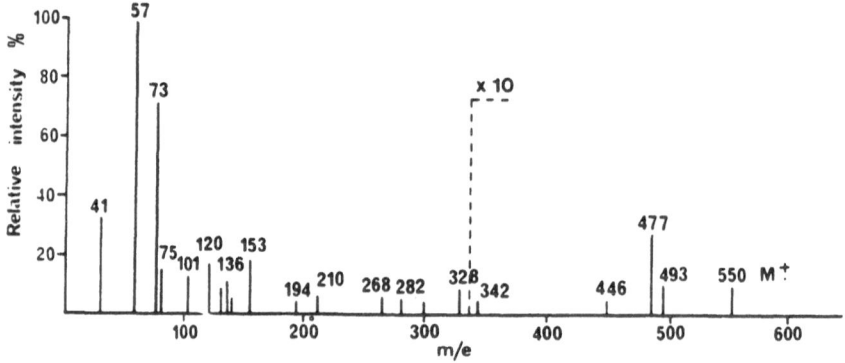

FIG. 17. Mass spectrum of trimethyl-silyl-hepta
fluoro butyryl derivative of ZAMI 1305.

The fragmentation pattern is very similar to that of the intact drug itself.

It is worth while noting that the ion at 282 a.m.u. is the molecular ion of ZAMI 1305, and that the ions at 268 a.m.u., 194 a.m.u. and 136 a.m.u. also appear in the spectrum of the intact drug.

The ion at 477 a.m.u. corresponds to the loss of the trimethylsilyl group from the molecule, and the ion at 493 a.m.u. to the loss of the ter butyl radical.

The quantitative determinations concern only the unchanged drug.

The results obtained by gas chromatographic analyses are summa-

TABLE 1.

BLOOD LEVELS (mcg/ml) of Z 1305 AFTER A SINGLE ORAL ADMINISTRATION

Each result (mean ± standard deviation) is the mean of six determinations

Specie	Dose per Kg. of body weight (mg)	Time after Z 1305 administration			
		30 min.	1 h	3 h	6 h
RAT	50	1.051±0.034	6.81±0.20	1.772±0.032	0.892±0.034
DOG	10	1.331±0.043	10.072±0.23	2.847±0.059	2.126±0.14

TABLE 2.

ESTIMATION OF Z 1305 IN URINE AFTER A SINGLE ORAL ADMINISTRATION (mcg. tot.)

Each result is the mean of six determinations

ANIMAL	Dose per Kg. of body weight	Z 1305		
	mg	0 - 6h	6 -12h	12 - 24 h
RAT	50	21.05	81.97	28.94
DOG	10	39.28	121.82	51.37

TABLE 3.

ESTIMATION OF FREE AND CONJUGATED FORMS OF Z 1305 IN ANIMAL URINE AND SERUM AFTER INGESTION OF DRUG					
ANIMAL	Initial dose per Kg. of body weight (mg)	URINE		SERUM	
		FREE	CONJUGATED	FREE	CONJUGATED
RAT	50	27%	73 %	32%	68%
DOG	10	24%	76 %	29 %	71%

rized in Table 1. From this table, it can clearly be seen that
ZAMI 1305 is rapidly absorbed in° the rat and the dog. The maximum
blood level is reached during the first hour after oral administration. Only a small amount of the administered drug is recovered
in urine in the form of the unmetabolized compound (Table 2).

Table 3 shows the percentage of the drug present as conjugated
forms, which are the glucoronic derivatives.

In conclusion, ZAMI 1305 is well-absorbed after oral administration. In the body the drug is extensively metabolized. In fact,
seven metabolites are found "in vivo", in the rat and the dog.

Finally, we have confirmed the difference between the biological fates of ZAMI 1305 in male and female animals and therefore
we can use this finding to help explain the divergent toxicological
effects in the two sexes.

REFERENCES

1) Italian Patent No. 41001 12/1/1977.
2) R. Longoni, P. Berntsson, N. Bild and M. Hesse, Helv. Chim.
   Acta, 1977, 60, 103.

ISOLATION AND IDENTIFICATION OF THE CHLOROFORM SOLUBLE URINARY

METABOLITES OF SULOCTIDIL IN MAN

R. Roncucci, W. Cautreels, M. Martens, C. Gillet,

K. Debast and G. Lambelin

Continental Pharma, Machelen, Belgium

INTRODUCTION

Suloctidil is a new vasoactive drug endowed with interesting biochemical, pharmacological and therapeutic properties. Its metabolic fate has been investigated in various animal species and in man (1). Fig. 1 lists the human urinary metabolites identified in previous studies.

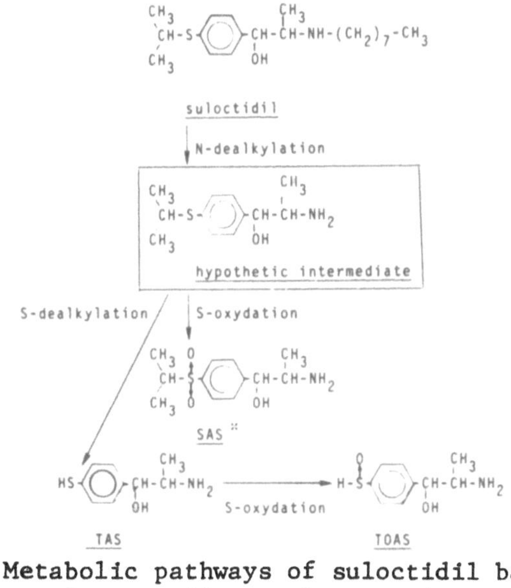

FIG. 1. Metabolic pathways of suloctidil based on previous studies on human urine.
*Abbreviation (see table IV for the entire compound name).

In those studies the metabolites present in urine have been divided
in chloroform soluble and water soluble fractions. The aim of this
paper is to complete the identification of the chloroform-soluble
metabolites in the 0-24 hours urine of man.

EXPERIMENTAL

Biological sampling. A healthy and fully informed volunteer
(male, 24 years old), with normal renal and hepatic functions
participated in this study. In order to facilitate· the localisation
of the metabolites throughout the isolation procedure a mixture
of ^{14}C-labelled, deuterium labelled and unlabelled suloctidil was
used (Fig. 2).

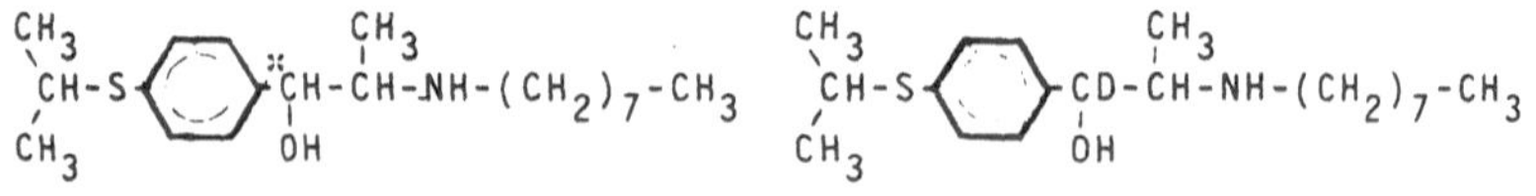

FIG. 2. Structures of ^{14}C-deuterium-labelled
suloctidil used in this study.

300 mg of the mixture was orally administered to the volunteer after
breakfast. Urine samples were collected during 24 hours after the
intake of the drug. The urine samples collected were mixed
thoroughly and stored at -20°C pending analysis.
    Isolation and separation of the chloroform soluble metabolites.
The urine collected was brought to pH 12 by means of a 5N sodium-
hydroxide solution and extracted with chloroform (twice, 3 : 1 v/v).
The combined chloroform layers were dried over anhydrous sodium
sulphate and evaporated to complete dryness under reduced pressure
at maximum 35°C. The residue obtained was redissolved in 5 ml of a
solvent mixture of ethyl acetate, absolute ethanol and ammonia
(25%) (3.5, 1.5, 0.1; v/v).

After complete dissolution 10 µl were taken for the determination
of the radioactivity by liquid scintillation counting. At the same
time the radioactivity was measured on the remaining aqueous layer.
The radioactivity extracted with chloroform represented 37% of the
radioactivity initially present in the urine.

The first separation was performed using a silicagel column

(Merck, cat.n° 10401, model B). The concentrated chloroformic extract
was injected onto the column and eluted with the solvent mixture
described above. The eluate was collected in 5 ml fractions. About
1000 ml was allowed to percolate through the column with a mean
flow rate of 0.3 ml per minute. From each fraction, 100 µl were
immediately sampled to determine the radioactivity present. The
chromatography was stopped when the total amount of radioactivity
eluted reached more than 95% of the radioactivity initially present
on top of the column. Radioactivity was plotted versus elution
volumes in order to determine the volume of each separated radioactive peak. Accordingly pools were prepared and their relative
radioactivity measured. Pools were then evaporated to dryness under
reduced pressure at 35°C. The residues obtained were redissolved
twice in 5 ml ethanol and quantitatively transferred into a centrifuge
tube. The elution pattern of the radioactivity peaks on the silica
column is represented in Fig. 3.

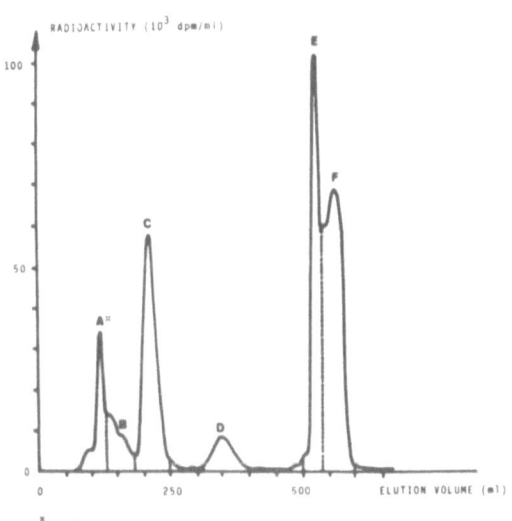

FIG. 3. Elution pattern of the chloroformic
extract on the silica gel column.

The percentages of the radioactivity of each pool relative to the
total radioactivity recovered are presented in Table 1.

    The remainder of the ethanolic solutions of the pools was
concentrated under reduced pressure and applied on a PLC-plate
(20 x 20 cm, Merck cat n° 5717) or a TLC-plate (5 x 20 cm, Merck
cat n° 5723). In the case of TLC, as many plates were used as
necessary to absorb the entire extract of the pool. Each pool was
chromatographed with the solvent selected during a preliminary
search on a small fraction of the pool. After chromatography the

TABLE 1
Percent in radioactivity of each pool eluted from
the column.

Pool	Elution fraction (ml)	%
A	80 → 130	9.5
B	135 → 180	5.7
C	185 → 250	23.2
D	315 → 400	5.7
E	510 → 535	21.6
F	540 → 590	34.2

separated radioactive spots were located on the TLC plates by means
of radioscanning (Packard TLC radioactivity scanner, model 2700).
In the case of PLC, silicagel fractions were scraped off every·half
cm and combusted in a sample oxidizer (Packard, model 306). The
silica of each radioactive zone was scraped off and transferred
into a large centrifuge tube. According to the solubility of each
metabolite present, the radioactive fraction was extracted with
water, hydrochloric acid (0.1 N) or methanol and back-extracted
with chloroform. The extraction recoveries throughout the chromatographic procedure were monitored by radioactive measurements. The
final extracts were concentrated and stored under nitrogen at -20°C
pending identification by G.C.-M.S.. The chromatographic separations
performed are summarized in Table 2.

Identification techniques. G.C.-M.S. analysis. - Derivatization
procedures: a) Persilylation: The residue of the isolated radioactive
spot is dissolved in 100 μl 1,4-dioxane and 100 μl BSTFA. One drop
of TMCS is added in order to catalyze the silylation. The mixture
is allowed to react overnight at room temperature prior to GC-MS
analysis. b) Persilylation with deuterated agents: The residue of
the isolated radioactive spot is dissolved in 100 μl pyridine and
100 μl of deuterated BSA (Supelco cat n° MD-1060). The mixture
is allowed to react overnight at room temperature prior to G.C.-M.S.-
-analysis. - GC-MS conditions: GC-MS analysis was performed using a
Varian 2700 gas chromatograph coupled to a Varian CH7 single-
-focusing mass spectrometer by means of a double stage Watson-
-Biemann separator. Data acquisition and data elaboration was
performed on-line using the MAT SS 101 data system. Typical GC-MS
conditions are given in Table 3. Complete mass spectra from m/e
20 to m/e 1000 were scanned manually at GC-peak maxima and stored
on memory disks for further processing.

MS-analysis by direct introduction. The residue of the isolated
radioactive spot was dissolved in 100 μl chloroform. From this
10 μl were brought into a quartz crucible and evaporated to dryness

TABLE 2
Separation by TLC and PLC of pools A, B, C, D,
E and F.

Pool	Fraction nr	Elution mixture used	Chroma- togr. system	Rf x 100	Extraction -solvent of the silica gel	% radio- activity relative to the radio- activity in urine prior to extraction
A	A.1	Ethyl Ac : 70	TLC	0	methanol	0.2
	A.2	n-butanol : 30		51		0.3
	A.3	Ac.acid : 2		65		2.1
	A.4			84		0.6
B	B.1	Ethyl.Ac : 70	TLC	2	methanol	0.1
	B.2	n-butanol : 30		30		0.1
	B.3	Ac-acid : 2		40		0.5
	B.4			46		0.3
	B.5			53		0.7
	B.6			60		0.1
C	C.1	Ethyl.Ac : 70	PLC	33	H_2O	8.2
	C.2	n.butanol : 30		56		0.1
		Ac.acid : 2				
D	D.1	Ethyl.Ac : 70	TLC	27	methanol	1.8
		n-butanol : 30				
	D.2	Ac.acid : 2		50		0.5
E	E.1	Ethanol :100	PLC	33	0.1 N HCl	5.7
	E.2	NH_3 (25%) : 2		48		0.2
	E.3			71		<0.1
F	F.1	Ethanol :100	PLC	19	0.1 N HCl	3.0
	F.2	NH_3 (25%) : 2		32		8.3
	F.3			43		0.2

TABLE 3
Typical GC-MS conditions.

G.C.-column : 1.5 meter, 3% SE-30 on supelcoport
              80-100 mesh
oven-temperature : 100°C up to 270°C at 10°C/min
He-flow : ± 30 ml/min
injector-temperature : 300°C
Vacuum : < 10^{-6} Torr
electron beam current : 300 μA
ionisation potential : 70 eV
electron multiplier high voltage : 2 kV

at 200°C. The crucible was mounted on the direct inlet probe of the
Varian CH7 mass spectrometer and introduced into the ion source.
Heating the probe up to 500°C within 500 seconds, mass spectra were
scanned on-line from m/e 20 to m/e 1000 using the MAT SS 101 data
system. Alternatively, mass spectra were recorded using an oscillo-
graphic UV-recorder in order to detect metastable ion peaks.

High resolution M.S.-analysis. High resolution mass spectrometric analysis was performed on a Jeol 01SG2-special double focusing mass spectrometer. The photoplates obtained were scanned using
a Jeol-densitometer, coupled on-line with a Jeol G.C.-6 data system.
Samples were dissolved in 50 µl 1,4-dioxane. From these, 10 µl
was brought into a quartz crucible. The crucible was mounted on the
direct inlet probe of the mass spectrometer and introduced into
the ion source. Mass spectral data were recorded on photoplates.

RESULTS AND DISCUSSION

The identification of the suloctidil metabolites present in the
different fractions started with the G.C.-M.S. analysis of a
persilylated aliquot of the fractions. Confirmation of the results
obtained was performed by G.C.-M.S. after persilylation with deuterated silylating agents and by direct introduction mass spectrometric analysis at both low and high resolution power. Several of
the metabolites identified were synthesized and similarly analyzed
in order to confirm unambiguously the structures proposed.

Table 4 lists the suloctidil metabolites identified in the
different radioactive fractions investigated during this study.
Several metabolites have been noticed to be present in more than
one fraction. This can be explained by incomplete separation during
column and thin layer chromatography. The mass spectrometric
identification of the aminopropanol compounds and of the dihydroxy-
propane compounds has been described in detail previously (1, 2).

Two new series of suloctidil metabolites have been identified
during this study: compounds containing an isopropylsulfoxide moiety
formed by S-oxidation of the isopropylthiol-moiety of suloctidil
and compounds containing a pyrrolidone moiety formed by oxidation
of the alkyl side chain.

Since the sulfoxide-moiety does not resist the overall
temperatures during G.C.-M.S. analysis, those compounds were
identified by direct introduction M.S. with low and high resolving
power. This will be discussed in detail elsewhere.

Fig. 4 shows the mass spectrum of one of the pyrrolidone-
-metabolites (SPS) identified after persilylation and G.C.-M.S.-
-analysis of fraction B4 (Table 4). The fragmentation pattern
proposed is shown in Fig. 5. The presence of this metabolite was

TABLE 4
Identified metabolites of suloctidil in the different radioactive fractions separated.

Fraction nr.	Structure of identified metabolites	Name	Abbreviation	%
A.1	–	–	–	0.2
A.2	–	–	–	0.3
A.3	(SO₂-Ø ...OH, OH)	1-(4-isopropylsulfonylphenyl)-1,2-dihydroxypropane	SSG	⎫
	(CH₃-S-Ø ... pyrrolidone OH)	1-(4-methylthiophenyl)-2-pyrrolidone propanol	MPS	⎪
	(S-Ø ... pyrrolidone OH)	1-(4-isopropylthiophenyl)-2-pyrrolidone propanol	PS	⎬ 2.1
	(SO₂-Ø ... pyrrolidone OH)	1-(4-isopropyl sulfonyl phenyl)-2-(x-hydroxypyrrolidone propanol	HSPS	⎭
A/4	(H-S-Ø ...OH, OH)	1-(4-thiophenyl) 1,2 dihydroxy propane	TGS	0.6

contd.

TABLE 4 (contd.)

Fraction nr.	Structure of identified metabolites	Name	Abbreviation	%
	(isopropyl)–S–∅–C(CH₃)(OH)–OH	1-(4-isopropylthiophenyl)1,2-dihydroxypropane	SG	
	(isopropyl)–SO₂–∅–C(OH)–OH	1-(4-isopropylsulfonyl phenyl)1,2-dihydroxypropane	SSG	0.1
B.1	–	–	–	
B.2	(isopropyl)–S–∅–C(=O)–OH	4-isopropylthio-benzoic acid	CAS	
B.3	(isopropyl)–S–∅–C(=O)–OH	4-isopropylthio-benzoic acid	CAS	0.5
	CH₃–SO₂–∅–(pyrrolidone)–OH	1-(4-methylsulfonylphenyl)2 pyrrolidone propanol	MSPS	
B.4	(isopropyl)–SO₂–∅–(pyrrolidone)–OH	1-(4-isopropylsulfonylphenyl)2-pyrrolidone propanol	SPS	0.3

contd.

TABLE 4 (contd.)

Fraction nr.	Structure of identified metabolites	Name	Abbreviation	%
E.5		1-phenyl-1,2-dihydroxy-propane	BSG	0.7
		1-(4-isopropylsulfoxy-phenyl)-1,2-dihydroxy-propane	SOSG	
		4-isopropylsulfoxybenzoic acid	SOCAS	
		1-(4-isopropylsulfonylphenyl)-2-pyrrolidone propanol	SPS	
B.6		1-(4-methylthiophenyl)-2-pyrrolidone propanol	MPS	0.1
		1-(4-isopropylsulfonylphenyl)-2-(x-hydroxypyrrolidone)-propanol	HSPS	

contd.

TABLE 4 (contd.)

Fraction nr.	Structure of identified metabolites	Name	Abbreviation	%
C.1		1-phenyl-2-pyrrolidone propanol	BPS	8.2
		1-(4-isopropylsulfoxyphenyl)-2-pyrrolidone-propanol	SOPS	
C.2	?	–	–	0.1
D.1	?		–	1.3
	?			
D.2	–	–	–	0.1
E.1		1-(4-isopropylsulfonylphenyl)-2-aminopropanol	SAS	5.7
E.2	–	–	–	0.2
E.3	–	–	–	<0.1
F.1		1-(4-isopropylsulfenylphenyl)-2-aminopropanol	SAS	3.0

contd.

TABLE 4 (contd.)

Fraction nr.	Structure of identified metabolites	Name	Abbreviation	% "
F.2		1-(4-isopropylsulfoxyphenyl)-2-aminopropanol	SOAS	8.3
		1-(4-thiophenyl)-2-aminopropanol	TAS	
		1-(4-isopropylsulfoxyphenyl)-2-aminopropanol	SOAS	
		1-(4-isopropylsulfonylphenyl)-2-aminopropanol	SAS	
F.3		1-(4-isopropyl-sulfonylphenyl)-2-aminopropanol	SAS	0.2

" % obtained comparing the radioactivity of the fraction to that present in the urine.

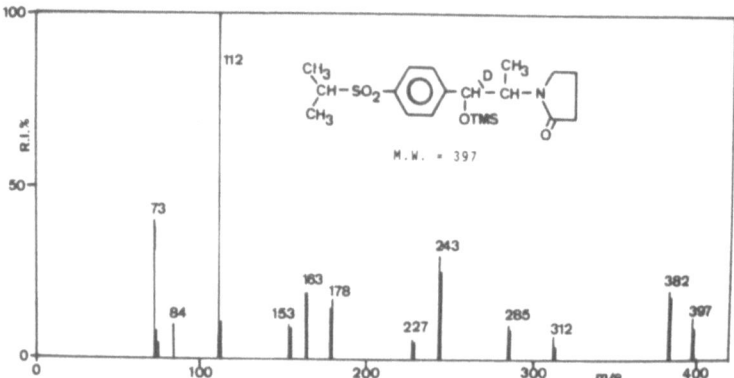

FIG. 4. Mass spectrum of silylated SPS present in
fraction B4.

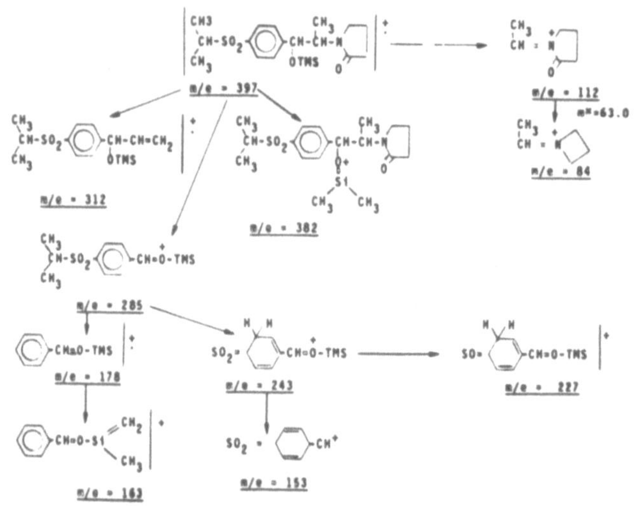

FIG. 5. Proposed fragmentation of silylated "SPS".

confirmed by analysis of the corresponding synthetic compound (3).

Since the fraction containing this metabolite (fraction B4)
was relatively pure a high resolution mass spectrometric analysis
by direct introduction was also performed. The structure of two
specific ions of the pyrrolidone-moiety (m/e 112 and 84) was
confirmed. Results are given in Table 5.

The proposed fragmentation of the pyrrolidone-moiety has been
further confirmed by the presence of a metastable ion peak at m/e
63.0 (=84^2/112), observed during a low resolution M.S.-analysis.
This fragmentation is presented in detail in Fig. 6.

TABLE 5
Results obtained after high resolution mass spectrometric
analysis of fraction B.4.

Experimental mass-value	Calculated mass-value	Error	Composition
84.0819	84.0813	+ 0.0006	$C_5^{12}H_{10}N$
	.0768	+ 0.0051	$C_4^{12}C^{13}H_9N$
112.0769	112.0762	+ 0.0005	$C_6^{12}H_{10}NO$
	.0717	+ 0.0051	$C_5^{12}C^{13}H_9NO$
	.0708	+ 0.0060	$C_6^{12}H_{12}$

$C_6H_{10}NO$ CH=N with CH$_3$ (pyrrolidone ring)   m/e 112

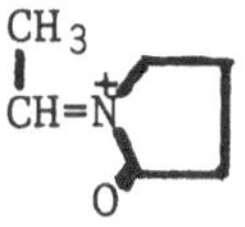

$C_5H_{10}N$ CH=N with CH$_3$ (pyrrolidine ring)   m/e 84

m/e = 112   m/e = 84

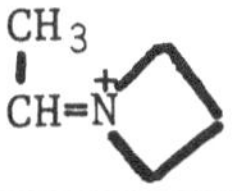

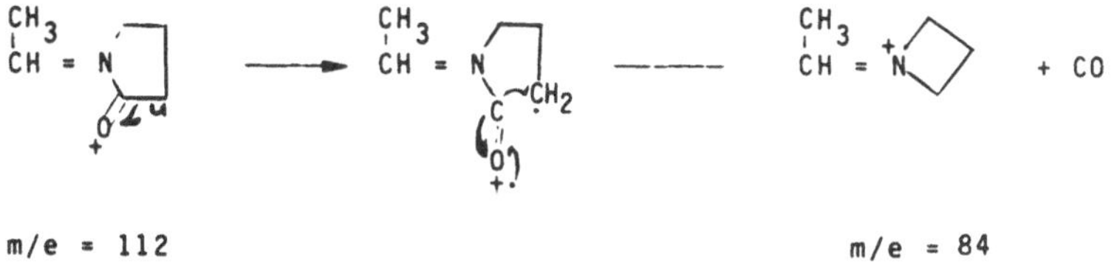

FIG. 6. Fragmentation of the pyrrolidone moiety
of the suloctidil metabolites.

All pyrrolidone-metabolites identified are characterized by
these fragments (m/e 84, 112) and by the corresponding (M-85) and
(M-112) fragments.

In every case the twin-peak effect caused by administration of
the 50/50 mixture of unlabelled and monodeuterated suloctidil is
clearly visible.

Fig. 7 shows the mass spectrum of HSPS, another pyrrolidone
metabolite identified in fraction A3 (Table 4). The mass spectra

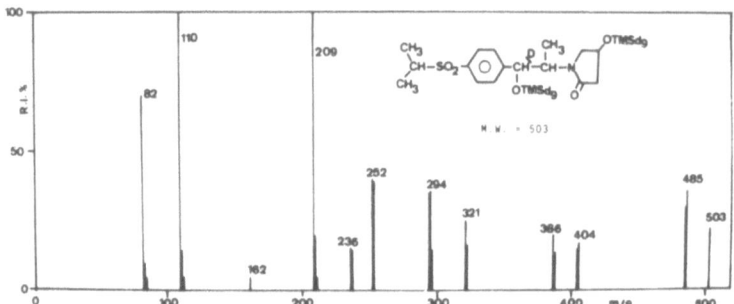

FIG. 7(a). Mass spectrum of HSPS after persilylation
        with BSA/d9.

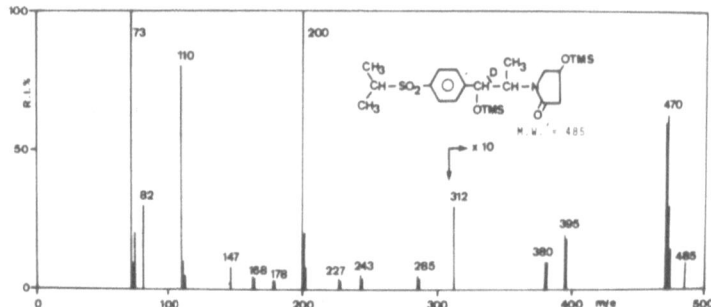

FIG. 7(b). Mass spectrum of persilylated "HSPS".

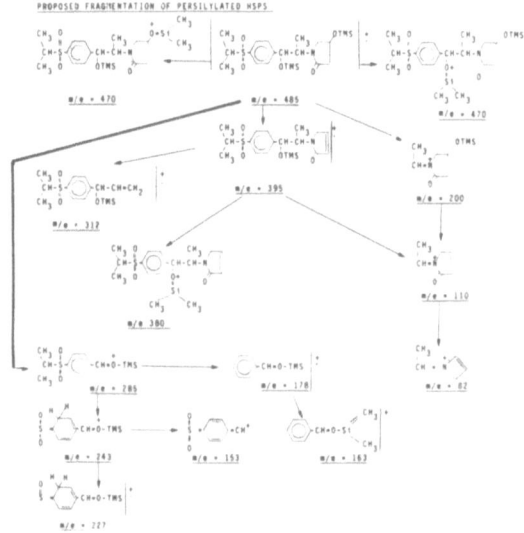

FIG. 8. Proposed fragmentation of persilylated HSPS.

figured were obtained after persilylation and d$_9$-persilylation. The fragmentation pattern proposed is shown in Fig. 8.

Since not all radioactive fractions have been identified yet and since some metabolites (thiol-and isopropylsulfoxide metabolites) cannot be volatilized quantitatively by G.C.-M.S., it is impossible at the present stage of our research to calculate the percent of each metabolite relatively to the total radioactivity initially present in the urine. Because of its very low abundance in 0-24 h urine (< 0.01% of the total radioactivity present) unchanged suloctidil was not detected during this study. However, unchanged suloctidil has been detected in human urine by mass fragmentographic analysis (4).

Structures presented in Table 4 enable us to propose a chart for suloctidil metabolic pathways taking into account only the structures isolated from the chloroformic extract of the urine.

The identification of the pyrrolidone-metabolites may be due to an artifact (5). If this should be the case, these cyclic metabolites have to be replaced in Fig. 9 by the corresponding open structures.

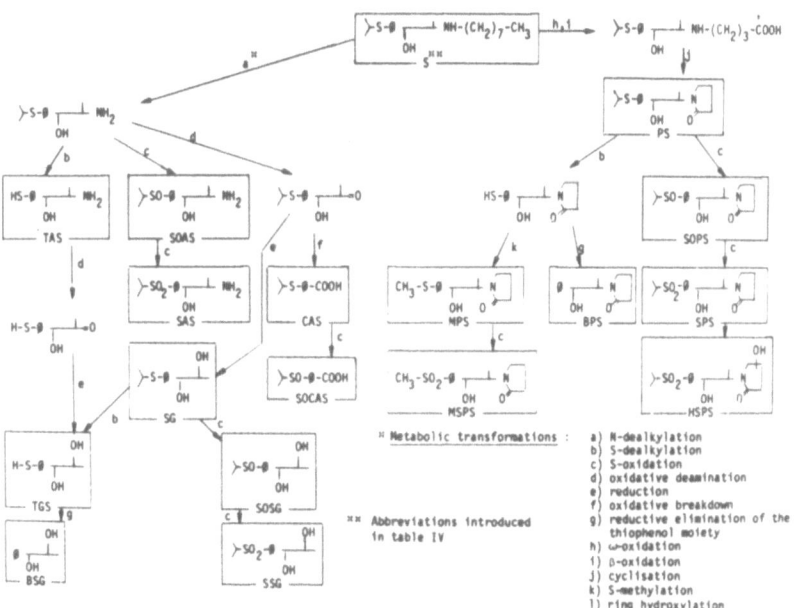

FIG. 9. Proposed metabolic pathway of suloctidil.

Until now the presence of these open structures was not
confirmed in the chloroform soluble fraction of the urinary
metabolites. However, recent results obtained for the analysis of
the water soluble fraction of the urinary suloctidil metabolites
indicated the presence of some of them.

CONCLUSIONS

Seventeen different suloctidil metabolites have been identified
in the chloroform soluble fraction of the urine of a human
volunteer. Using the data obtained a metabolic pathway of suloctidil
in humans has been proposed.

Two major pathways for the metabolic degradation of suloctidil
may be defined: one starts with the N-dealkylation of the octyl
side-chain, the other originates from ω-oxidation and subsequent
β-oxidations of the alkyl side chain. The latter results in the
formation of a series of pyrrolidone-compounds. Further research
will be necessary to confirm the existence of the corresponding
open structures (aminocarboxylic acid-like) for the pyrrolidone
compounds identified.

ACKNOWLEDGEMENTS

Dr. E. Esmans and Ing. J. Verreydt (laboratories of organic
chemistry, University of Antwerp (RUCA), dir. prof. Dr. F.
Alderweireldt) are thanked for their assistance in running the
high resolution MS analyses.

This work was supported by a grant of IRSIA (Institut pour la
Recherche Scientifique dans l'Industrie et l'Agriculture, Convention
n° 2813).

REFERENCES

1) R. Roncucci, M.-J. Simon, K. Debast, R. Vandriessche, C. Gillet
   and G. Lambelin, in "Advances in Mass Spectrometry in Biochemistry
   and Medicine", Vol. 2, A. Frigerio, Ed., Spectrum, New York,
   1977, p. 37.
2) R. Roncucci, M.-J. Simon, M. Martens, K. Debast and G. Lambelin,
   "Abstract of the Symposium Metabolisme van geneesmiddelen",
   Wilrijk, Belgium, March, 1977.
3) G. Gillet, to be published.
4) M. Martens, R. Roncucci, K. Debast and G. Lambelin, in "Recent
   Developments in Mass Spectrometry in Biochemistry and Medicine",
   Vol. 1, A. Frigerio, Ed., Plenum Press, New York, 1978, p. 133.
5) J. De Graeve, P. Kremers, J.E. Gielen, Thérapie, 1977, 32, 195.

ARE STABLE SULFENIC ACIDS POSSIBLE METABOLITES OF SULOCTIDIL?

W. Cautreels, M. Martens, R. Roncucci, C. Gillet, K.

Debast and G. Lambelin; Continental Pharma, Research

Laboratories, Machelen, Belgium

INTRODUCTION

In earlier work on the identification of suloctidil metabolites in urine several sulfenic acids have been hypothesized (1) (Table 1).

Sulfenic acids are known as very unstable compounds. Only three stable sulfenic acids have been described in the literature (2-4). In all of them the sulfenic acid moiety is stabilized by formation of intramolecular hydrogen-bridges. The structure of suloctidil and its sulfenic acid metabolites does not permit this kind of stabilisation. For this reason their detection has been considered as a possible artifact.

Shelton et al. (5) described the thermolysis of t-butylsulfoxide at temperatures of 80°C. Using I.R.-absorption as a monitor the formation of t-butylsulfenic acid was confirmed (Fig. 1).

Since overall temperatures during GC-MS analysis largely exceed 80°C (injector temperature 300°C), thermolysis of sulfoxides during the injection of the sample may not be excluded.

Since all identification work on the metabolism of suloctidil was performed by means of GC-MS, it is plausible that the sulfenic acids detected in the mass spectrometer may be pyrolysis products of the corresponding sulfoxy compounds. In this paper the identification of the sulfoxide metabolites of suloctidil will be described and their detection as sulfenic acids as an artefact will be discussed.

TABLE 1
Sulfenic acid metabolites proposed.

$$\left. \begin{array}{l} \underset{CH_3}{\overset{CH_3}{\diagdown}}CH-S-\!\!\left\langle\bigcirc\right\rangle\!\!-\underset{OH}{\overset{CH_3}{\underset{|}{CH}}}-CH-NH-(CH_2)_7-CH_3 \end{array} \right\}$$ unchanged drug

"suloctidil"

↓ metabolisation

$$\underset{H-S}{\overset{O}{\uparrow}}-\!\!\left\langle\bigcirc\right\rangle\!\!-\underset{OH}{\overset{CH_3}{\underset{|}{CH}}}-CH-OH$$

$$\underset{H-S}{\overset{O}{\uparrow}}-\!\!\left\langle\bigcirc\right\rangle\!\!-C\overset{\diagup\,O}{\underset{\diagdown\,OH}{\Vert}}$$

$$\underset{H-S}{\overset{O}{\uparrow}}-\!\!\left\langle\bigcirc\right\rangle\!\!-\underset{OH}{\overset{CH_3}{\underset{|}{CH}}}-CH-N$$

$$\underset{H-S}{\overset{O}{\uparrow}}-\!\!\left\langle\bigcirc\right\rangle\!\!-\underset{OH}{\overset{CH_3}{\underset{|}{CH}}}-CH-NH_2$$

sulfenic acid metabolites
proposed in earlier work.

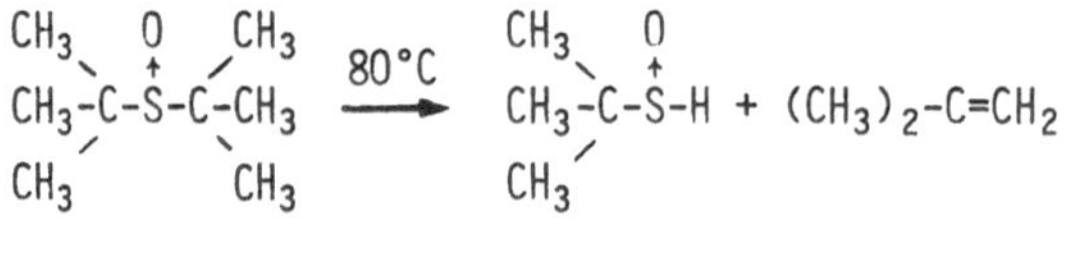

$$\underset{CH_3}{\overset{CH_3}{\diagdown}}CH_3-\underset{CH_3}{\overset{O}{\underset{\diagup}{C-S-C}}}-CH_3 \xrightarrow{80°C} \underset{CH_3}{\overset{CH_3}{\diagdown}}CH_3-\overset{O}{\underset{\diagup}{C-S-H}} + (CH_3)_2-C=CH_2$$

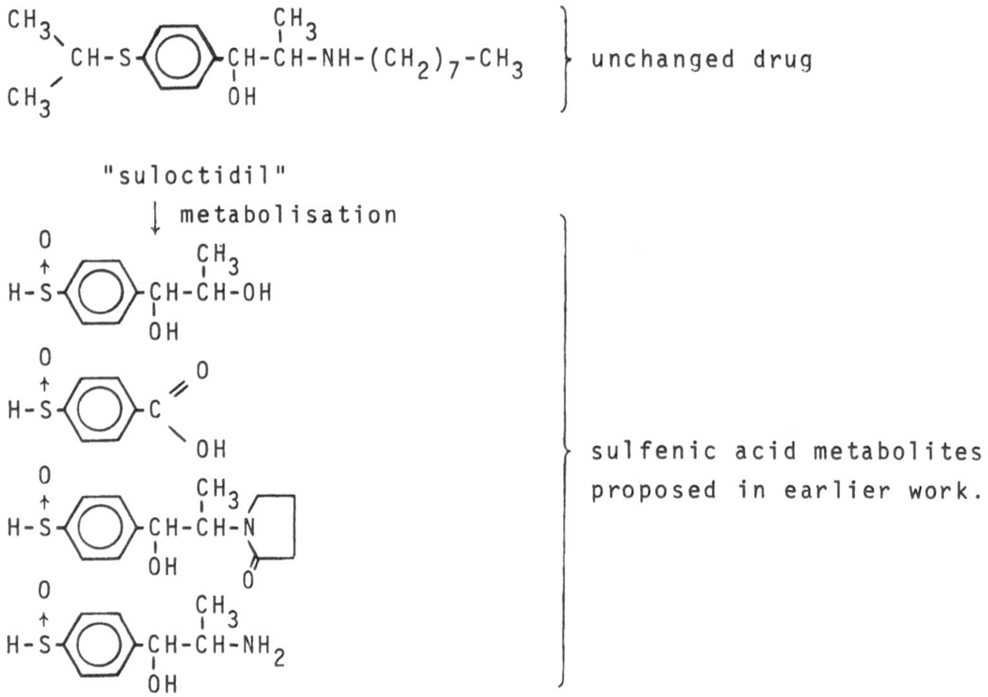

FIG. 1. Thermolysis of t-butylsulfoxide.

EXPERIMENTAL

Biological sampling. Urine sampling. A healthy and fully informed volunteer, with normal renal and hepatic functions participated in this study. In order to facilitate the localisation of the metabolites throughout the isolation procedure a mixture of ^{14}C-labelled, monodeuterated and unlabelled suloctidil was administered orally to the volunteer at a dose of 300 mg (Fig. 2).

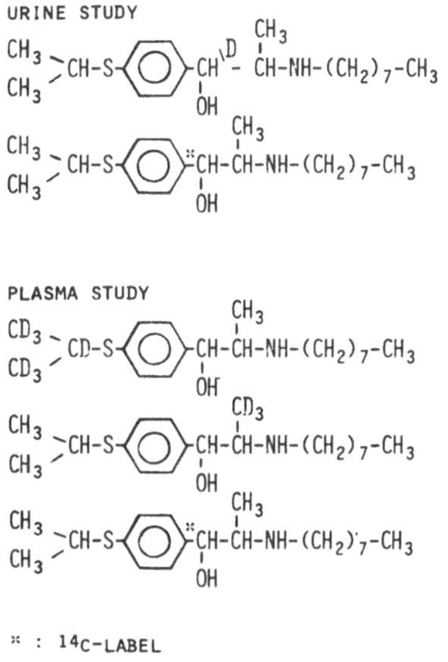

FIG. 2. Labelling of suloctidil.

Urine samples were collected during 24 hours after the intake of the drug. Blood sampling. A healthy and fully informed volunteer with normal hepatic and renal functions participated in this study. In order to facilitate the localisation of the metabolites throughout the isolation procedure a mixture of ^{14}C-labelled, trideuterated, heptadeuterated and unlabelled soluctidil was administered orally to the volunteer at a dose of 300 mg (Fig. 3). 350 ml of blood was collected 4 hours after the intake of the drug.

Isolation and separation of the metabolites. Metabolites were isolated and separated successively by solvent extraction, column chromatography and thin layer chromatography. The different metabolites are monitored during the separation procedure measuring the radioactivity present in the fractions obtained: for the urine and blood samples by Martens et al. (6).

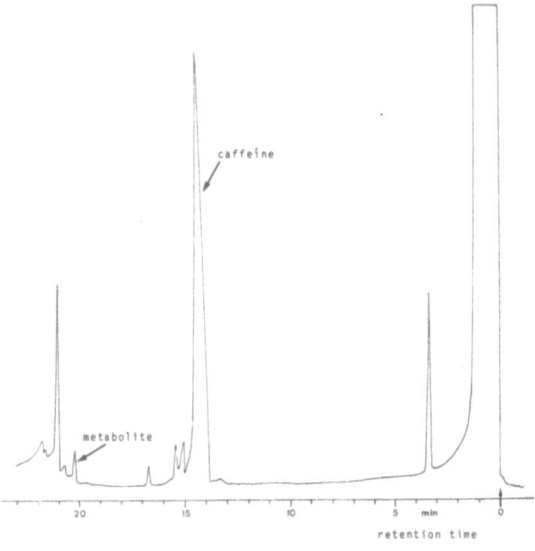

FIG. 3. TIC obtained after persilylation of the
radioactive fraction.

Derivatization procedures for GC-MS analysis. Persilylation.
The residue of the isolated radioactive fraction was dissolved in
100 μl 1,4-dioxane and 100 μl BSTFA. One drop of TMCS was added in
order to enhance silylation. The mixture was allowed to react
overnight at room temperature pending GC-MS analysis. Oxidation
with hydrogenperoxide. In order to oxidize sulfoxides into the
corresponding sulfones, the following derivatization technique was
used:

$$
\begin{array}{ccc}
\begin{array}{c} O \\ \uparrow \\ -S- \end{array} \Big\} &
\begin{array}{c} H_2O_2\ (30\%) \\ \xrightarrow{\hspace{2cm}} \\ 60°C/1\ h \end{array} &
\Big\} \begin{array}{c} O \\ \uparrow \\ -S- \\ \downarrow \\ O \end{array}
\end{array}
$$

The residue of the isolated radioactive fraction was dissolved in
1 ml methanol (if the fraction was already silylated) and evaporated
to dryness using a rotavapor. To the residue obtained 1 ml of H_2O_2
(30%) was added and the mixture was allowed to react at 60°C. After
one hour the mixture was carefully evaporated to dryness and
silylated using the method described above.

GC-MS analysis. GC-MS/EI analysis. GC-MS/EI analysis was
performed on a Varian 2700 gas chromatograph coupled to a Varian
CH 7 single focusing mass spectrometer by means of a double stage
Watson-Biemann separator. Data acquisition and data elaboration was

performed on-line using the MAT SS 101 data system. GC-MS/CI analysis.
GC-MS/CI analysis was performed using a Ribermag 1010B GC-MS system
with ammonia as reagent gas. Data acquisition and data elaboration
was performed on-line using the Riber 400 data system. Gas chromatographic conditions. Typical GC-conditions are: GC-column: -2 meter,
3% SE-30 on Supelcoport, 80-100 mesh; oven temperature: -200°C up
to 270°C at 10°C/min, -270°C isotherm; he-flow: 25 ml/min; injector
temperature: 300°C.

Direct introduction mass spectrometric analysis. Low resolution
MS analyses were run on a Varian CH 7 single focusing mass spectrometer. The direct introduction probe was programmed from room
temperature to 500°C within a time delay of 500 sec. Mass spectral
data were recorded on-line using a MAT SS 101 data system.

High resolution MS analyses were run on a Jeol 01SG2 special
double focusing mass spectrometer (resolution ± 15,000). Mass
spectral data were recorded on photoplates. The photoplates
obtained were scanned using a Jeol-densitometer, coupled on-line
with the Jeol GC-6 data system.

RESULTS AND DISCUSSION

Identification of suloctidil metabolites in human urine. The
chloroform soluble urinary suloctidil metabolites were separated
by subsequent column and thin layer chromatography. Only two of
the fractions obtained (fraction I and fraction II), containing
respectively 8 and 3% of the radioactivity initially present in
the urine, will be considered here.

The total ion current chromatogram (TIC) obtained for fraction
I after silylation is shown in Fig. 3. By preparative gas chromatography it was proved that peak present was not radioactive.
Interpretation of the mass spectrum of this peak pointed to the
presence of caffeine. Only one of the minor GC-peaks in the TIC
showed a twin peak effect due to the presence of deuterium, thus
representing a suloctidil metabolite. The corresponding mass spectrum
is shown in Fig. 4 and the fragmentation pattern is shown in Fig. 5.
Because of the low response in the TIC, this peak could not stand
for all the radioactivity present in this fraction. This discrepancy
either could be incomplete silylation of the sulfenic acid moiety
or the instability of these compounds at the GC-temperatures used.

This phenomenon was explained by direct introduction mass
spectrometric analysis of an aliquot of this fraction. The mass
spectrum obtained is shown in Fig. 6 and the corresponding fragmentation pattern in Fig. 7. This time MS analysis revealed the presence
of a sulfoxy metabolite in the extract. The structure of this

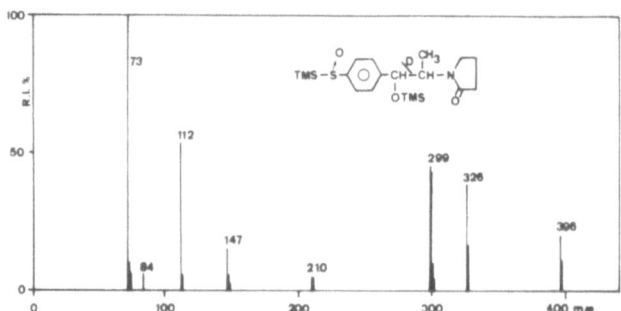

FIG. 4. Mass spectrum of the persilylated suloctidil
metabolite.

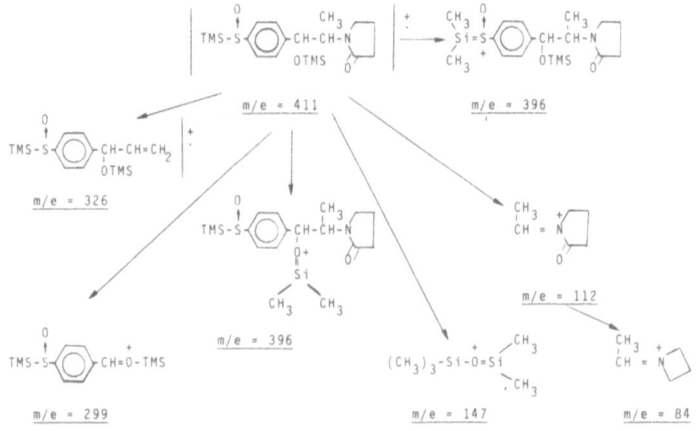

FIG. 5. Proposed fragmentation of the persilylated
suloctidil metabolite.

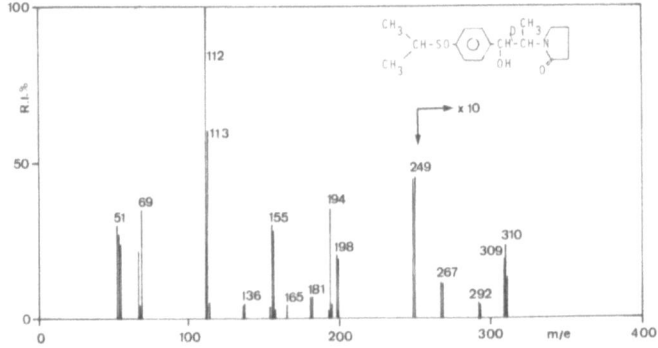

FIG. 6. Mass spectrum of the radioactive fraction
after direct introduction.

metabolite was also confirmed by comparison with the mass spectrum
of the synthetic metabolite. The presence of this compound was
further confirmed by GC–MS after oxidation of the fraction with
hydrogen peroxide and subsequent silylation. The TIC obtained is
shown in Fig. 8.

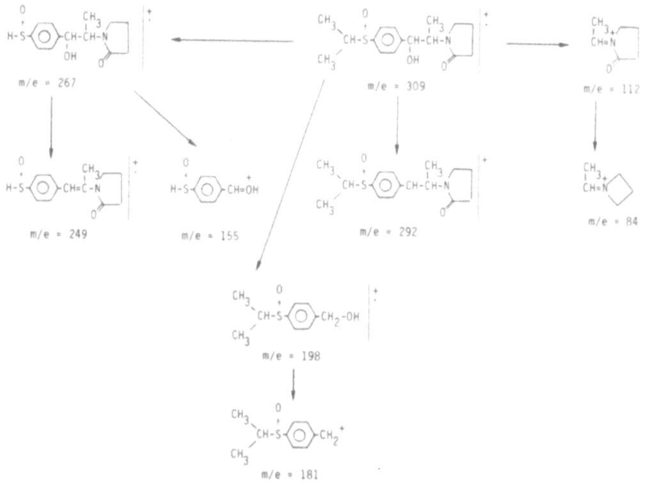

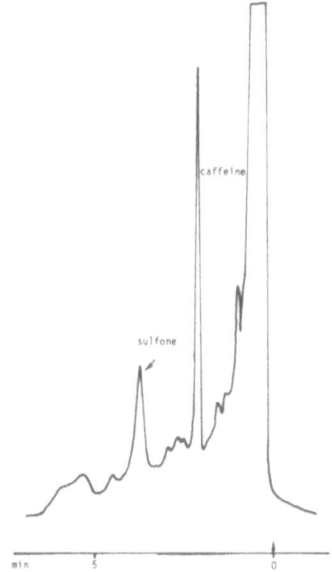

FIG. 7. Fragmentation of the sulfoxide metabolite.

FIG. 8. Total ion current chromatogram of the
radioactive fraction after H_2O_2-oxidation
and silylation.

MS analysis revealed the presence of caffeine in the first peak and
the sulfone analogue of the sulfoxide metabolite identified by
direct introduction MS in the second peak (Fig. 9).

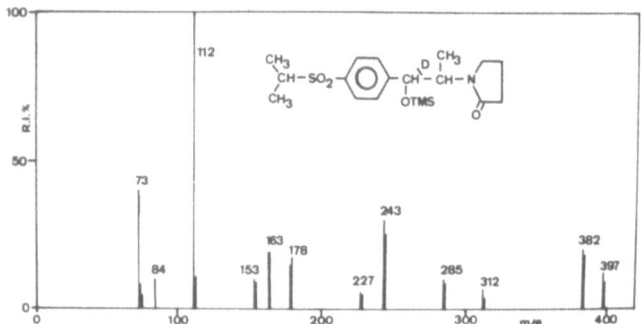

FIG. 9. Mass spectrum of the silylated sulfone-
compound.

Thus, in contrast to sulfoxide compounds, sulfones seem to be stable
during GC-MS analysis.

The total ion current chromatogram obtained for fraction II
after silylation only shows a minor peak with a twin peak effect
in the corresponding mass spectrum (Fig. 10).

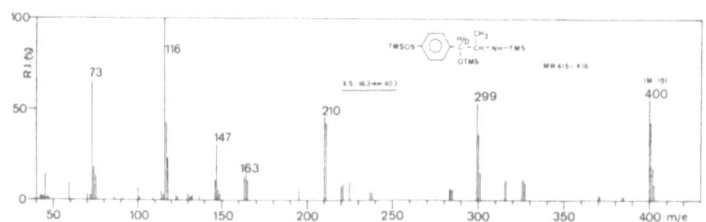

FIG. 10. The sulfenic acid metabolite identified
in the second radioactive fraction.

The direct introduction mass spectrum showed the presence of the
corresponding sulfoxide compound in this fraction (Fig. 11)

Identification of suloctidil metabolites in human plasma. One
radioactive fraction obtained after purification and separative
steps and accounting for 20% of the total radioactivity initially
present in the plasma was considered here. The TIC obtained by

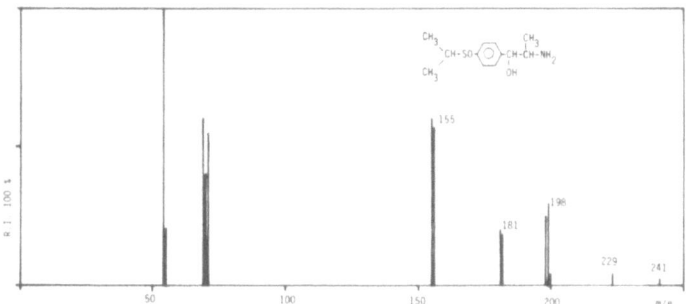

FIG. 11. Mass spectrum of the aminoalcohol sulfoxide
metabolite.

GC-MS after silylation did not show relevant peaks. However, the
direct introduction mass spectrometric analysis suggested the
presence of the sulfoxide-pyrrolidone compound already identified
in human urine (see Figs. 6 and 8). Since the concentration level
of this compound in the sample was rather low its presence was
confirmed by GC-MS/CI analysis after oxidation (H_2O_2) and silylation.
Fig. 12 shows the total ion current chromatogram obtained and the
mass chromatograms for m/e 398 (M+1) and m/e 415 (M+18), the quasi
molecular ions obtained with ammonia as CI-reagent gas.

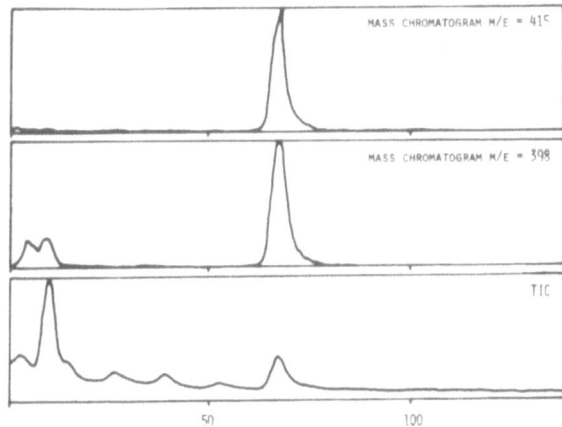

FIG. 12. GC-MS/CI (NH_3) analysis of the human
plasma extracts.

Fig. 13 shows the CI-mass spectra of both the corresponding synthetic
compound and the plasma metabolite. Important enough is the presence
of the deuterium labelling (d3 and d7) which furthermore confirms

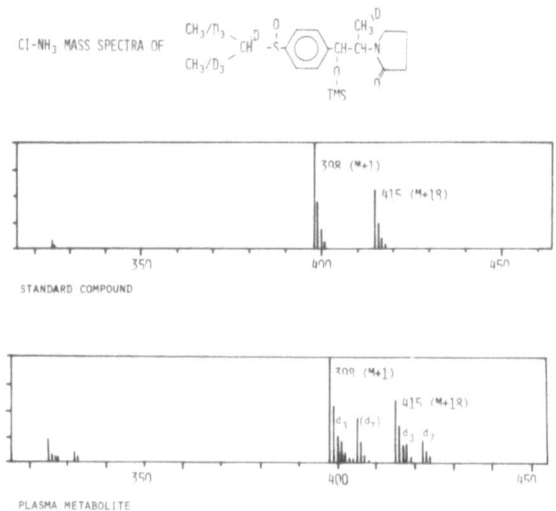

FIG. 13. CI-NH$_3$ mass spectra.

the structure proposed for this metabolite.

High resolution MS-analysis of a synthetic sulfoxide metabolite.
The direct introduction (low resolution) mass spectrometric analysis
for the standard sulfoxide compound and for the metabolites proposed
shows some characteristic ion fragments at m/e 198, 181 and 155.
The structure of these ion fragments was already proposed in Fig. 7.
In order to confirm these ion structures a high resolution mass
spectrometric analysis was performed on the corresponding synthetic
compound. The results obtained are shown in Table 2.

Due to the low concentration levels of the metabolites in the
different fractions only the most abundant ion fragment of the
sulfoxide moiety (m/e 155) has been confirmed by a high resolution
MS analysis.

CONCLUSIONS

During previous studies on the metabolic fate of suloctidil
several sulfenic acid compounds were proposed. The results obtained
in this study clearly demonstrate that the detection of sulfenic
acid metabolites by GC-MS is an artifact, which consists in the
pyrolysis of sulfoxy compounds into the corresponding sulfenic
acids at temperatures normally used in GC. The detection of
silylated sulfenic acids in the mass spectrometer results from the
on-column silylation of the pyrolysis products formed (Fig. 14).

Oxidation of the sulfoxides into the corresponding sulfones

TABLE 2
High-resolution MS-analysis of the synthetic
sulfoxide compound.

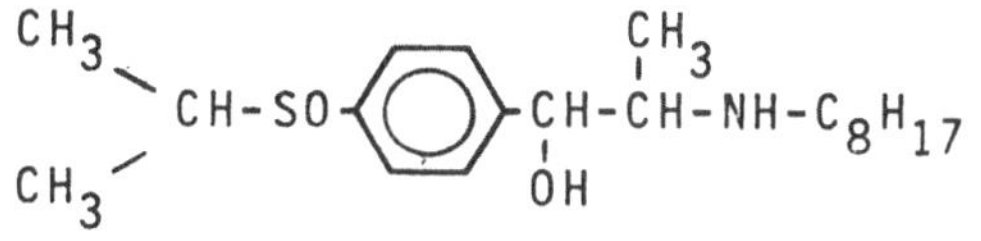

$$CH_3\text{-CH-SO-}\bigcirc\text{-CH-CH-NH-}C_8H_{17}$$

m/e measured	formula	structures proposed
198.0716	$C_{10}H_{14}O_2S$	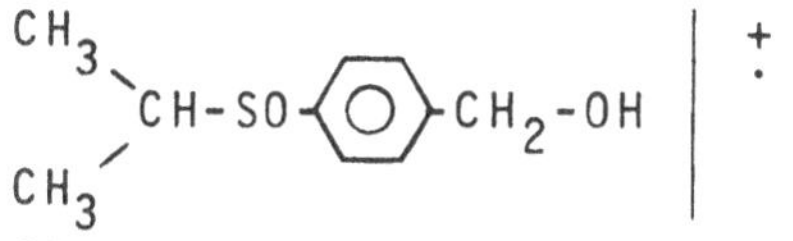 CH_3-CH-SO-⟨O⟩-CH_2-OH
181.0708	$C_{10}H_{13}OS$	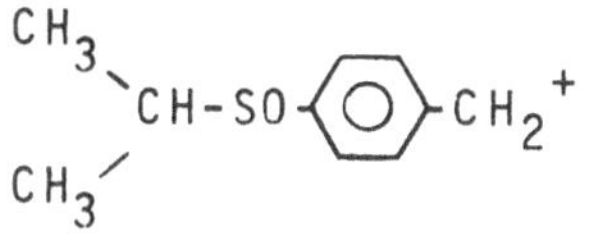 CH_3-CH-SO-⟨O⟩-CH_2⁺
155.0141	$C_7H_7O_2S$	 H-SO-⟨O⟩-CH=ỚH

FIG. 14. Detection of sulfenic acids by GC-MS.

by hydrogenperoxide permits their detection by GC-MS. This may be
the basis of a new method for their quantification in several body
fluids.

    The artifact described above is of course not limited to the
suloctidil metabolism but should be kept in mind for all drugs
containing an alkylthio group.

ACKNOWLEDGEMENTS

Dr. E. Esmann and Ir. J. Verreydt (Laboratories of Organic Chemistry, University of Antwerp (RUCA), Dir. Prof. Dr. F. Alderweireldt) are thanked for their assistance in running the high resolution MS-analyses.

The application Laboratories of Ribermag (Paris, France) are thanked for running the GC-MS/CI analyses.

This work was supported by a grant of IRSIA (Institut pour la Recherche Scientifique dans l'Industrie et l'Agriculture, Convention n° 2830).

REFERENCES

1) R. Roncucci, M.-J. Simon, M. Martens, K. Debast and G. Lambelin, "Abstract of a Communication presented at the Symposium on Metabolisme van geneesmiddelen", Wilrijk, Belgium, March 17, 1977.
2) K. Fries, Berichte, 1912, 45, 2965.
3) T.C. Briuce and R.T. Harkin, J. Am. Chem. Soc., 1957, 79, 3150.
4) B.C. Pal, M. Uziel, D.G. Doherty and W.E. Cohn, J. Am. Chem. Soc., 1969, 91, 3834.
5) J.R. Shelton and K.E. Davis, J. Am. Chem. Soc., 1976, 89, 718.
6) M. Martens, R. Roncucci, K. Debast and G. Lambelin, in "Recent Developments in Mass Spectrometry in Biochemistry and Medicine", Vol. 1, A. Frigerio, Ed., Plenum Press, New York, 1978, p. 133.

THE ELECTRON IMPACT FRAGMENTATION OF PRODUCTS RELATED TO N-(ALLYL)-
-3,3-DIPHENYLPROPYLAMINE, A METABOLITE FROM THIOPROPAMINE ®

V. Borzatta, M. Cristofori and A.G. Giumanini

Laboratori Alfa Farmaceutici and Università di Bologna,

Bologna, Italy

During the study of the metabolism of the new oral drug
THIOPROPAMINE ® (REDDEN), prepared in Alfa Farmaceutici Laboratories,
we identified, among others, a metabolite in rat and human urines,
whose acetyl derivative exhibited intense interrelated ions at m/e
113, 72, 70, the origin of which appeared to be linked to the
presence of the amide function, since the corresponding amine gave
three related peaks respectively at m/e 71, 30 and 70 (Figs. 1 and
2).

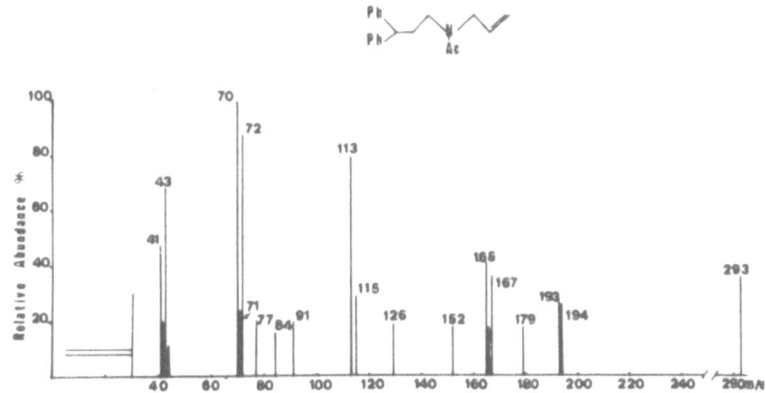

FIG. 1. EI-mass spectrum of N-(allyl)-N-(3,3-diphenylpropyl)
        acetamide (GLC inlet, ion En. 70 eV).

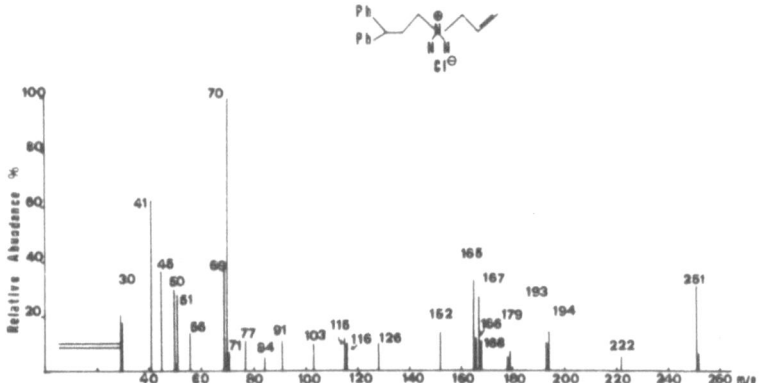

FIG. 2. EI-mass spectrum of N-(Allyl)-3,3-diphenylpropylamine
        hydrochloride (direct inlet, 20°C, ion En.
        70 eV).

A molecular peak at m/e 293 and 251, the peaks for an intact
benzhydryl group at m/e 167, 165 and 152 indicated that, if the
compound had an insaturation, as seemed likely in the aliphatic
chains, it had to be present in the less substituted alkyl chain.
Thermal as well as electron impact induced eliminations from
propanolamine structures were discarded. Unlikely highly efficient
(so to erase $(M-1)^+$ and M^+ ions even at low ionization energy) were
also ruled out by the synthesis of the saturated compound N-(Propyl)-
-3,3-diphenylpropylamine and its N-acetamide (Figs. 3 and 4).

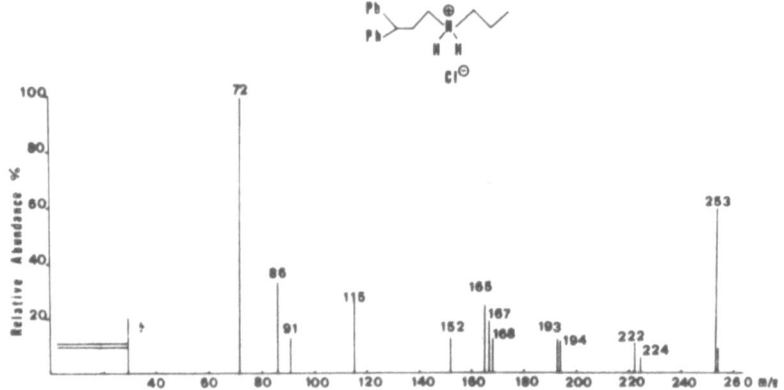

FIG. 3. EI-mass spectrum of N-(Propyl)-3,3-diphenylpropylamine
        hydrochloride (direct inlet,T   20°C,Ion.En. 70 eV).

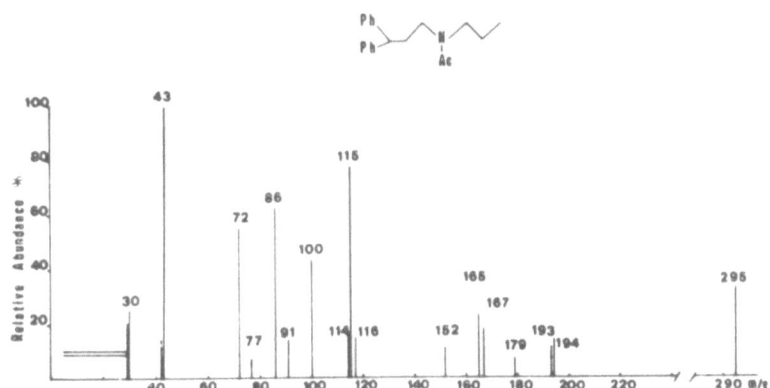

FIG. 4. EI-mass spectrum of N-(Propyl)-N-(3,3-diphenyl-
propyl) acetamide (GLC Inlet, ion En. 70 eV).

Interestingly, three out of the four compounds mentioned
showed the oftentimes strong peaks corresponding to an apparent
formula

$$H_3C-N\begin{array}{l} R' \\ R \end{array}$$

where R=H or Ac and R'= allyl or propyl, and its decay products.

Another compound, a possible metabolite which turned out not
be present, the N-methyl derivative of N-(Allyl)-3,3-diphenylpropyl-
amine presented the same behaviour (Fig. 5). In the saturated amine
N-(Propyl)-3,3-diphenylpropylamine the by far prevalent fragmentations are the α-β bond cleavage (m/e 222 and 72) and the "benzyl"
cleavage (m/e 167), followed by elimination of two hydrogen atoms
(m/e 166 and 165) or methyl group expulsion (m/e 152). The latter
is a common characteristic for all amines with a benzhydryl group
in the structure. In fact in another compound prepared for
comparison's sake, N-(Propyl)-3,3-diphenylallylamine (Fig. 6) and
its acetamide (Fig. 7) did not yield this triade, although it
made us aware of the danger of too quick guesses, when we found a
peak at m/e 194 and the related peak at m/e 115 for both of them.
In fact the composition of m/e 115 was found to be partly $C_9H_7^+$ by
high resolution MS.

The other part was again $H_3C-N-Ac$ ⌐ + . The peak at m/e 194

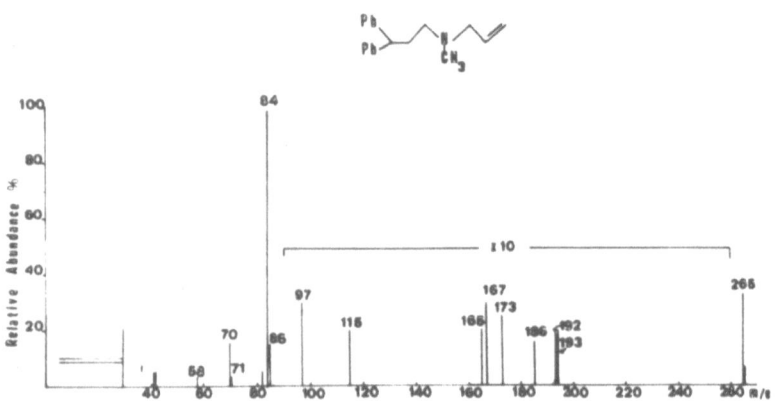

FIG. 5. EI-mass spectrum of N-(Allyl)-(methyl) N-3,3-diphenyl
propylamine (Direct inlet, T   20°C, ion
En. 70 eV).

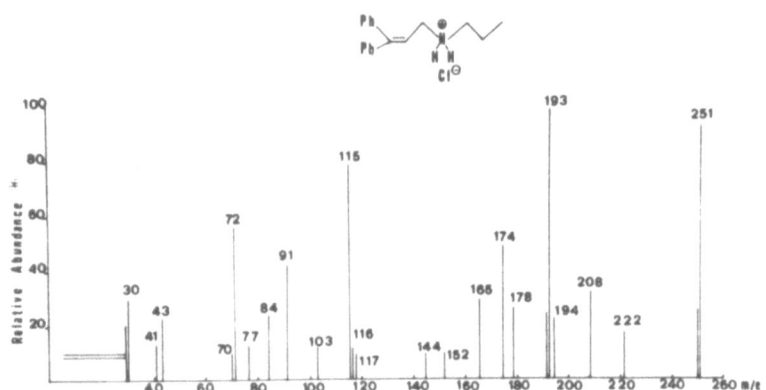

FIG. 6. EI-mass spectrum of N-(Propyl)-3,3-diphenylallyl-
amine hydrochloride (Direct inlet, T   20°C,
ion En. 70 eV).

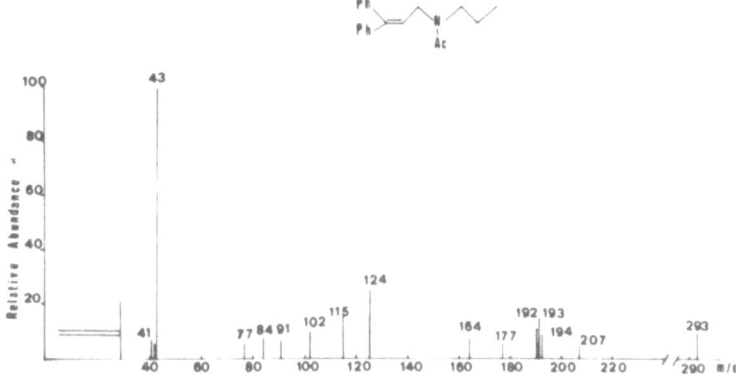

FIG. 7. EI-mass spectrum of N-(Propyl)-N-(3,3-diphenylallyl) acetamide (GLC inlet, ion En . 70 eV).

and that at m/e 115 containing oxygen pointed to a rearrangement of the original structure to that of allylic isomer under electron impact.

The presence of m/e 194, on the other hand, was a common feature of all thses spectra and together with it comes its decomposition peak at m/e 115. The expected high stability of the 3,3-diphenylallyl carbonium ion, perhaps even by phenyl anchimeric assistance, as hinted by the elimination of two hydrogen atoms, showed up with a base peak at m/e 193 in the spectrum of the corresponding amine.

How does the peak at m/e 113 form, and, above all, what is its composition?

High resolution showed it to be 100% $C_6H_{11}N$ O, as we said, formally corresponding to

H_3C-N ⟍ Ac

Analogously, we found m/e 115 to be (in part) $C_6H_{13}N$ O from the saturated acetamide. It is remarkable that the sum of the relative abundances of these ions and their decay products may be very high (well above 100%), just when, e.g., the "usual" α-β bond cleavage is close to zero.

The species precursor to these ions is in any case the parent ion, as often confirmed by the appropriate metastable ion. This obviously implies the transfer of a hydrogen atom to the splitting moiety of the parent ion.

Whence the question: where does it come from?

To this end we prepared (Fig. 8) the deuterated analogs, i.e., N-(Allyl)-3,3-diphenylpropylamine-3-d and its acetamide (Figs. 9 and 10).

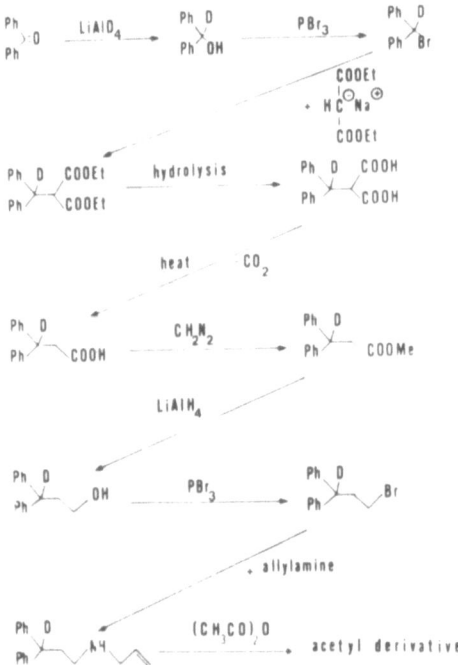

FIG. 8. Synthesis of N-(Allyl)-3,3-diphenylpropyl-
amine-3-d and its acetamide.

The rationalization of the observed fragmentations can be seen in Chart 1 (amine) and Chart 2 (amide). It appears that both the D from the methine group and, even more surprisingly, one of the hydrogens of the adjacent methylene were transferred to nitrogen containing moiety. A reasonable pathway for this unprecedented rearrangement is shown in Chart 3, where the "onium" structures are possibly expected to partly rearrange to the amine (amide) tautomer. A few other interesting features could be noticed. The fortunate

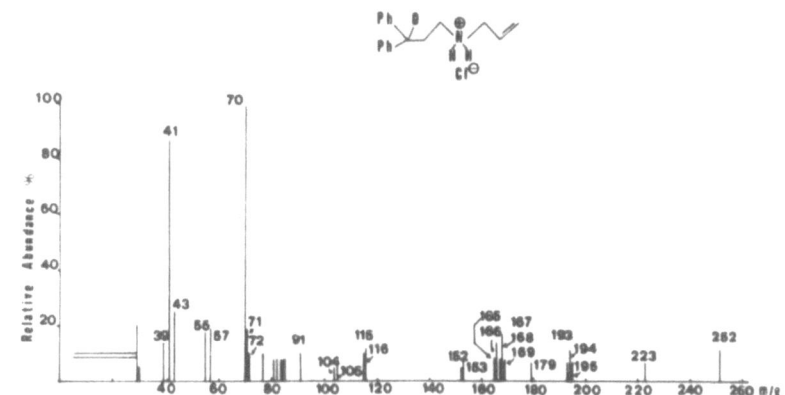

FIG. 9. EI-mass spectrum of N-(Allyl)-3,3-diphenylpropyl-
amine-3-d hydrochloride (Direct inlet, T   20°C,
ion En. 70 eV).

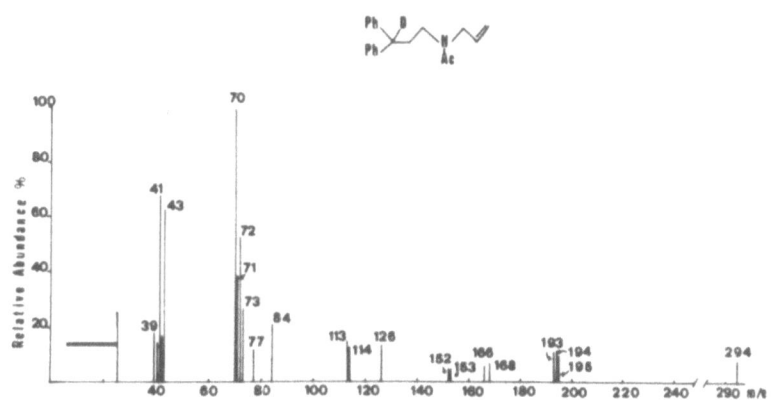

FIG. 10. EI-mass spectrum of N-(Allyl)-N-(3,3-diphenyl-
propyl-3-d) acetamide (Direct inlet, T   20°C,
ion En. 70 eV).

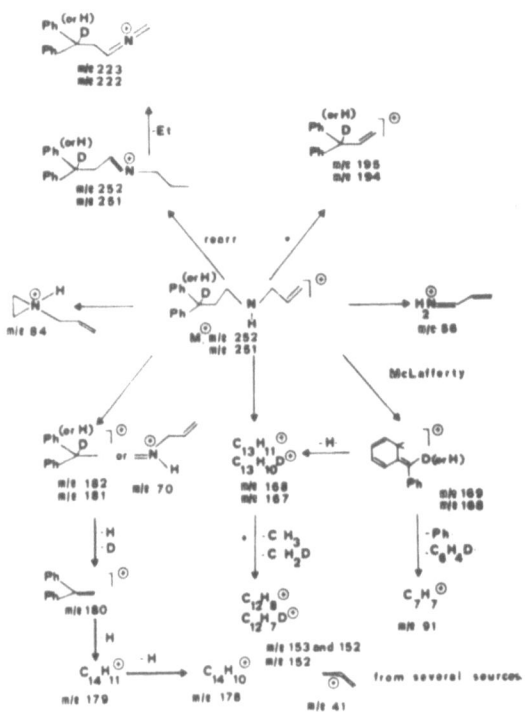

CHART 1. Fragmentation
of N-(Allyl)-3,3-
diphenylpropylamine and
N-(Allyl)-3,3-diphenyl-
propylamine-3-d
hydrochloride under EI.

CHART 2.
Fragmentation of
N-(Allyl)-N-(3,3-
diphenylpropyl-3-d)
acetamide under EI.

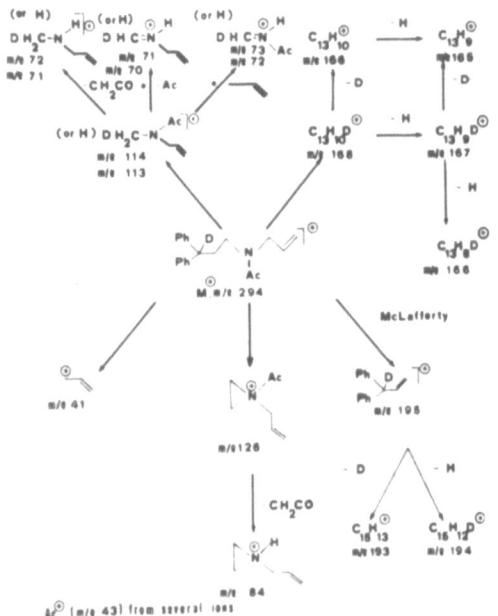

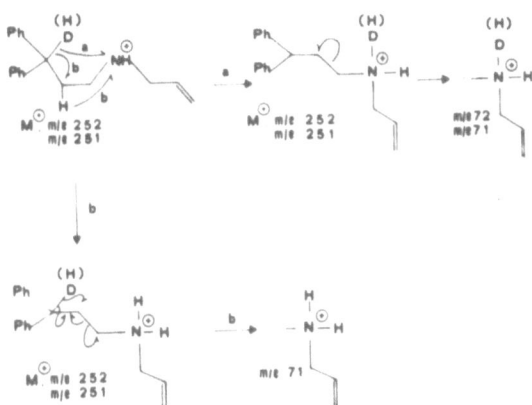

CHART 3. Fragmentation of N-(Allyl)-3,3-diphenyl-
propylamine and N-(Allyl)-3,3-diphenyl-
propylamine-3-d hydrochloride under EI.

coincidence of the absence of the charged hydrocarbon fragment from
the rearrangement involving hydrogen migration in the spectrum of
the amine (m/e 181) made it possible to conclude that the charged
hydrocarbon fragment ion at m/e 182 (D) (m/e 181 with H) formed
by the α-β bond cleavage retained its original structure under
further breakdowns; in fact, it lost only a deuterium atom (not
an H atom) in the next decomposition. Deuterium migration would
have caused H-D scrambling in this step.

A McLafferty rearrangement in the amine (m/e 169 for the D-
-derivative) involving one of the phenyl groups looks like an ion
which selectively lost only hydrogen, as deduced by the comparison
of the intensity ratios of the rearrangement peak with that at one
mass unit less, but scrambling certainly occurred in the following
methyl group expulsion (m/e 153 and 152) and, for the D-amide,
also in the loss of two "hydrogens".

The McLafferty rearrangement involving the amide function gave
an ion (m/e 195 for the D-amide) which gave scrambling, because it
also lost a hydrogen atom.

Methine deuteration finally made it possible to trace back the
several H and D rearrangements and losses, the cyclization and
phenyl losses of the ion at m/e 195 (194 in the H-derivative) on
its way to m/e 115 and 116 (115 in the H-derivative). Chart 4 shows
this detective work.

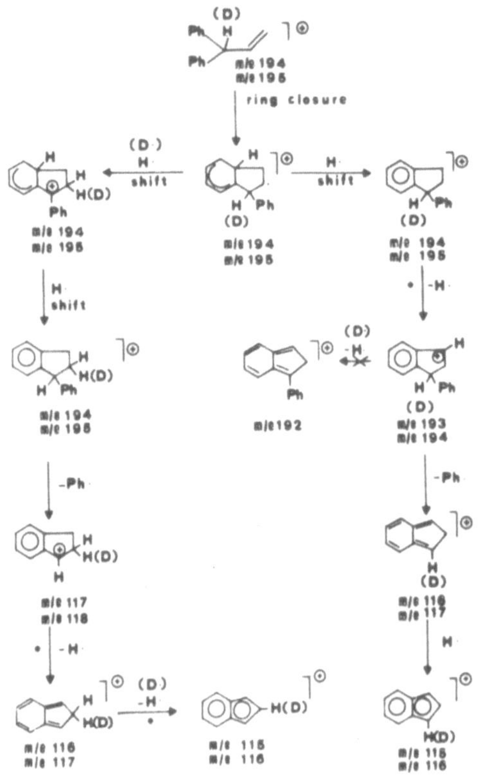

CHART 4. Fragmentation of N-(Allyl)-3,3-diphenyl-
propylamine and N-(Allyl)-3,3-diphenyl-
propylamine-3-d hydrochloride under EI.

After this picture of the unexpected fragmentations of these
compounds, it might appear astonishing that the saturated amine
N-(Propyl)-3,3-diphenylpropylamine (Fig. 3) behaved as a classical
textbook example.

N-Deuteromethylation of the amine made it possible to reach
certainty about the sole elimination of either an allylic or
methyl hydrogen from the rearrangement ion at m/e 88 with no
incursion from the methyl deuterium atoms; the transition was
confirmed by the observation of the corresponding metastable ion.

URINARY METABOLITES FROM (2-ETHYL-2,3-DIHYDRO-5-BENZOFURANYL)

ACETIC ACID

C.Casalini°,G.Cesarano°,G.Mascellani°,G.Tamagnone^ and

A.G.Giumanini¨; °Alfa Farmaceutici, Bologna; ^Schiapparelli,

Torino; ¨University of Bologna, Bologna, Italy

The identification of urinary metabolites of 2-ethyl-2,3-dihydro-benzofuranyl acetic acid (1a) (prepared by Schiapparelli Research Laboratories (1, 2)) was undertaken as a consequence of discovery of its good antiinflammatory activity and low general systemic toxicity. The search for metabolites was preliminarily restricted to acidic compounds in the reasonable assumption that the acidic function originally present in 1a could not be destroyed. In fact from an acidified urine chloroform extract of treated rats four compounds related to Ia (Fig. 1) could be separated by tlc.

They were the result of biological hydroxylation of 1a at different sites; 5a appeared to be a product of subsequent dehydratation. Incidentally, 1a was present in the urine only in tiny traces. All of the metabolites were first methylated with diazomethane to the methylesters, then acetylated with acetic anhydride at the free alcohol hydroxy group. Tlc techniques made it possible to effect a preliminary separation of 1a-5a, which was very efficiently carried out after the two esterification steps on the esters 1b - 5b (Fig. 1).

The former separation was rather poor for the pair 2a and 4a. Gas chromatography on 1b - 5b confirmed the efficiency of their tlc separations.

Introduction of 1a - 5a via the direct inlet system and 1b - 5b (also via the GC inlet) into the ion source of a mass spectrometer gave important structural information besides the molecular weights for all the isomers, 5a and 5b.

The two latter compounds in a very pure state exhibited UV

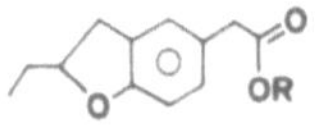

1 a: R = H
b: R = Me

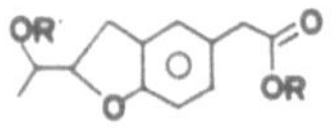

2 a: R, R' = H
b: R = Me; R' = Ac

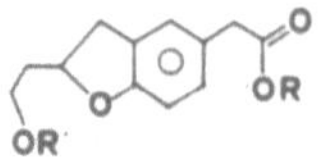

3 a: R, R' = H
b: R = Me; R' = Ac
c: R = Me; R' = H

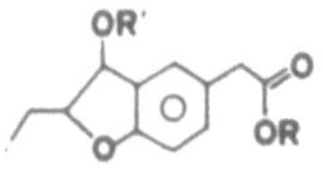

4 a: R, R' = H
b: R = Me; R' = Ac

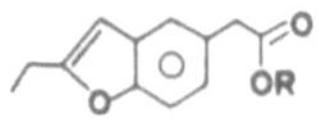

5 a: R = H
b: R = Me

FIG. 1. The drug, its metabolites and derivatives thereto related.

spectra quite different from those of 1a and 1b, yielding the original hint for a fully conjugated cyclic structure (benzofuran), whose UV spectral characteristics are definitely different. The mass spectra of 5a (Fig. 2) showed the prominent losses of a methyl and a carboxyl group in this and the reversed sequence, as confirmed by the appropriate metastable ions; the ion at m/e 144 eventually lost carbon monoxide and a hydrogen atom. The spectral behaviour of 5b (Fig. 3) was exactly repetitive of that of 5a (Scheme 1).

Compound 5a, though, might not be the product of a biochemical transformation, but either an artifact of the acidic treatments in the course of its separation or simply the result of a facile chemical elimination of water in vivo from 4a, of which an authentic specimen was actually found very sensitive under this respect to acidic conditions when a careful work up of the urine with avoidance of strong heating or long contact with acid was carried out, 4a neatly predominated over 5a. Hydroxy compound 4a lost water under electron impact (the process was practically the only one observed below 20 eV); its mass spectrum (Fig. 4) showed the same fragmentations as that of 5a. In addition, one could observe the loss of the ethyl group, the carboxy group and, as the result of a rearrangement, C_3H_7 radical (Scheme 2). An attempt at deuteration of the free acid 4a with D_2O - THF directly in the quartz probe of the direct inlet

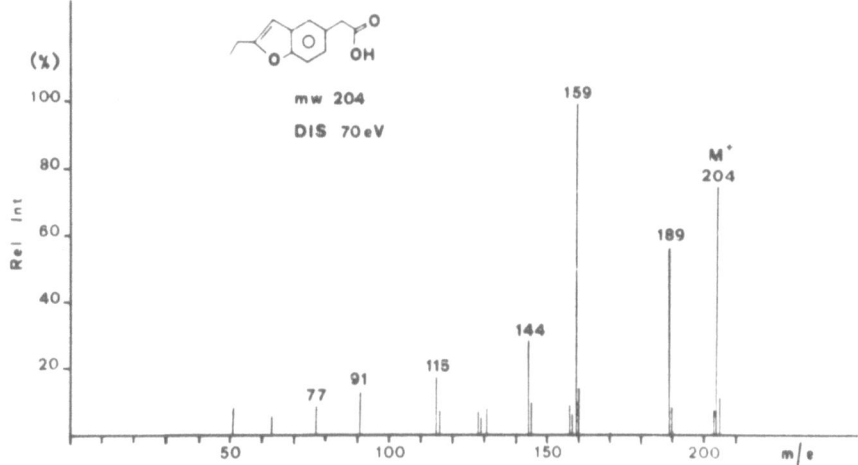

FIG. 2. EI-mass spectrum of 5a.

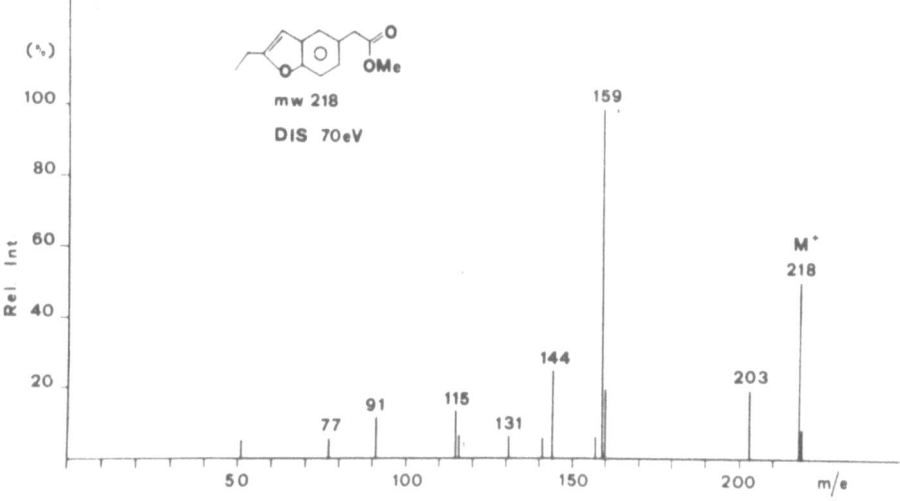

FIG. 3. EI-mass spectrum of 5b.

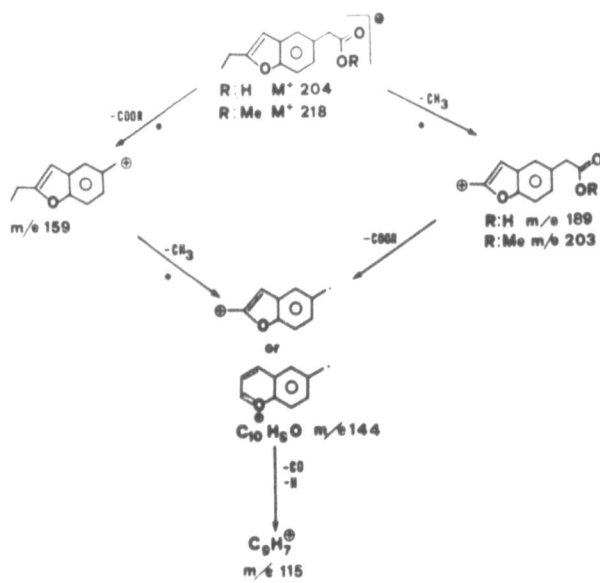

SCHEME 1. Mass fragmentation of 5a (R=H) and
           5b (R=Me).

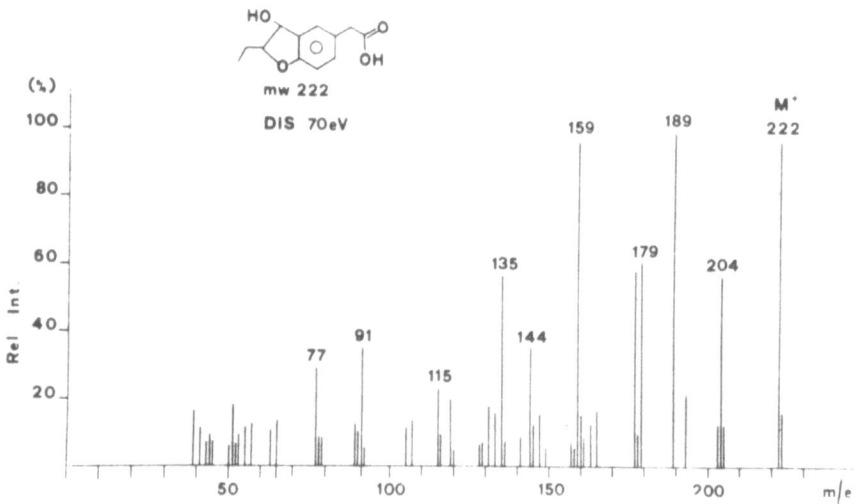

FIG. 4. EI-mass spectrum of 4a.

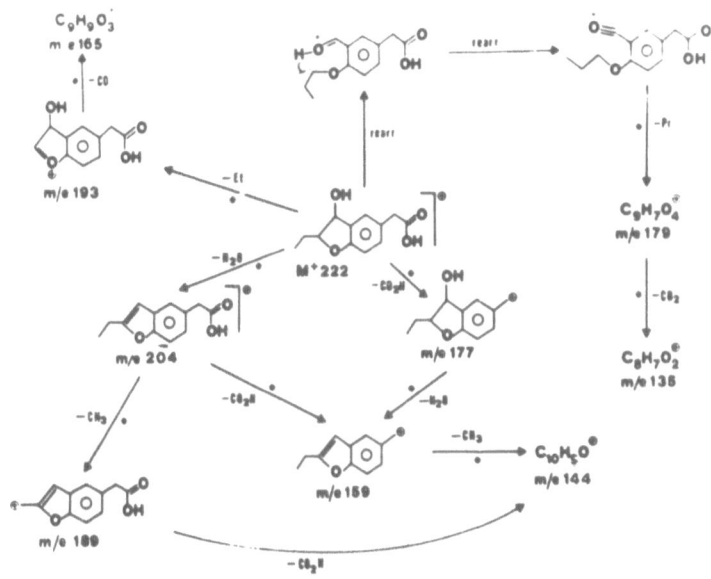

SCHEME 2. Mass fragmentation of <u>4a</u>.

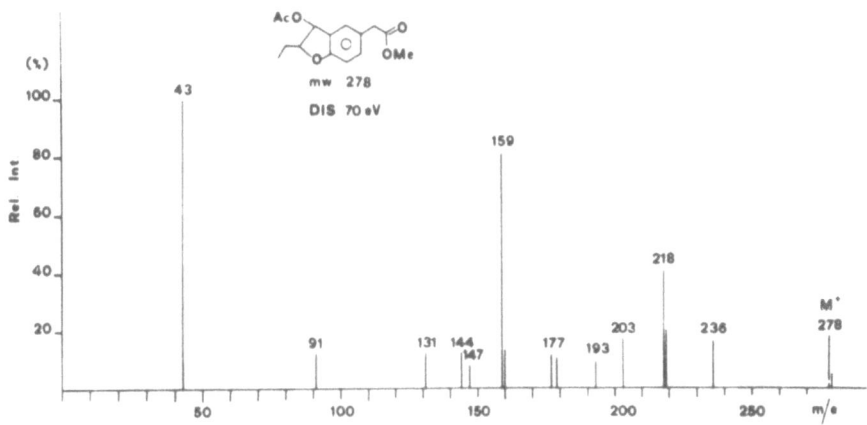

FIG. 5. Ei-mass spectrum of <u>4b</u>.

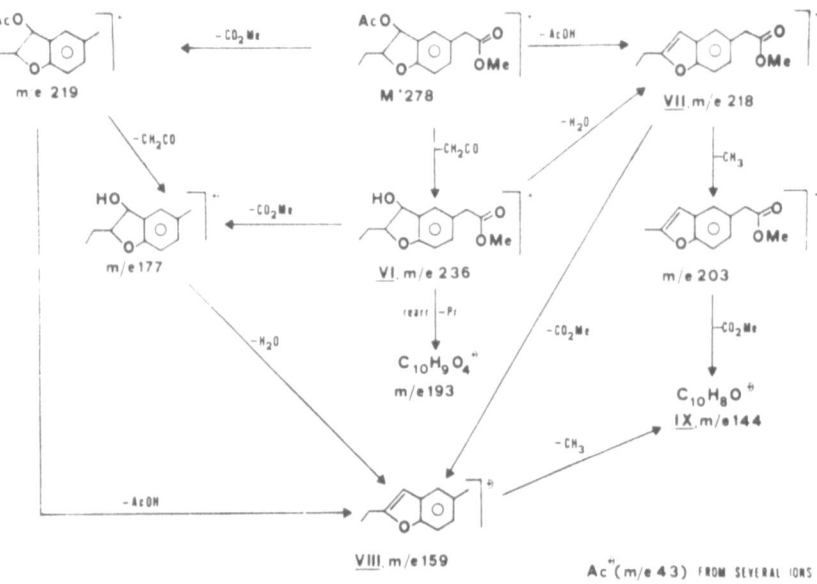

SCHEME 3. Mass fragmentation of 4b.

system, followed by evaporation of the solvent at 40° and 760 torr
led to water elimination. Diester 4b confirmed the structure of 4a
(Fig. 5).

As shown in Scheme 3, three different routes took to ion VIII
from the molecular precursor ion, which can undergo initial ketene,
carboxymethyl radical and acetic acid expulsion. Ion VI undergoes
the same type of rearrangement cleavage observed in the spectrum
of 4a: the elimination of a C_3H_7 fragment. Methyl elimination from
the side chain could be triggered at higher electron energy only
from ion VII to give an ion (m/e 203) ending up at ion IX, which is
also the terminal for VIII.

Uncommon and interesting spectral features were exhibited by
the metabolite oxidized in the β-position of the side chain, 3a
(Fig. 6) and the two esters therefrom derived, 3b (Fig. 7) and 3c
(Fig. 8). A common primary process was the cleavage of the carboxyl
(or methyl carboxylate) function followed by water (or acetic acid)
and finally ethene to yield the apparently very stable ion at m/e
131 (Scheme 4). Other secondary processes (Scheme 5) played an
important role in the fragmentation of the diester. After loss of
the COOMe group, either a hydrogen atom (m/e 218) or, surprisingly,
methyl acetate could be lost, just to follow up with the discarded
alternative and end up with an ion at m/e 144. The second primary

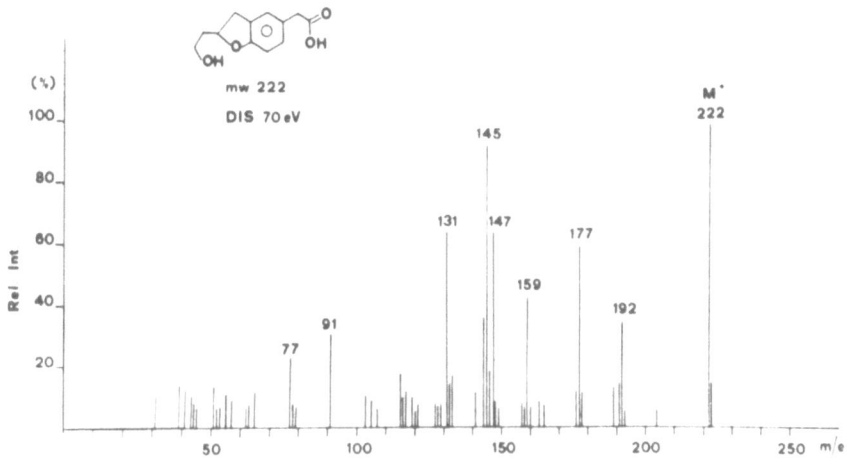

FIG. 6. EI-mass spectrum of 3a.

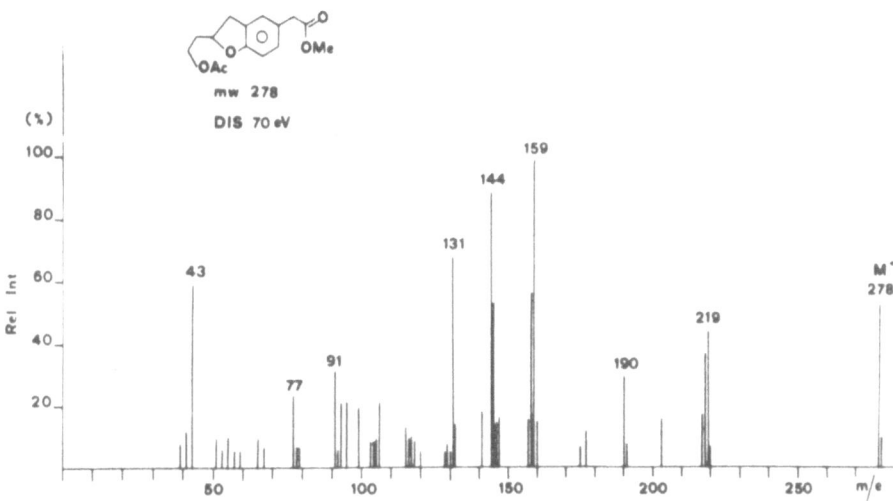

FIG. 7. EI-mass spectrum of 3b.

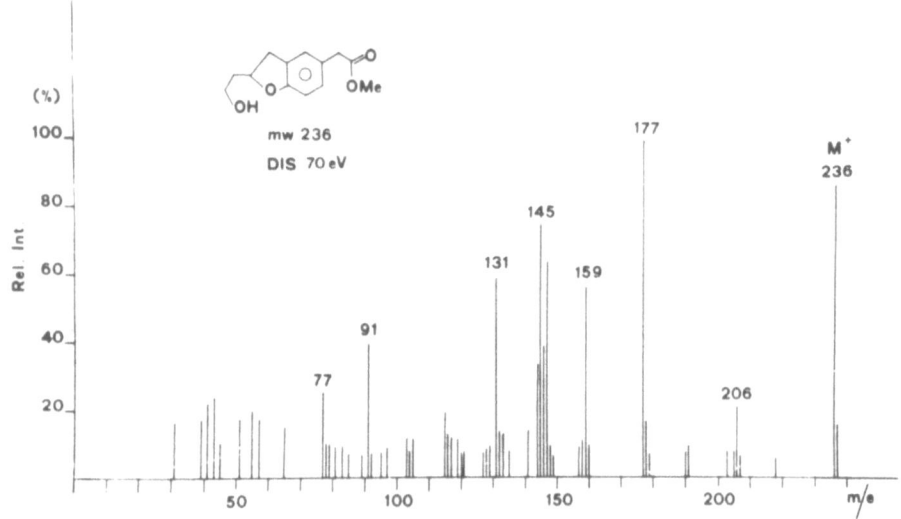

FIG. 8. EI-mass spectrum of 3c.

SCHEME 4. Mass fragmentation of 3a (R=H) and 3c (R=Me).

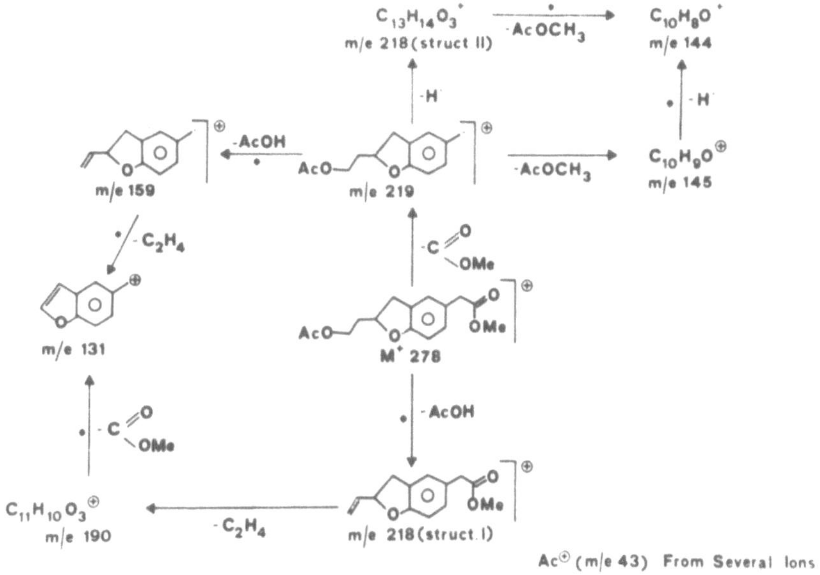

$C_{13}H_{14}O_3^{\cdot}$
m/e 218(struct II) $\xrightarrow{-AcOCH_3}$ $C_{10}H_8O^{\cdot}$
m/e 144

$\xrightarrow{-AcOH}$ m/e 159   AcO—   m/e 219 $\xrightarrow{-AcOCH_3}$ $C_{10}H_9O^{\oplus}$
m/e 145

$\cdot | -H$

$-C_2H_4$

m/e 131

$-C\overset{\nearrow O}{\underset{\searrow OMe}{}}$

AcO— M^+ 278

$\cdot | -C\overset{\nearrow O}{\underset{\searrow OMe}{}}$

$C_{11}H_{10}O_3^{\oplus}$
m/e 190 $\xleftarrow{-C_2H_4}$ m/e 218(struct. I)

$\cdot | -AcOH$

$Ac^{\oplus}$ (m/e 43) From Several Ions

SCHEME 5. Mass fragmentation of 3b.

process to be operative was strongly influenced by the internal
H-bonding of the OH function with the ether oxygen of the dihydro-
furane ring. The diester 3b, in fact, lost acetic acid, most probably
via a McLafferty rearrangement, followed by ethene expulsion and
finally by methyl carboxylate cleavage to reach, by an alternative
pathway, the ubiquitary ion at m/e 131. Interestingly, this sequence
implies the formation of an ion at m/e 218 of different structure
from the one previously cited. On the other hand, the ion of the
oxyester 3c, as well as the oxyacid 3a took quite a different course,
which suppressed both the expected onium cleavage and water
elimination. They eliminated formaldehyde (to ions at m/e 206 and
192, respectively) with great efficiency, then the carboxyl (methyl
carboxylate) function and a cascade of hydrogen atoms.

The metabolism of the drug 1a gave rise to a third metabolite
2a, where the alcohol function was introduced into the side chain
in a β-position with respect to the ether linkage. The mass spectrum
(Fig. 9) of this compound, dominated by a peak at m/e 131, is not
very different in a qualitative fashion from those of its isomers:
it is therefore a much easier way for ions of the same composition
to decay to the ion at m/e 131, the common dead end of many pathways
of all side chain oxy compounds inspected. Its origin (Scheme 6)
may be twofold: from m/e 176 (or 132) and m/e 159 with relative
contributions which may depend on the precursors structure, as the

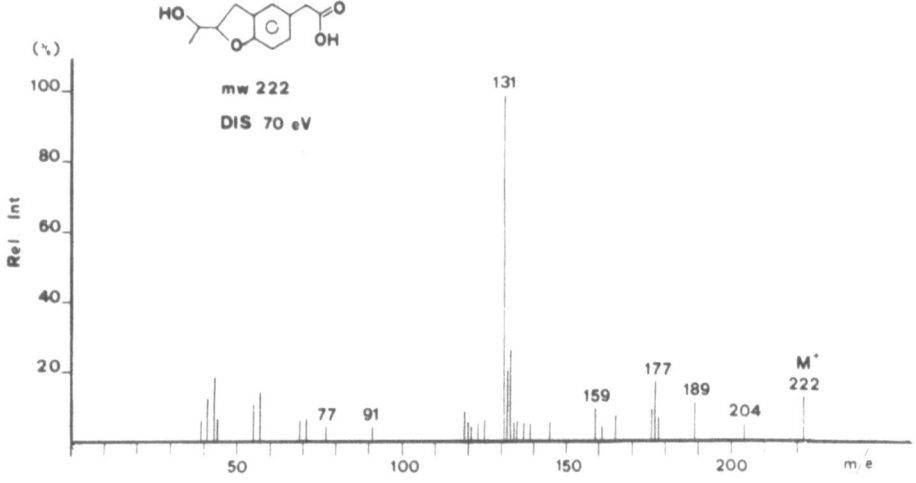

FIG. 9. EI-mass spectrum of 2a.

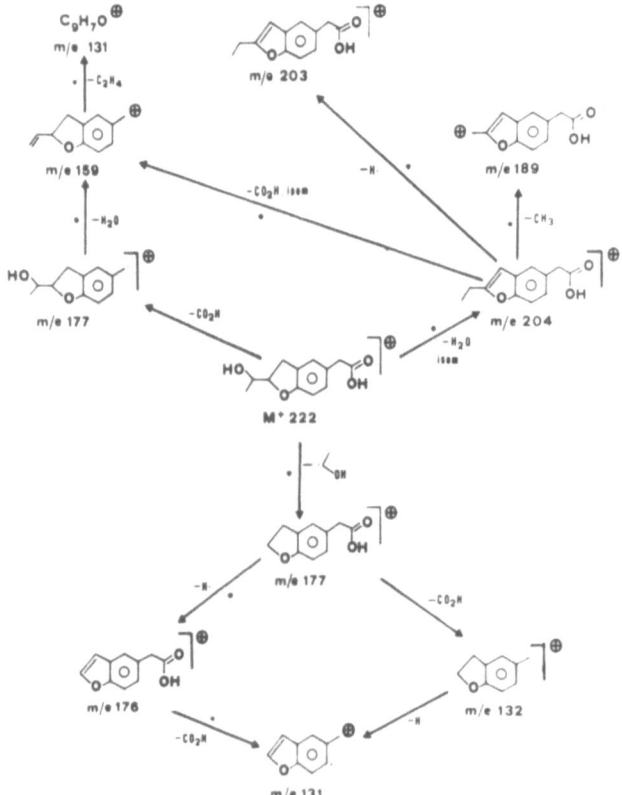

SCHEME 6. Mass fragmentation of 2a.

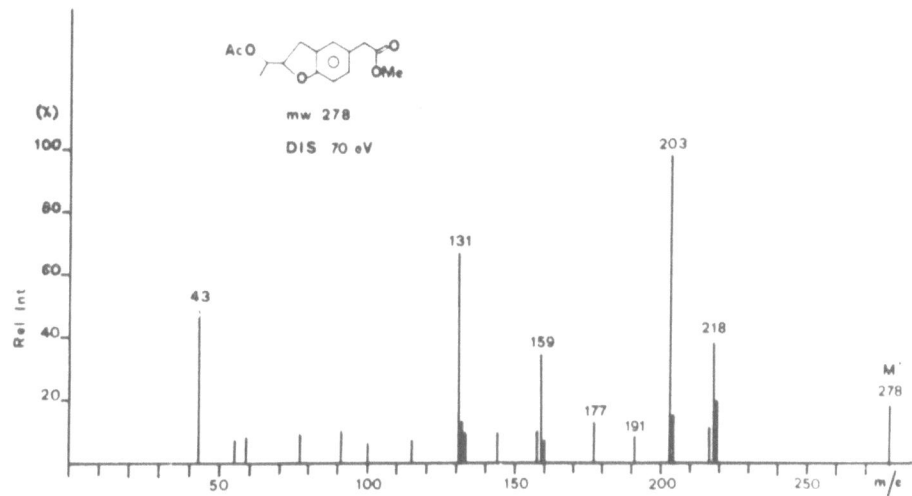

FIG. 10. EI-mass spectrum of <u>2b</u>.

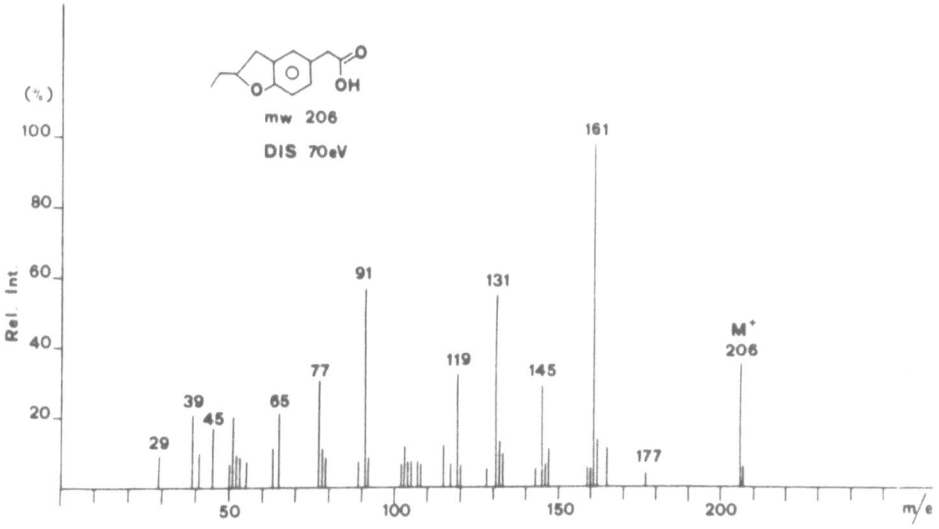

FIG. 11. EI-mass spectrum of <u>1a</u>.

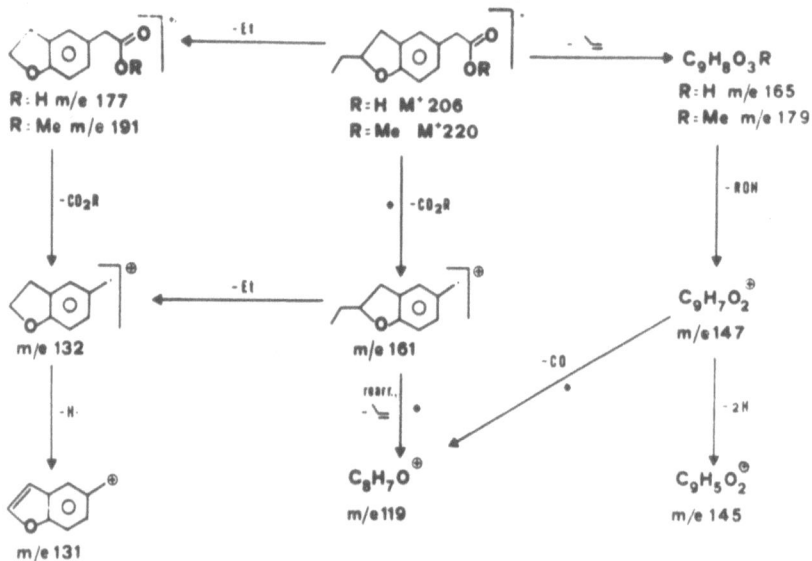

SCHEME 7. Mass fragmentation of <u>1a</u> (R=H) and <u>1b</u> (R=Me).

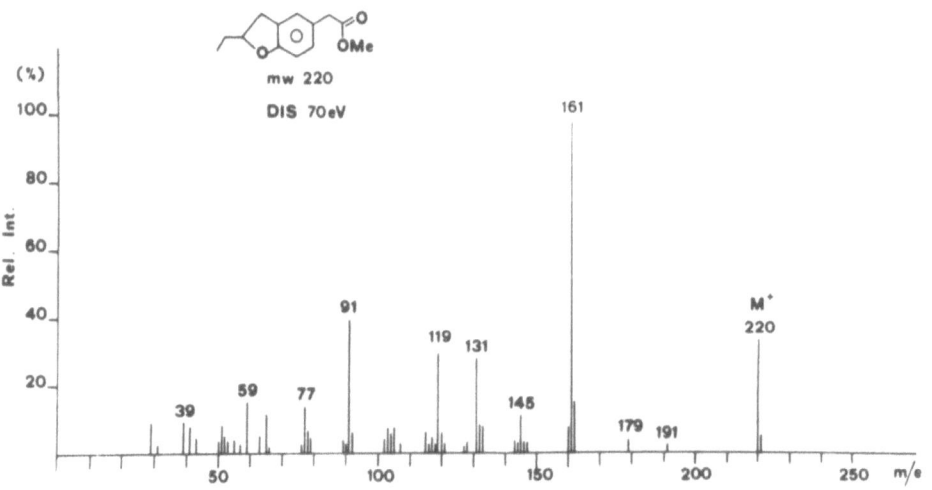

FIG. 12. EI—mass spectrum of <u>1b</u>.

variable intensity of the corresponding transitions are marked by more or less intense metastable ions.

The fragmentation routes of diester 2b closely paralleled those of the oxy acid 2a (Fig. 10).

The mass spectrum of the original drug 1a (Fig. 11) neatly showed the effect of the lack of the oxygen entered into the structure by metabolism. In fact, 1a lost under electron impact (Scheme 7) the carboxyl group (and its methyl ester 1b the ester function (Fig. 12)) to yield the base peak at m/e 161, which can in turn decay by loss of an ethyl radical and a hydrogen atom to the ever present ion at m/e 131. The usual alternative route to m/e 131 asks for loss of the alkyl chain from the parent ion, followed by loss of the carbomethoxy function and a hydrogen atom. Both routes rule out an oxygenless composition for the ion at m/e 131.

This fact in connection with the absence of metastable ions for loss of CO for m/e 159 (except for 3b) and the very low intensity of the ion at m/e 131 in the spectra at 5a and 5b, where the possible precursor at m/e 159 was the base peak, is taken as an indication that this route has a low importance if any at all. This deduction was promptly confirmed by high resolution mass spectrometry, for which we are indebted to Varian, Bremen. Interesting losses of 41 and 42 mu showed up twice in the mass spectrum of both 1a and 1b; they are the result of rearrangement of the initial structures.

REFERENCES

1) G.F. Tamagnone and F. De Marchi, Brit. Pat. 26078 (June 12, 1974); U.S. Pat. 4,029,811 (June 14, 1977).
2) G.F. Tamagnone, M.U. Torrielli and F. De Marchi, in "Simposio scientifico nell'Industria Farmaceutica in Italia", Rome, October 2-4, 1975.

THE MAJOR METABOLITES OF NICLOSAMIDE: IDENTIFICATION BY MASS

SPECTROMETRY

L.A. Griffiths and V. Facchini

Department of Biochemistry, University of Birmingham,

Birmingham, England

INTRODUCTION

Despite the widely held view that the extensively used intestinal taeniacide, niclosamide (2',5-dichloro-4'-nitro salicylanilide) is poorly absorbed from the mammalian gastrointestinal tract (1-3) the only experimental support for such a view is provided by an early study by(4),who employing a colorimetric method of limited specificity assessed the levels of niclosamide remaining in the gastrointestinal tract at selected intervals of time following oral administration. Synthesis of a $/^{14}C/$niclosamide and its employment in metabolic studies has now shown that this drug is not only absorbed in considerable amounts but is extensively metabolized in the rat. The major metabolites have been characterized by chromatography and mass spectrometry.

EXPERIMENTAL

Animals. Groups of three rats (200-300 g) of the Wistar strain were employed in each experiment under conditions and on a diet previously described (5).
   Compounds. Non-labelled niclosamide was kindly donated by Bayer AG, Wuppertal, W._Germany. The $/^{14}C/$niclosamide (2',5-dichloro--4'-nitrosalicyl/_carbonyl-^{14}C_/anilide) was prepared by the condensation of $/$carboxy-^{14}C_/-5-chlorosalicylic acid (6) with 2-chloro-4-nitro-aniline in the presence of PCl_3 to give a compound shown to be 97% radiochemically pure.

2',5-Dichloro-4' amino salicylanilide was obtained by the

121

reduction of niclosamide with aq. TiCl$_3$.

Dosage. Each animal received a single dose of either $/^{-14}C_/$ niclosamide (20 mg; 26.7 µCi/m.mole) for the radiometric investigations or 50 mg of unlabelled niclosamide for the ms. studies.

Chromatography. Three solvents systems were employed for the characterization of metabolites including solvent A (Table 1). Whatman paper No. 3 MM was employed for both qualitative and preparative chromatography.

Enzymic hydrolysis. For the hydrolysis of isolated conjugates, treatment with β-glucuronidase (Sigma, E. coli Type II) was as described by Griffiths and Smith (5).

Mass spectrometry. Mass spectra were determined in an AEI MS902 with direct probe insertion. A temperature of 220° was employed. The electron energy was 70 eV.

Radiometric measurement. Measurement of the radioactivity of both biological fluids or isolated metabolites were carried out in a Philips scintillation counter, under standard conditions previously described (7).

Biliary-cannulation procedures. These were also as previously described (7).

Microfloral incubations. Niclosamide was incubated under N$_2$ in a glucose-yeast   peptone broth with an inoculum derived from the rat caecum by a method previously described (5). The appropriate uninoculated controls were run concurrently.

Pretreatment with oral antibiotics. Partial suppression of the rat intestinal microflora in vivo was achieved by administering 200 mg of a neomycintetracycline–Bacitracin mixture (100:50:50 by weight) for two days prior to administration of niclosamide and for seven days after (8).

RESULTS

Oral administration of $/^{-14}C_/$niclosamide to a group of rats was shown to result in the excretion of considerable radioactivity in urine over 144 h (Table 2). Evidence was also obtained of considerable biliary excretion following oral administration to biliary cannulated rats.

The major biliary metabolite: NB1, (Table 1) was shown to be hydrolysed by β-glucuronidase to a product which showed similar chromatographic and mass spectrometric characteristics to niclosamide. The molecular ion was seen at m/e 326 and the base peak at m/e 155. The postulated fragmentation pathway is shown in Fig. 1. Niclosamide (unconjugated) was also found as the major metabolite in faeces.

The major metabolite of urine was shown by chromatography and mass spectrometry (Table 1) to be 2',5-dichloro-4'-amino-salicyl-anilide. The fragmentation (Fig. 2) was observed to follow a similar route to that of niclosamide. Additionally, several labile conjugates

TABLE 1
The metabolites of niclosamide

Metabolites	R_F in solvent A^	% of dose excreted in non-cannulated animals	Identification by chromatography and m.s.
Bile			
NB 1	0.49	–	Niclosamide-O-glucuronide
NB 2	0.79	–	Niclosamide
Urine			
NU 1	0.07	1.63	Unknown
NU 2	0.19	2.58	Conjugates of 2',5-dichloro 4' amino-salicyl-anilide
NU 3	0.32	4.07	
NU 4	0.46	6.85	
NU 5	0.65	30.83	2'5-Dichloro-4'-aminosalicylanilide
NU 6	0.84	4.99	Niclosamide
Faeces			
NF 1	0.30	1.82	Unknown
NF 2	0.57	10.41	2',5-dichloro-4'-aminosalicylanilide
NF 3	0.71	34.64	Niclosamide

^ Solvent A: n-butanol-NH$_3$-ethanol-water (10:1:10:4).

Metabolites were detected by use of modified Ehrlich reagent alone or following reduction of metabolite with TiCl$_3$/CH$_3$COOH spray reagent.

TABLE 2.
Excretion of radioactivity as % of dose following oral
administration of $[^{14}C]$niclosamide (20 mg; 19.5 µCi/
m.mole)

	Normal rats (over 144h)	Biliary-cannulated rats (over 48h)
Urine	51.5%	10.9%
Faeces	47.4%	2.3%
Bile	–	14.5%

Values are means of 3 experiments.

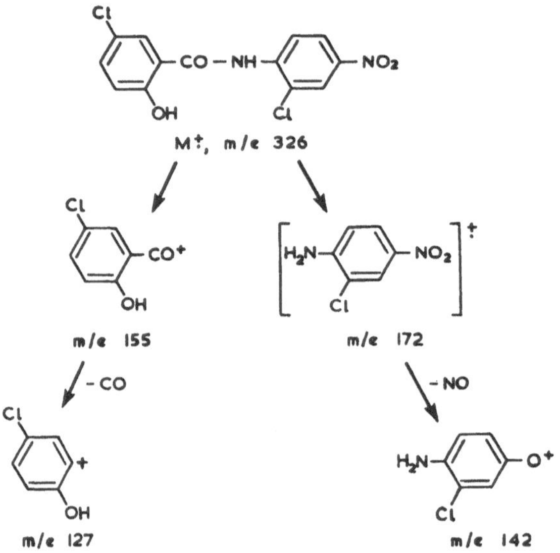

FIG. 1. Fragmentation of 2',5-dichloro-4-nitro-
salicylanilide (niclosamide), the aglycone
of the biliary conjugate.

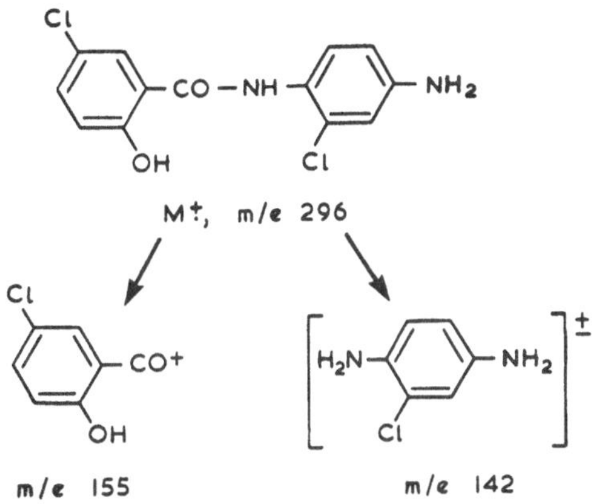

FIG. 2. Fragmentation of the faecal metabolite,
2',5-dichloro-4'-aminosalicylanilide.

NU_2, NU_3 and NU_4 were detected in urine. NU_2 was shown to be an N-
-glucuronide of 2',5-dichloro-4'-amino-salicylanilide being
chromatographically identical with a product obtained by direct
interaction of the amino-derivative of niclosamide with glucuronic
acid in phosphate buffer at pH 6.4.

The major metabolite of faeces was shown to be unchanged
niclosamide although considerable amounts of 2',5-dichloro-4'-
-salicylanilide were also present (Table 1). The latter metabolite
was shown to be formed on incubation with the intestinal microflora
in vitro under anaerobic conditions. Partial suppression of
dichloro-amino-salicylanilide formation was shown in vivo when
niclosamide was administered to antibiotic-treated rats. It is
thought that the limited formation of dichloro-amino-salicylanilide
observed was due to incomplete suppression of the microflora by
the antibiotics used.

On intraperitoneal administration to biliary-cannulated rats
where contact between the drug and the intestinal microflora is
excluded, it was notable that no reduction of the niclosamide
occurred in the tissues. This indicates that in the intact animal
despite the demonstrated absorption of niclosamide, tissue enzymes
do not significantly contribute to the formation of dichloro-amino-
-salicylanilide.

DISCUSSION

Contrary to the often stated view that the drug niclosamide is not significantly absorbed from the mammalian gastrointestinal tract, the current investigation has shown that considerable absorption of the drug and its reduced metabolite, 2',5-dichloro--4'-amino salicylanilide does occur. This is evidenced by the high level of excretion of metabolites in urine and bile. Despite rapid uptake from the gut, it is evident that the therapeutically-active levels maintained in the lower intestine are due not only to non--absorbed niclosamide but also to niclosamide released from the biliary conjugate by the β-glucuronidase of the intestinal microflora, since no conjugated niclosamide was found to be eliminated in the faeces of animals with an intact bile duct.

Niclosamide is known to be a very safe drug for the treatment of human intestinal cestode infections and this freedom from toxicity has been attributed to low absorption from the gut (4). It is however evident that other explanations for low mammalian toxicity must now be sought.

REFERENCES

1) L.S. Goodman and A. Gilman, Eds., "The Pharmacological Basis of Therapeutics", Macmillan, London, 1975, p. 1024.
2) Martindale, "The Extra Pharmacopoeia", N.W. Blacow, Ed., Pharmaceutical Press, London, 1973, p. 151.
3) P.G.C. Douch and H.M. Gahagan, Xenobiotica, 1977, 5, 301.
4) R. Strufe and R. Gonnert, Arzneimittel-Forsch., 1960, 10, 886.
5) L.A. Griffiths and G.E. Smith, Biochem. J., 1972, 128, 901.
6) A. Leulier and L. Pinet, Bull. Soc. Chim., 1927, 41, 1363.
7) A. Barrow and L.A. Griffiths, Xenobiotica, 1972, 2, 575.
8) R. Gingell, J.W. Bridges and R.T. Williams, Xenobiotica, 1971, 1, 143.

THE METABOLISM OF DEUTERIUM-LABELLED ANALOGUES OF Δ^1-,Δ^6-, and Δ^7-

TETRAHYDROCANNABINOL AND THE USE OF DEUTERIUM LABELLING

D.J. Harvey and W.D.M. Paton

University Department of Pharmacology

Oxford, England

INTRODUCTION

The _in vivo_ and _in vitro_ metabolism of Δ^1-tetrahydrocannabinol (Δ^1-THC,I), the main psychoactive principle of _Cannabis sativa L._, has been extensively studied in several species and to date some

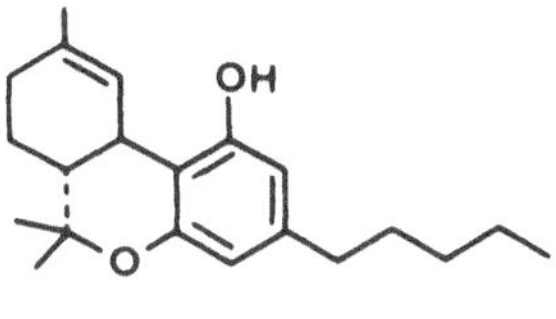

Δ^1-THC

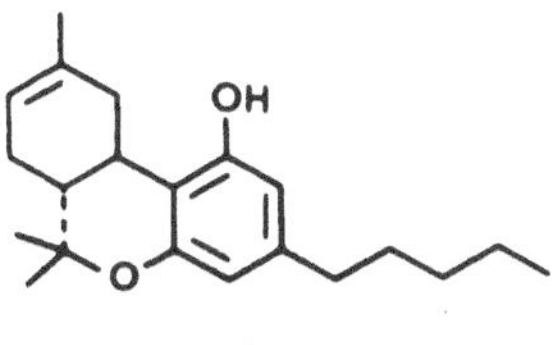

Δ^6-THC

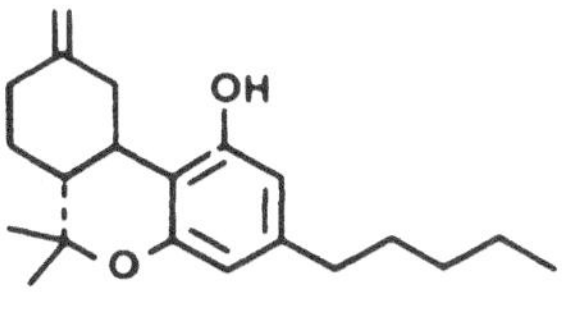

Δ^7-THC

Structures.

127

forty to fifty metabolites have been identified (1-4). Fewer studies
have been made with the isomeric Δ^6-THC (II), but from reported
work it would appear that similar metabolic routes are involved
(5, 6). The metabolism of the inactive Δ^7-THC (III) does not seem
to have been reported. The major biotransformation pathways of Δ^1-
and Δ^6-THC involve allylic and aliphatic hydroxylations, oxidation
to acids and ketones, β-oxidation of the pentyl side-chain, epoxide
formation and hydrolysis, reduction of the double bond and conjugation
with glucuronic acid. Metabolites containing up to three metabolically
introduced groups have been reported. Many of these metabolites are
present in only very small amounts (1-10 ng/g of tissue) and thus
their isolation in quantities sufficient for the recording of NMR
or IR spectra is difficult. GC-MS although having the sensitivity
to record spectra of such small amounts of material, does not
always yield sufficient information for full structure elucidation.
Although information on the nature of functional groups can be gained
quite easily by the variation of the derivatives usually employed
in GC-MS work, determination of position and stereochemistry of
functional groups is more difficult, particularly as some isomeric
compounds have very similar mass spectra.

In order to exploit the sensitive GC-MS technique fully we have
been investigating methods for obtaining the maximum structural
information from the spectra and in this paper we discuss various
ways in which deuterium labelling can be used to examine the
metabolism of Δ^1, Δ^6-, and Δ^7-THC.

EXPERIMENTAL

Materials. Δ^1-THC was obtained from the National Institute for
Drug Abuse.

1",1",2",2"-$[^2H_4]\Delta^1$-THC. 3,5-Dimethoxy-1-n-(1',1',2',2'-
$[^2H_4]$pentyl) benzene was prepared by α-exchange of the 2-hydrogens
of 1(3,5-dimethoxy)-1-pentanone followed by hydrogenation using
deuterium and a Pd/C catalyst. This was demethylated to $[^2H_4]$-
-olivetol and condensed with (+)-trans-mentha-2,8-dien-1-ol to give
1",1",2",2"-$[^2H_4]$-Δ^1-THC (7) of isotopic purity: $[^2H_0]$, 2.9%;
$[^2H_1]$, 2.9%; $[^2H_2]$, 7.2%; $[^2H_3]$, 17%; $[^2H_4]$, 70%. Full
details will be published elsewhere.

3-$[^2H_1]\Delta^6$-THC. Verbenone was reduced with lithium aluminium
deuteride to $[^2H_1]$-verbenol (8) and this was condensed with
olivetol (9) to give 3-$[^2H_1]\Delta^6$-THC which was purified by chromatography on Florisil.

3-$[^2H_1]\Delta^1$-THC. 3-$[^2H_1]\Delta^6$-THC was isomerised to 3-$[^2H_1]\Delta^1$-
-THC in 45% yield with zinc chloride followed by potassium t-amyl
oxide (10).

3-$/^-{^2}H_1$ $/\Delta^7$-THC. Irradiation of 3-$/^-{^2}H_1$ $/\Delta^6$-THC with a Hg vapour lamp for 1 hour (11) followed by chromatography on Florisil gave 3-$/^-{^2}H_1$ $/$-Δ^7-THC in 50% yield. All three 3-$/^-{^2}H_1$ $/$ THC's contained 83% deuterium enrichment. Full details of their preparation will be published elsewhere.

Dosage of animals. Nine male Charles River CD-1 mice (23-25 g) were used to study metabolism of each THC. They were divided into three groups of three and treated i.p. with the drug suspended in Tween-80 and isotonic saline at a concentration of 100 mg/kg as follows:

Group a   THC
Group b   THC and $/^-{^2}H$ $/$THC (1 : 1 molar)
Group c $/^-{^2}H$ $/$ THC.

The animals were killed after one hour and the livers were removed.

Extraction and separation of metabolites. Livers were homogenized in isotonic saline (10 ml) and the metabolites and non-polar lipids were extracted with ethyl acetate (3 x 20 ml) and separated on Sephadex LH-20 with chloroform and chloroform : methanol mixtures as described previously (3, 12). Five fractions were obtained as shown in Table 1. Extracts from the three mice in each group were pooled. Aliquots representing the extract from about 200 mg of liver were converted into trimethylsilyl (TMS), $/^-{^2}H_9$ $/$ TMS, methyl ester-TMS, methyloxime-TMS or alkylboronate derivatives or were reduced with lithium aluminium deuteride as described (3).

GC-MS. Low resolution GC-MS was performed with a VG Micromass 12B mass spectrometer interfaced to a VG 2040 data system and via a glass jet separator to a Varian 2400 GLC. The chromatograph was fitted with a 2 m x 2 mm glass column packed with either 3% SE-30 or 3% OV-17 on 100-120 mesh Gas Chrom Q (Applied Science Laboratories Inc., State College, PA, USA). The column was temperature programmed from 170° to 300° at 2°/min, the injector, separator and ion source temperatures were 300°, 250° and 260° respectively and data acquisition was started when the column reached 190°. Spectra were recorded at 25 eV, continuously throughout the elution of the chromatogram using a scan speed of 3 sec/decade and an inter-scan delay of 2 sec. The mass spectrometer was operated with an accelerating voltage of 2.5 KV and a trap current of 100 μA. Ion chromatograms were plotted using the "Mass Max" technique (13).

High resolution GC-MS and direct probe mass spectra were recorded with a VG Micromass 70/70F mass spectrometer interfaced to the data system and a similar chromatographic system to that described above. The mass spectrometer was operated with a resolution of 5000, an accelerating voltage of 4 KV, a trap current of 200 μA and an electron energy of 70 eV.

TABLE 1.
Fractionation on Sephadex LH-20 of liver extracts from THC treated mice

Fraction	Solvent	Vol.	Contents		
			Δ^1-THC	Δ^6-THC	Δ^7-THC
1	CHCl$_3$	18 ml	triglycerides	triglycerides	triglycerides
2	CHCl$_3$	8	Δ^1-THC	Δ^6-THC,epoxide	Δ^7-THC
3	CHCl$_3$	10	fatty acids	fatty acids	fatty acids
4	CHCl$_3$	35	mono-OH,keto-OH	mono-OH	mono-OH
5	20%/MeOH/CHCl$_3$	50	polar metabs.	polar metabs.	polar metabs.

TABLE 2.
Summary of the major hydroxy, ketone and acid
metabolites of Δ^1-THC produced by the mouse.

No	R^1	R^2	R^3	R^4	R^5
1a	CH$_2$OH	H	H	H	H
1b	H	α-OH	H	H	H
1c	H	β-OH	H	H	H
1d	CH$_2$OH	α-OH	H	H	H
1e	CH$_2$OH	β-OH	H	H	H
1f	CH$_2$OH	H	OH	H	H
1g	CH$_2$OH	H	H	OH	H
1h	CH$_2$OH	H	H	H	OH
1i	CH$_2$OH	α-OH	OH	H	H
1j	CH$_2$OH	α-OH	H	OH	H
1k	CH$_2$OH	α-OH	H	H	OH
1l	COOH	H	H	H	H
1m	COOH	α-OH	H	H	H
1n	COOH	H	OH	H	H
1o	COOH	H	H	OH	H
1p	COOH	H	H	H	OH
1q	COOH	α-OH	OH	H	H
1r	COOH	α-OH	H	OH	H
1s	COOH	α-OH	H	H	OH
1t	CH$_2$OH	=O	H	H	H
1u	H	=O	OH	H	H
1v	H	=O	H	OH	H
1w	H	=O	H	H	H

Multiple ion monitoring was performed with the VG Micromass 70/70F mass spectrometer fitted with an 8 channel peak selector.

RESULTS AND DISCUSSION

In order to reduce losses of metabolites by adsorption or degradation and to make full use of the GC–MS technique, pre–GLC clean–up procedures were kept to a minimum and GLC was relied on to achieve separation of the metabolites. Ethyl acetate was used to extract the metabolites but this solvent also extracted the high molecular weight neutral lipids such as triglyceride and cholesterol. The latter, present in large amounts, were separated by gel filtration on Sephadex LH–20 as outlined in Table 1. This method gave satisfactory results with the production of fractions containing large numbers of metabolites but few tissue constituents. Further fractionation of fraction 5 using graded elution with chloroform : methanol mixtures separated many of the more polar metabolites into groups of similar compounds such as diols, but small quantities of the more polar compounds were sometimes lost and were thus best observed in the more complex fraction 5. Separation could usually be achieved by varying the GLC phase.

Most of the metabolites of Δ^1-THC identified in fractions 4 and 5 were found to contain from 1 to 3 additional groups at positions 2"-, 3"-, 4"-, 6α-, 6β-, or 7-. All positions were hydroxylated and, additionally oxidation to a ketone or an acid occurred at positions 6- and 7- respectively (Table 2). Δ^1-THC-7-oic acid and its mono- and di-hydroxylated derivatives were the most abundant metabolites (Fig. 1). Δ^6-THC was metabolised in a similar way (Fig. 2) with allylic hydroxylation at positions 5α-, 5β-, and 7-, and side-chain hydroxylation again at positions 2"-, 3"-, and 4"- (Table 3). Δ^7-THC metabolism was somewhat different (Fig. 3) as the 7-position could not undergo allylic hydroxylation, the process which produced tha major metabolites with the other isomers. Instead the metabolites could best be rationalized in terms of initial epoxidation of the double bond, a mechanism which occurred to a minor extent with Δ^6-THC. However the intermediate epoxide was not observed. Hydrolysis of this epoxide gave two isomers 1,7- -diols IIIh (Scheme 1) which were then hydroxylated further in the side chain, again at positions 2"-, 3"-, and 4"-. Rearrangement of the epoxide gave two 7-oxo-HHC's (Scheme 1) again not isolated and these, in turn were either oxidized to a pair of HHC-7-oic acids, one of which was present as a major metabolite or reduced to 7- -hydroxy-HHC's. Again side-chain hydroxylation of these metabolites was observed. Allylic hydroxylation of Δ^7-THC was a minor metabolic pathway.

Much information on the nature of these metabolites was obtained

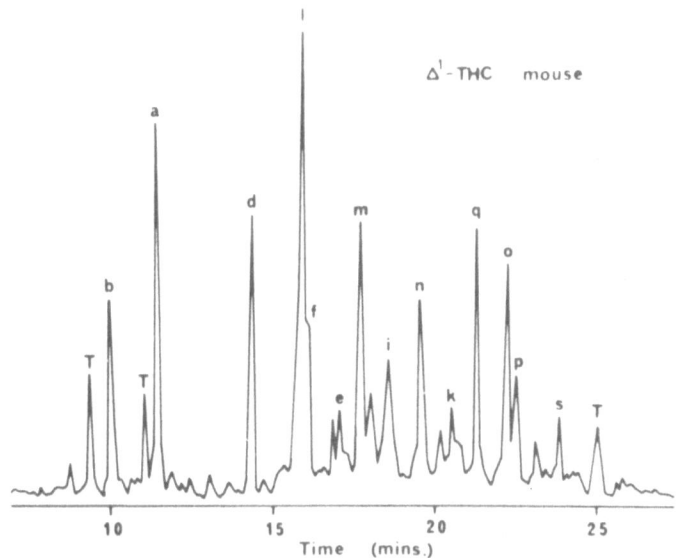

FIG. 1. Limited ion (m/e 300-700) chromatogram produced using the Mass Max technique of the metabolites of Δ^1-THC (TMS derivatives) extracted from mouse liver. Compound identification is given in Table 2. Separation was with a 3% SE-30 column using the conditions given in the experimental section. Peaks marked T are tissue constituents.

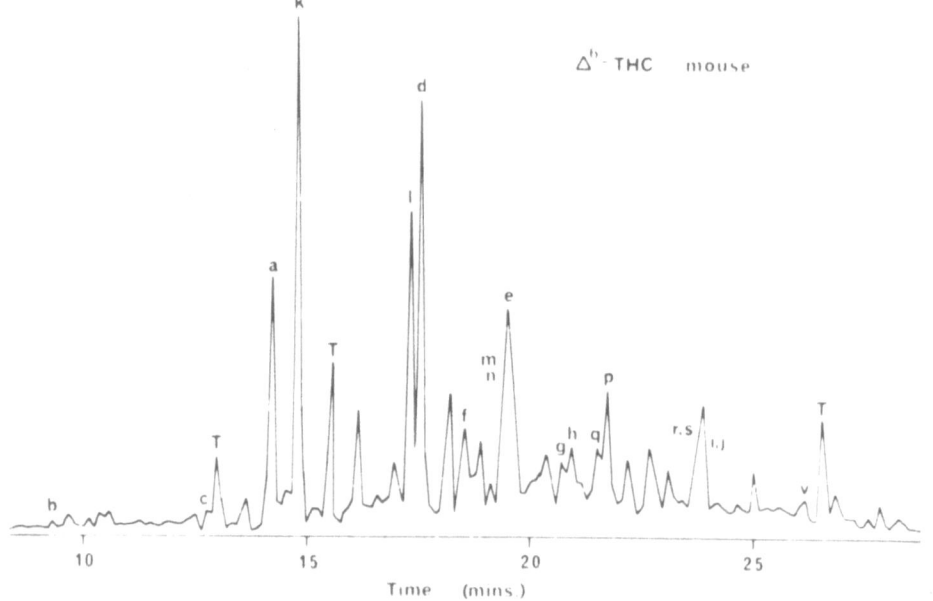

FIG. 2. Mass Max chromatogram (m/e 300-700) of the
Me-TMS derivatives of the metabolites of
Δ^6-THC extracted from mouse liver.
Separation conditions are as for Fig. 1,
peak identification is given in Table 3.

TABLE 3
Summary of the major metabolites of Δ^6-THC produced
by the mouse (See Fig. 2).

No.	R^1	R^2	R^3	R^4	R^5
IIa	CH_2OH	H	H	H	H
IIb	H	α-OH	H	H	H
IIc	H	β-OH	H	H	H
IId	CH_2OH	α-OH	H	H	H
IIe	CH_2OH	β-OH	H	H	H
IIf	CH_2OH	H	OH	H	H
IIg	CH_2OH	H	H	OH	H
IIh	CH_2OH	H	H	H	OH
IIi	CH_2OH	α-OH	H	OH	H
IIj	CH_2OH	α-OH	H	H	OH
IIk	COOH	H	H	H	H
IIl	COOH	α-OH	H	H	H
IIm	COOH	β-OH	H	H	H
IIn	COOH	H	OH	H	H
IIo	COOH	H	H	OH	H
IIp	COOH	H	H	H	OH
IIq	COOH	α-OH	OH	H	H
IIr	COOH	α-OH	H	OH	H
IIs	COOH	α-OH	H	H	OH
IIt	COOH	β-OH	OH	H	H
IIu	COOH	β-OH	H	OH	H
IIv	COOH	β-OH	H	H	OH

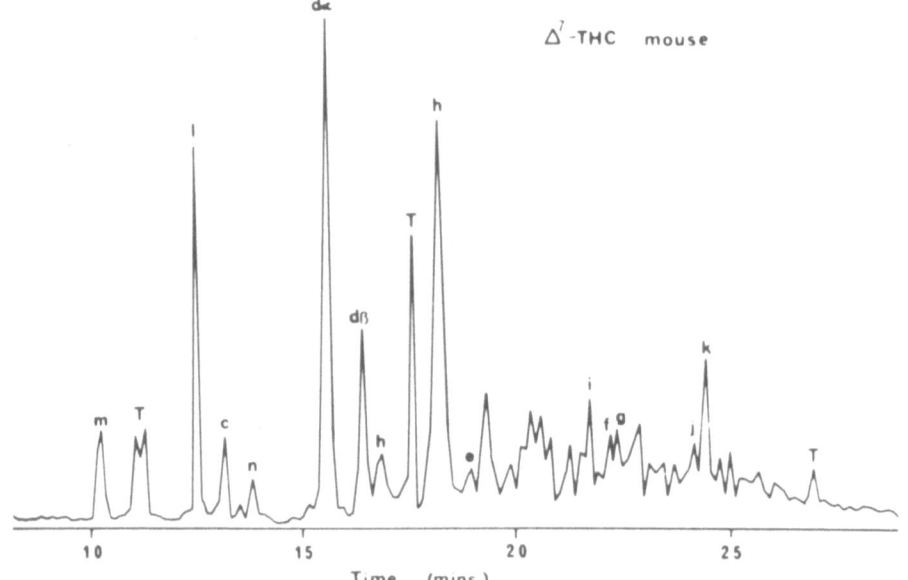

FIG. 3. Mass Max chromatogram (m/e 300-700) of the
metabolites of Δ^7-THC (TMS derivatives)
extracted from mouse liver. Separation
conditions are given in Fig. 1, peak
identification is as shown in Scheme 1 with
α and β to signify the different isomers
where stereochemistry is known. Peaks
marked T are tissue constituents.

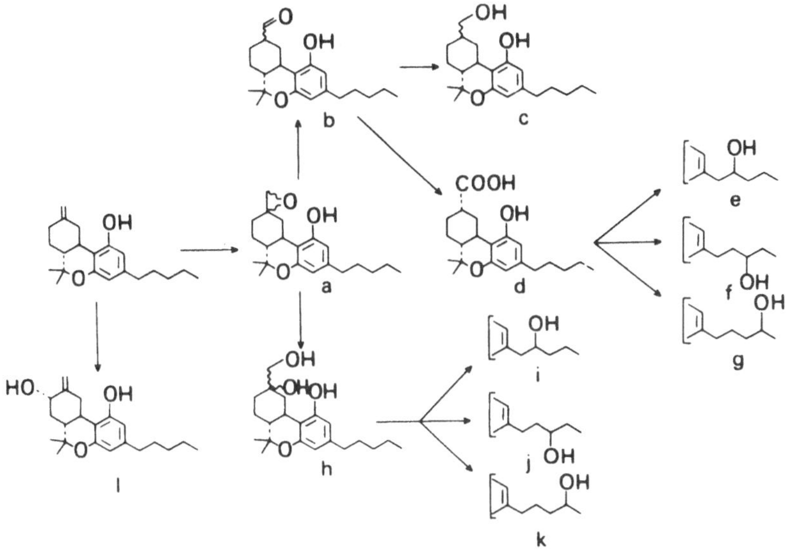

SCHEME 1. Outline of Δ^7-THC metabolism. Each of
the metabolites a - k is accompanied by
its C - 1 epimer.

by the use of deuterium labelling in the following ways: (a) The
isotope doublet technique; (b) Use of deuterated derivatives; (c)
Reduction with lithium aluminium deuteride; (d) Metabolism of
labelled cannabinoids: i. displacement of deuterium (metabolic
switching), ii. mass spectral fragmentation involving loss of
deuterium.
  (a) Isotope doublet technique. Chromatography of the liver
extracts on Sephadex LH-20 gave reasonably clean metabolic fractions.
However as separation was mainly on the basis of gel filtration,
residual liver constituents in the metabolic fractions generally
had similar molecular weights and polarities to the metabolites.
Gas chromatography on non-polar phases such as SE-30 did not always
produce good separation of these compounds and in some cases these
were easily confused with metabolites. A vivid example of this was
the similarity in retention time and mass spectra of the TMS
derivatives of 7-hydroxy-Δ^1-THC and monopalmitin from the liver.
Both compounds formed bis-TMS derivatives with the same molecular
weight (474) and both fragmented by loss of CH_2OTMS to give the
base peak at m/e 371. In the absence of reference spectra it would
be easy to confuse these compounds and as THC and its metabolites
in general have similar molecular compositions to many neutral
lipids, there is a real danger that other tissue constituents could
be confused with metabolites. The isotope doublet technique (14, 15)

is thus valuable in any study of cannabinoid metabolism as a method
for the rapid identification of GLC peaks produced by metabolites.

The metabolism of Δ^1-THC was examined using two deuterated
analogues, the 1",1",2",2"-$/^{-2}H_4\_/$- and 3-$/^{-2}H_1\_/$ - species. The
compounds were administered individually and as a 1 : 1 molar
mixture for the isotope doublet study. Of the two, 1",1",2",2"-
-$/^{-2}H_4\_/$ Δ^1-THC gave the clearest results as the doublet peaks were
separated by 4 amu. With the 1 amu separation of Δ^1-THC and 3-$/^{-2}H_1\_/$
Δ^1-THC it was sometimes difficult to find the doublets in complex
chromatograms where many of the GLC peaks had several components.
Fig. 4 shows the spectra of a 1 : 1 molar mixture of the unlabelled
and labelled analogues of the TMS derivative of Δ^1-THC-7-oic acid.

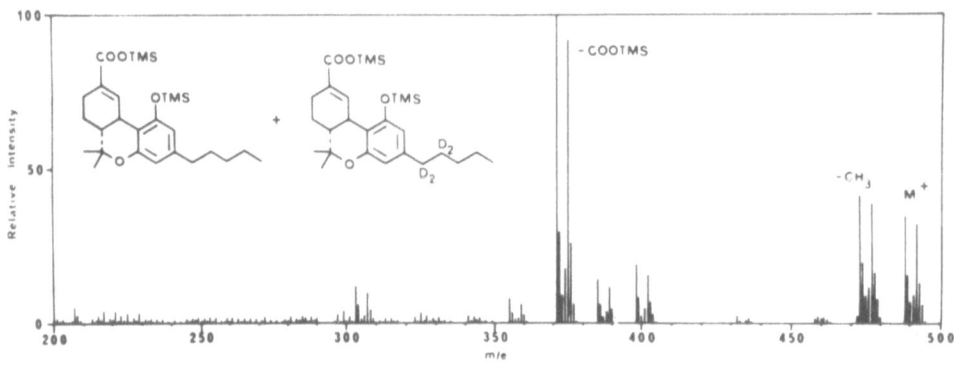

FIG. 4. 25 eV Mass spectrum of the TMS derivative
of Δ^1-THC-7-oic acid produced by metabolism
of a 1 : 1 molar mixture of Δ^1-THC and 1",
1",2",2"- $/^{-2}H_4\_/$ Δ^1-THC.

It is important with the isotope doublet technique that the deuterium
is sited at a position from which it is not lost during metabolism
and for this reason 3-$/^{-2}H_1\_/$ Δ^1-THC had the advantage as no 3-
-substituted metabolites of Δ^1-THC are known. 2"-Hydroxylation of
Δ^1-THC by the mouse leads to major metabolites (3) and thus the
isotope doublet technique using 1",1",2",2"-$/^{-2}H_4\_/\Delta^1$-THC could not
be used as effectively to identify metabolites substituted in this
position.
     (b) Use of deuterated derivatives. The first stage in the
identification of the metabolites involved the determination of the
number and nature of the functional groups. This was achieved by
the preparation of several derivatives such as, for example, TMS,
$/^{-2}H_9\_/$ TMS, methyl ester-TMS, and methyloxime TMS derivatives. The
difference between the molecular weights of the TMS and $/^{-2}H_9\_/$TMS
(16) derivatives gave the total number of hydroxyl and carboxylic

acid groups; the latter could be identified by the formation of
methyl esters. Methyloximes were used to identify ketones although
in the case of the 5-oxo-Δ^6-THC metabolites, steric factors prevented
their formation and identification had to be based on reduction to
the corresponding alcohol with aluminium deuteride and determination
of the deuterium content as described below.

(c) Reduction with lithium aluminium deuteride.Having established
the number and nature of the functional groups, the next stage was
to establish their position. TMS groups tended to direct the
fragmentation of THC metabolites to produce, in many cases, base
peaks which were characteristic of the position of hydroxyl
substitution. For example the side-chain hydroxyl group produced
ions at $\sqrt{\text{M-C}_4\text{H}_9}$ $\sqrt{}^+$, m/e 145, $\sqrt{\text{M-C}_4\text{H}_7\text{OTMS}}$ $\sqrt{}^{+\cdot}$, and m/e 117 when
the hydroxy group was in the 1"-, 2"-, 3"-, or 4"- positions
respectively (17, 18). Ring hydroxylation led to less specific ions
as substituents such as the 6α-OTMS group of 6α-hydroxy-Δ^1-THC and
the 5α group in 5α-hydroxy-Δ^6-THC both give abundant $\sqrt{\text{M-TMSOH}}$ $\sqrt{}^{+\cdot}$
ions. Losses of CH_2OTMS could be interpreted in terms of a 7-, 9-,
or 10- hydroxy metabolite. A convenient way to reduce this pattern
was by the interconversion of metabolites or their conversion into
known compounds. Advantage was taken of the fact that the
metabolites of Δ^1-THC fall into several series, the members of which
contain substitution in equivalent positions as outlined in Table
2. Thus the diols containing 7-hydroxylation together with the
second hydroxyl group at positions 6α-, 2"-, 3"- or 4"- were
accompanied by the corresponding 7-oic acids. Reduction of the
metabolic fraction with lithium aluminium hydride or deuteride
reduced all keto and acid metabolites to mono-, di-, or tri-
-hydroxy species. By using lithium aluminium deuteride, the
proportion of the alcohol which had been produced by reduction of
a ketone, or acid metabolite in which equivalent positions were
substituted, could readily be determined by the incorporation of
one or two deuterium atoms respectively. This technique provided
a good method for correlating the structures of related metabolites
with known compounds. With Δ^1-THC metabolites for example, Δ^1-THC-
-7-oic acid was reduced to 7-hydroxy-Δ^1-THC and 6α-hydroxy-Δ^1-THC-
-7-oic acid gave the 6α,7-diol; standards were available for both
of these compounds. Fig. 5 shows the spectrum of the TMS derivative
of 5β,7-dihydroxy-Δ^6-THC produced by metabolism and by reduction
of various oxygenated species.

(d) Metabolism of labelled cannabinoids. Another method for the
determination of the position of substitution, applicable to several
metabolites simultaneously, was to administer a deuterium labelled
analogue of the compound and look for metabolites in which the
deuterium had been replaced.

i. Displacement of deuterium. The metabolism of deuterated
drugs has been investigated quite extensively in recent years (19)
and it is apparent that where biotransformation involves the
breaking of a carbon-deuterium bond, the primary kinetic isotope

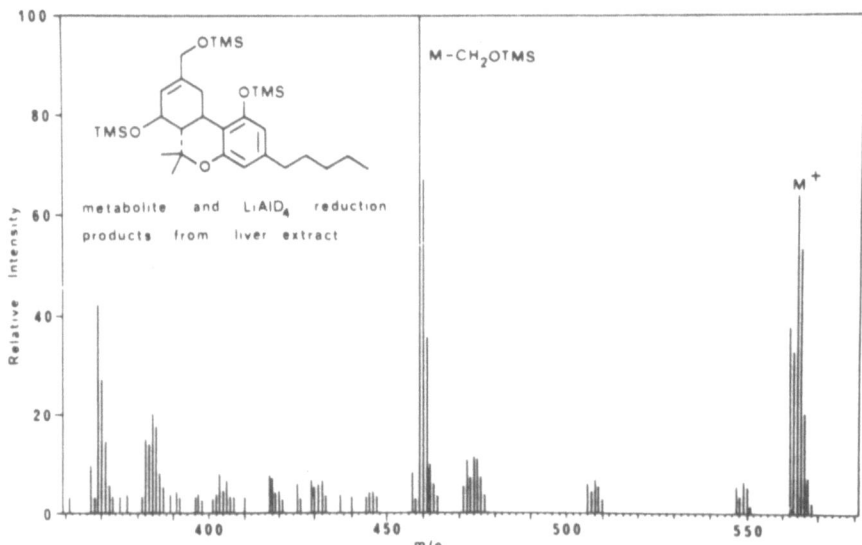

FIG. 5. 25 eV Mass spectrum (m/e 360-570) of the TMS
derivative of 5β,7-dihydroxy-Δ⁶-THC produced
by reduction of the mouse liver extracts with
lithium aluminium deuteride. The molecular
ion shows the diol metabolite (m/e 562) and
ions containing one, two or three deuterium
atoms produced by reduction of a hydroxy-
-ketone, a hydroxy-acid and a keto-acid
respectively. The loss of two of the
derivative atoms in the $\underline{/}$ M-CH-OTMS $\underline{/}^{+}$· ion
shows that the acid was in the 7-position.

effect causes a reduction in the rate of the reaction. This can be
illustrated, for example, by the prolongation of sleeping time
produced by 3'-$\underline{/}$ ^{2}H$_2$ $\underline{/}$ butethal over that produced by the unlabelled
drug (20). In this instance the presence of deuterium slows
metabolic hydroxylation and increases the plasma half life. In cases
where the drug is metabolised by several alternate pathways,
incorporation of deuterium at one of the sites of attack will
divert the metabolism in favour of the other pathways. This effect,
know as metabolic switching (21), has been observed for drugs such
as caffeine and antipyrine (21). Because metabolic replacement of
deuterium is of obvious use in identifying sites of metabolic
hydroxylation we investigated the metabolism of 1",1",2",2"-
-$\underline{/}$ ^{2}H$_4$ $\underline{/}$Δ¹-THC, originally synthesised as an internal standard for
quantitative work, as a model. 2"-Hydroxylation by the mouse is a
major metabolic route but 1"-hydroxylation has not been observed.
Any secondary isotope effect produced by the 1"-deuterium atoms
would not be expected to alter the hydroxylation of the 2"-position

significantly. Pronounced metabolic switching was observed when the
metabolites of 1",1",2",2"-$[^2H_4]$-Δ^1-THC were examined. The abundance
of the two major 2"-hydroxy-containing metabolites, 2"-hydroxy-Δ^1-
-THC-7-oic acid and 2",6α-dihydroxy-Δ^1-THC-7-oic acid were reduced
to 10% and 20% of their original values respectively as recorded
by multiple ion monitoring. The same effect is expected to occur
with deuterium substitution at other sites of metabolism. One
advantage of this technique was that it enables the identification
in one GC-MS run of all metabolites containing substitution at the
site of deuteration. Reduction of the concentration of major
metabolites simplified chromatograms and facilitated identification
of minor constituents with similar retention times. Another possible
use of this method would be to provide information on the relative
pharmacological importance of hydroxylated metabolites by alterations
in the activity of the drug.

As a side-line to this study, the numbers of deuterium atoms
in the fragment ions derived from cleavage of the side-chain were
all found to be consistent with the proposed mechanisms for their
formation (17, 18). Thus cleavage between C-1" and C-2" took place
with the 2"- and 3"-hydroxy metabolites to give ions at m/e 145
and $[M-144]^{+\cdot}$ respectively. It has already been confirmed by
deuterium labelling that formation of the $[M-144]^{+\cdot}$ ion involves
transfer of a hydrogen from C-3" to the cannabinoid nucleus (3).
    ii. Mass spectral fragmentations involving loss of deuterium.
With several of the metabolites, the stereochemistry of the
substituents was difficult to determine. In some cases, for example,
the spectra of pairs of isomers were similar, in other cases the
only difference was the presence in the spectrum of one of a pair
of isomers, of a rearrangement ion, the formation of which could be
interpreted in terms of either isomer. Rearrangement ions, involving
interaction of groups from different parts of the molecule are
sensitive to stereochemical changes. Thus if the atom or group with
which the metabolic substituent interacts, is known, deductions can
be drawn about the substituent's stereochemistry.

Many mass spectral fragmentations involve the interaction of
a group such as TMSO with hydrogen and the expulsion of TMSOH as a
neutral molecule. In sterically rigid systems such as cyclohexane
rings, such eliminations require that the two groups come to within
bonding distance providing that ring cleavage does not precede
fragmentation, and usually involve 1 : 3 diaxial interactions (22).
However, with interactions such as the formation of the m/e 147 ion
(TMSO$^+$ = SiMe$_2$) from bis-trimethylsilyloxy steroids (23), changes
in ring conformation can occur and the abundance of the resulting
ion can be correlated with the position of substitution only when
that conformation which brings the reacting groups into closest
proximity is considered. As many of the hydroxylated metabolites of
the THC'S contain hydroxylation in a cyclohexene ring and as many
of the fragmentations involve the elimination of the derivatized

hydroxyl group as TMSOH, it was hoped that by labelling the
abstracted hydrogen, information on the stereochemistry of the
hydroxyl group could be gained. Such information would of course
be applicable to all metabolites containing hydroxylation at
corresponding positions providing similar losses of TMSOH were
produced. Since Δ^1- and Δ^7-THC were readily prepared directly from
Δ^6-THC, the preparation of a deuterium substituted Δ^6-analogue
readily afforded deuterated analogues in the other series.

The 3-position was chosen as the best site for deuterium for
three reasons: – it is axial (α-face) to several positions on the
terpene ring (5α, 1α in the chair conformation; 6α and 2α in the
boat form); – it is benzylic and thus fairly active; and 3-
hydroxylation has not been observed in cannabinoids and thus
metabolic replacement or metabolic switching could be used to
identify such a new series of compounds.

Loss of TMSOH from several known hydroxylated metabolites of
Δ^6 and Δ^1-THC were examined and good correlation was found between
the ability of the hydroxy group to abstract the 3-deuterium and
the stereochemistry of the hydroxy group (Table 4). Thus in the
Δ^1-THC series the TMS derivative of the 6α-hydroxy metabolite,
whose mass spectrum was dominated by the $/\overline{M-TMSOH}\,/^{+\cdot}$ ion, showed
quantitative elimination of the 3-deuterium. The same was true of
all the more highly substituted metabolites containing 6α-hydroxyl-
ation. The 6α-hydroxy group is in the α-axial position in the
half-boat conformation and suitably placed to abstract the deuterium.
No abstraction of the 3-deuterium was observed from metabolites
with 6β substitution. Similar behaviour was noted for the pair of
5-substituted Δ^6-THC'S.

Use of this method enabled several structural assignments
to be made. For example, metabolic reduction of Δ^6-THC-7-oic acid,
Δ^1-THC-7-oic acid (1), or oxidation of 7-oxo-HHC (from rearrangement
of the epoxy metabolites of Δ^7-THC) produced the same pair of
HHC-7-oic acids, IIIdα and IIIdβ

IIIdα IIIdβ

TABLE 4
Abstraction of the 3-hydrogen atom in fragmentations
involving loss ofTMSOH or HCOOTMS.

Cannabinoid	Substituent Eliminated	Other	Orientation[a]	Abstraction of 3-H
Δ^1-THC	6α	–	axial	+
"	6β	–	equatorial	–
"	6α	7-OH	axial	+
"	6α	7-COOH	axial	+
"	6β	7-OH	equatorial	–
Δ^6-THC	5α	–	axial	+
"	5β	–	equatorial	–
"	5α	7-OH	axial	+
"	5β	7-OH	equatorial	–
"	5α	7-COOH	axial	+
"	5β	7-COOH	equatorial	–
"	9	–	axial	+
HHC	1α	6α	axial	+
"	1α	6β	axial	+
"	1β	6β	equatorial	+ b
"	1α	7	axial	+ b
"	1β	7	equatorial	+ b
"	α-COOH	–	axial	+
"	β-COOH	–	equatorial	–
"	α-COOH	2"-OH	axial	+
"	α-COOH	3"-OH	axial	+
"	α-COOH	4"-OH	axial	+

a) Most stable conformation. The 1β groups, although axial in the
   boat forms project towards the β-face i.e. on the opposite side
   to the 3-$/^2$H$_/$.
b) Abstraction not expected, presumably preceded by ring opening.

whose compositions were confirmed by high reduction mass spectrometry. Their TMS derivatives both fragmented in a similar way
except for the presence in the spectrum of compound IIIdα of a
moderately abundant (29%) ion at m/e 372 (Fig. 6).

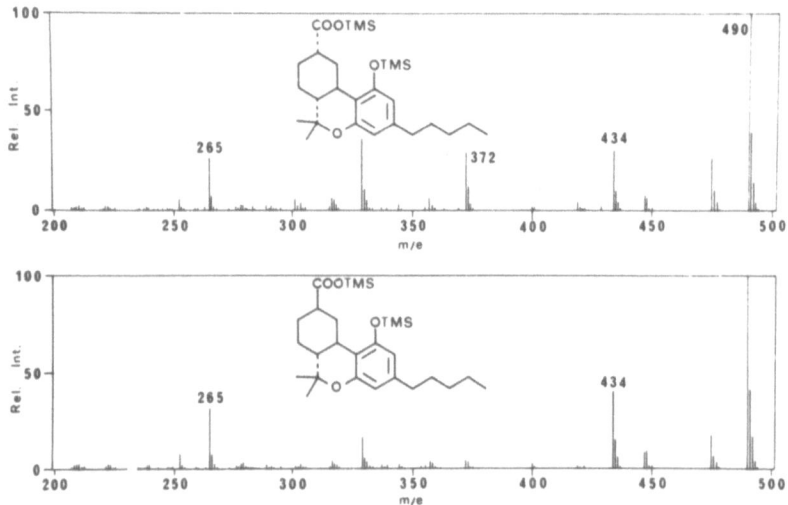

FIG. 6. 25 eV Mass spectrum (m/e 200-500) of the
TMS derivatives of the epimeric 3-$\sqrt{\ ^2H_1\ }$
HHC-7-oic acids.

Preparation of the $\sqrt{\ ^2H_9\ }$ TMS derivative showed this to be produced
by elimination of the COOTMS group together with a hydrogen. The
quantitative abstraction of the 3-deuterium in the spectrum of the
deuterated analogue clearly showed this isomer to be the axial (α)
compound (Fig. 7). This was subsequently confirmed chemically (1).
The other isomer could conceivably abstract either the 4β or 5β
hydrogens when in the boat form, thus the presence of the ion at
m/e 372 could be explained in terms of either structure if the
source of the reacting hydrogen was not known. Side-chain hydroxylated
analogues of these acids were also produced metabolically and
behaved in a similar way (Table 4).

     Unfortunately this method did not work with metabolites
hydroxylated in the 1-position. 1α,6β-Dihydroxy-HHC produced
metabolically by opening of the epoxide ring of 1β,6β-epoxy-HHC,
a metabolite of Δ^6-THC (24), formed a bis-TMS derivative under mild
conditions and a tris-TMS derivative under stronger conditions.
The former derivative had a free 1α-hydroxy group which was
eliminated as H_2O with abstraction of the 3-deuterium as expected,

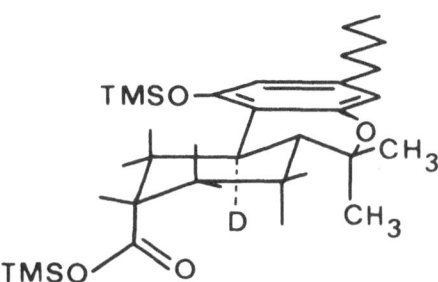

FIG. 7. Stereochemistry of the TMS derivative of the
α-isomer of 3-$\underline{/}$ $^2H_1$ $\underline{/}$HHC-7-oic acid.

and the tr̲i̲s̲-TMS derivative likewise eliminated TMSOH (elimination
of the 1α̲-̲T̲M̲S̲ group was confirmed by the preparation of the mixed
6β-TMS, 1α-$\underline{/}$ $^2H_9$ $\underline{/}$ TMS derivative (25)). A synthetic sample of
1α,6α-dihydroxy-HHC, prepared by the action of alkaline permanganate
on Δ^6-THC acetate, behaved similarly. However, a sample of 1β,6α-
-dihydroxy-HHC also showed the same behaviour with elimination of
the 1β group (checked by mixed TMS/$\underline{/}$ $^2H_9$ $\underline{/}$ TMS derivative). This
suggests prior ring cleavage and other ions in the spectrum of the
1β,6α-isomer ($\underline{/}$ M-131 $\underline{/}^+$ and $\underline{/}$ M-131-56 $\underline{/}^+$) tended to support this.
Likewise the TMS derivatives of 1α,7- and 1β,7-HHC, produced both
synthetically and metabolically from Δ^7-THC gave weak ions
corresponding to $\underline{/}$ M-TMSOH $\underline{/}^{+ \cdot}$ which showed loss of the 3-deuterium
from both isomers. Again this suggests ring cleavage with loss of
conformation and this was supported by the pronounced similarity
of their spectra; these were identical except for the presence of
a weak molecular ion from one compound.

The elimination of deuterium by the metabolic group provides
a valuable method for determining the stereochemistry of that
group providing that ring cleavage does not precede fragmentation.
Of the substituents in the 1,5, and 6 positions studied, only the
1-hydroxy substituents showed any significant sign of ring cleavage;
good correlations were found between deuterium abstraction and
functional group position for the other compounds. For structural
studies of unknown metabolites, therefore, it will probably be
necessary to observe the behaviour of the deuterium in each of the

pair of isomers in order to be sure that the elimination reaction proceeds with retention of the stereochemical relationship between the groups.

CONCLUSIONS

The examples described above illustrate some of the many ways in which deuterium can be used to obtain qualitative information on the structure of THC metabolites. These are by no means exhaustive; other experiments currently under way included the determination of the position of ketone group substitution by counting the number of α-hydrogens following replacement by deuterium. Deuterated THC'S are of course valuable internal standards for quantitative work and this will be the subject of further communications.

ACKNOWLEDGEMENTS

We thank the Medical Research Council for a programme research grant, Dr. E.W. Gill and Mr. M. Slater for the preparation of the deuterium labelled cannabinoids, and Miss J. Hughes and Mr. D. Perrin for expert technical assistance.

REFERENCES

1) D.J. Harvey, B.R. Martin and W.D.M. Paton, J. Pharm. Pharmacol., 1977, 29, 495.
2) D.J. Harvey, B.R. Martin and W.D.M. Paton, in "Mass Spectrometry in Drug Metabolism", A. Frigerio and E.L. Ghisalberti, Eds., Plenum, New York, 1977, p. 403.
3) D.J. Harvey, B.R. Martin and W.D.M. Paton, in "Recent Developments in Mass Spectrometry in Biochemistry and Medicine", Vol. 1, A. Frigerio, Ed., Plenum, New York, 1978, p. 161.
4) S. Agurell, M. Binder, K. Fonseka, J.-E. Lindgren, K. Leander, B. Martin, I.M. Nilsson, M. Nordqvist, A. Ohlsson and M. Widman, in "Marihuana; Chemistry, Biochemistry and Cellular Effects", G.G. Nahas, W.D.M. Paton and J.E. Idänpään-Heikkila, Eds., Springer-Verlag, New York, 1976, p. 141.
5) H.D. Christensen, R.I. Freudenthal, J.T. Gidley, R. Rosenfeld, G. Boegli, L. Testino, D.R. Brine, C.G. Pitt and M.E. Wall, Science, 1971, 172, 165.
6) M.E. Wall, Ann. N.Y. Acad. Sci., 1971, 191, 23.
7) R.K. Razdan, H.C. Dalzell and G.R. Handrick, J. Amer. Chem. Soc., 1974, 96, 5860.
8) R. Mechoulam, P. Braun and Y. Gaoni, J. Amer. Chem. Soc., 1972, 94, 6159.

9) A. Focella, S. Teitel and A. Brossi, J. Org. Chem., 1977, 42, 3456.
10) T. Petrzilka, W. Haefliger and C. Sikemeier, Helv. Chim. Acta, 1969, 52, 1102.
11) J.L.G. Nilsson, I.M. Nilsson, S. Agurell, B. Akermark and I. Lagerlund, Acta Chem. Scand., 1971, 25, 768.
12) D.J. Harvey, B.R. Martin and W.D.M. Paton, J. Pharm. Pharmacol., 1976, 29, 482.
13) P. Powers, M.J. Wallington, J.A.V. Hopkinson and G.L. Kearns, Paper presented at the 23rd Annual Conference on Mass Spectrometry and Allied Topics, American Society for Mass Spectrometry, Houston, Texas, Proceedings, 1975, p. 499.
14) D.R. Knapp, T.E. Gaffney and R.E. McMahon, Biochem. Pharmacol., 1972, 21, 425.
15) D.R. Knapp, T.E. Gaffney, R.E. McMahon and G. Kiplinger, J. Pharmacol. Exp. Ther., 1972, 180, 784.
16) J.A. McCloskey, R.N. Stillwell and A.M. Lawson, Anal. Chem., 1968, 40, 233.
17) M. Binder, S. Agurell, K. Leander and J.-E. Lindgren, Helv. Chim. Acta, 1974, 57, 1626.
18) M. Binder, in "Marihuana; Chemistry, Biochemistry and Cellular Effects", G.G. Nahas, W.D.M. Paton and J.E. Idänpään-Heikkilä, Eds., Springer-Verlag, New York, 1976, p. 159.
19) M.I. Blake, H.L. Crespi and J.J. Katz, J. Pharm. Sci., 1975, 64, 367.
20) M. Tanabe, D. Yasuda, S. LeValley and C. Mitoma, Life Sci., pr. I, 1969, 8, 1123.
21) M.G. Horning, J.-P. Thenot, K. Haegele, J. Nowlin, M. Stafford and K.R. Sommer, Fed. Proc., 1976, 35, 488.
22) J. Karliner, H. Buzikiewicz and C. Djerassi, J. Org. Chem., 1966, 31, 710.
23) S. Sloan, D.J. Harvey and P. Vouros, Org. Mass Spectrom., 1971, 5, 789.
24) D.J. Harvey and W.D.M. Paton, J. Pharm. Pharmacol., 1977, 29, 498.
25) P. Vouros and D.J. Harvey, Anal. Chem., 1973, 45, 7.

KINETICS OF DEGENERATE NUCLEOPHILIC EXCHANGE OF C(3)-HYDROXY GROUP IN 1-METHYL-3-HYDROXY-5-PHENYL-7-CLORO-2H-1,4-BENZODIAZEPIN-2-ONE

D.Srzić°, L.Klasinc°, B.Belin", F. Kajfež" and V. Šunjić"

°Institute "Rudjer Bošković", Zagreb, Yugoslavia

"CRC, Chemical Research Company, San Giovanni Natisone, Italy

INTRODUCTION

Recently (1-3) we studied configurational stability of the chiral centre and mechanisms of the acid-catalysed solvolytic racemization in C(3)-oxy 5-phenyl-7-cloro-2H-1,4-benzodiazepin-2- -ones 1-3 - see general formula.

	R	R_1
1	CH_3	OCH_3
2	H	OCH_3
3	CH_3	OH

The two different mechanisms were proposed (1-3) as potential routes of their racemization in acid-catalysed solvolysis: (a) ring- -chain tautomery and (b) degenerate nucleophilic exchange (identity reaction, i.e. exchange of the same groups).

A ring-chain tautomery, postulated as the possible route of fast racemization for C(3)-hydroxy 1,4-benzodiazepin-2-ones (Scheme

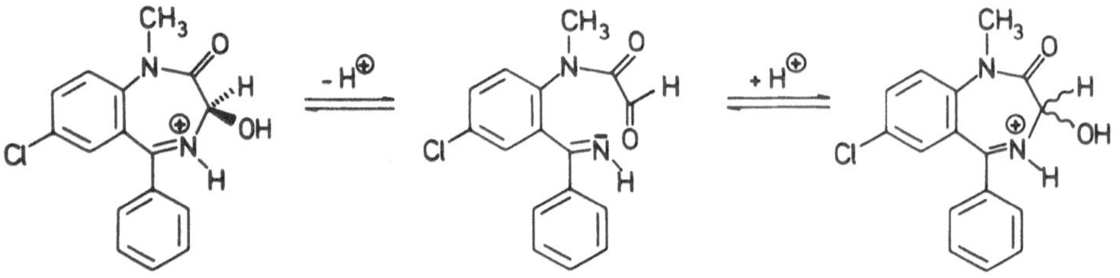

SCHEME 1. The acid catalysed ring-chain tautomery of 3.

1), is well known in the field of sugars and some nonaromatic five-
and six-membered heterocycles (4, 5). We presume that the ring-chain
tautomery has precluded resolution of the enantiomers of the compound
3 via diastereomeric esters with camphanic acid (6).

Observation that nucleophilic degenerate exchange of the C(3)-
-methoxy group in the acid methanolic solution of 1 and 2 occurred
easily, led to the assumption that the same type of reaction in acid
aqueous solution for 3-hydroxy derivatives may be one of the possible
ways of their racemization. In order to verify this hypothesis
exchange experiments with H_2O^{18} were performed (Scheme 2).

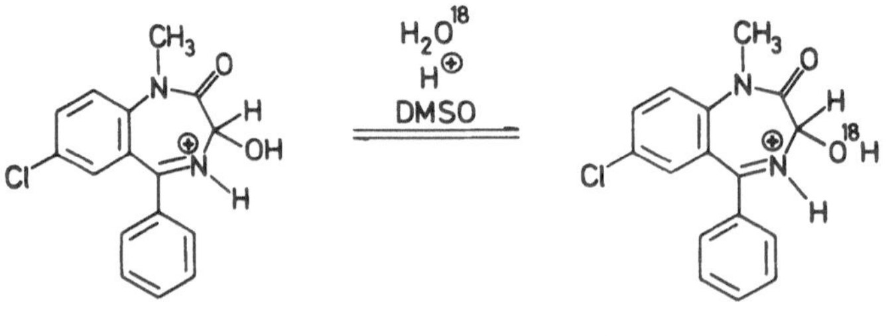

SCHEME 2. Degenerate nucleophilic exchange
3 $\longrightarrow$ 3-C(3)-^{18}OH

The acid catalysed replacement of C(3) hydroxy group was studied
kinetically to obtain a better insight into the mechanism of degenerate

nucleophilic exchange. The time course of exchange was followed by mass spectrometry.

EXPERIMENTAL

Racemic 1-methyl-3-hydroxy-5-phenyl-7-chloro-2H-1,4-benzodiazepin--2-one was dissolved in a mixture of DMSO and H_2O^{18}-H_2SO_4 (26.5% H_2SO_4). The samples, taken at regular time intervals from the thermostated solutions, were quenched by adding to 5% aqueous sodium acetate. After extraction of 3 into chloroform, evaporation and drying, the samples were directly injected into mass spectrometer.

The kinetic measurements have been performed by determination of the **relative** enhancement of the M+2 ion peak with time.

All mass spectrometric measurements were obtained on a Varian MAT CH-7 instrument, using a direct inlet system. Operating conditions were: electron energy 70eV, ionising current 100 µA, ion accelerating voltage 3kV.

Calculations of pseudo-first order rate constants and of thermodynamic magnitudes were carried out by computer using a non--linear least-squares program as already described (1, 3).

RESULTS AND DISCUSSION

The kinetic and thermodynamic data derived from mass spectro-metric measurements are given in Table 1.

TABLE 1.
Rate constants k_e and activation parameters for exchange reaction $3 \xrightarrow{e} 3$-C(3)-^{18}OH

Run no.	$t/°C$	k_e x 10^3/sec
1	25.2 ± 0.1	0.95 ± 0.02
2	35.0 ± 0.1	1.90 ± 0.02
3	49.9 ± 0.1	7.57 ± 0.05
4	59.8 ± 0.1	12.0 ± 0.1

$$H^{\neq} = 12.2 \pm 1.9 \text{ kcal/mol}$$
$$S^{\neq} = -27.2 \pm 5.6 \text{ e.u.}$$

The experimental investigations indicate that configurational instability of the chiral centre C(3) in the compound 3 is due to the exchange reaction. The total racemization rate of 3, however, should consist of racemisation via degenerate exchange on the ring, as well as via ring-chain tautomery. Of course, 3 could racemise after addition of O^{18} labelled water on the open aldehyde form as presented in Scheme 3.

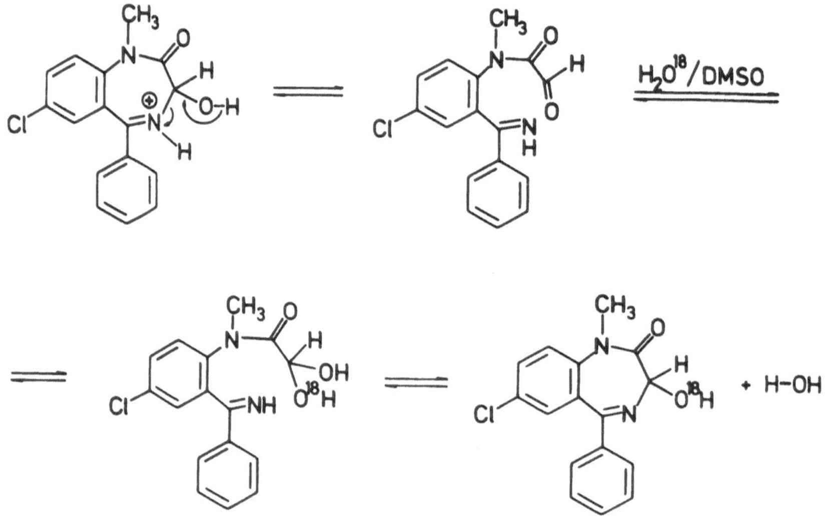

SCHEME 3. Possible mechanism of racemisation of 3.

In order to discriminate these possibilities and to get complete insight into the processes of racemization for this class of pharmacologically important compounds, we are presently engaged in determining separately the kinetics of the ring-chain tautomery.

REFERENCES

1) M. Štromar, V. Šunjić, T. Kovać, L. Klasinc and F. Kajfež, Croat. Chem. Acta, 1974, 46, 267.
2) V. Šunjić, R. Dejanović, A. Palković, L. Klasinc and F. Kajfež, Tetrahedron Lett., 1976, p. 4493.
3) V. Šunjić, A. Lisini, T. Kovać, B. Belin, F. Kajfež and L. Klasinc, Croat. Chem. Acta, 1977, 49, 505.
4) D. Thon and W. Schneider, Chem. Ber., 1976, 109, 2743.

5) G. Toth, G. Hornyak and K. Lempert, Chem. Ber., 1977, 110, 1492.
6) V. Šunjić, F. Kajfež, D. Kolbah and N. Blažević, Croat. Chem.
   Acta, 1971, 43, 205.

ON THE ANALYSIS OF FERMENTATION BROTH DURING BIOSYNTHESIS OF

CEPHALOSPORIN ANTIBIOTICS

M. Kač, J. Marsel and M. Pokorny°

University of Ljubljana, Ljubljana, Jugoslavija;

° KRKA, Novo mesto, Jugoslavija

INTRODUCTION

Process control of cephalosporin biosynthesis requires a fast and specific analytical method for the determination of the β-
-lactam antibiotic concentrations, and the estimation of its degradation products and precursors.

Recently, several analytical methods have been applied either for determination of cephalosporin C and its decomposition products in fermentation broth or for identification of metabolites of different β-lactam antibiotics. Beside the microbiological test (1), spectrophotometry (2), thin layer chromatography (3), the hydroxy-lamine (4) and idrometric (5) methods were mainly used for routine determination of β-lactam derivatives in fermentation broth. Besides these methods high performance liquid chromatography (HPLC) gained major attention in recent years due to its ability to separate similar compounds, i.e. its relatively high specifity and high sensitivity for the substances in question. By this method it was not only possible to control the fermentation process (6), it also allowed in vitro identification of different degradation products of cephalosporin C (7, 8) and metabolites of related β-lactam antibiotics in human urine (9) and serum (10).

In spite of this, HPLC alone cannot give an unequivocal answer if an unknown degradation product and/or metabolite is to be identif-ied. For this reason mass spectroscopy is a method which could also be promising in this field. However, this method was until now applied only for identification of two degradation products of cephalexin (7-(D-2-amino-2-phenylacetamido)-3-methyl-3-cephem-4-carboxylic

acid) (8). The lack of mass spectrometric data in the field of β-
-lactam and other antibiotics originates from the low volatility and
thermal instability of these compounds. These properties make the
electron impact (EI) ionisation technique impossible. For this
reason field desorption (FD) seems to be one of the few ionisation
techniquies applicable for identification of cephalosporin C and
chemically closely related compounds.

MATERIALS AND METHODS

TLC was performed on SiO_2 60 F 254 (Merck) precoated on
aluminium sheets with a mixture of n-butanol-acetic acid-water
(60 : 15 : 25) as mobile phase (11). Three different modes of
detection were used; UV at 254 nm, and two spraying agents: 0.2%
ninhydrine in ethanol and chloroplatinic acid (3).

The fermentation broth was extracted with ethyl acetate. First
we extracted the filtrate at the pH about 7 (adjusted with NH_4OH)
and then we repeated the extraction of the water phase at pH 2
(adjusted with 10% H_2SO_4). The extracts were dried with Na_2SO_4.

HPLC analyses were performed on an hp 1010 A chromatograph
equipped with a syringe loading injector and an hp 1030 B variable
wavelength UV detector, operating at 254 nm. A prepacked LiChrosorb
® RP 8 (10 μ) column (length 20 cm, i. 2.5 mm) was used with water
as mobile phase. The flow rate was 1 ml/min, the pressure at these
conditions was about 500 PSI.

Electron impact mass spectra were scanned with a CEC 21-110C
instrument at 6-kV accelerating voltage, 70 eV and 100 μA ionizing
current. Direct inlet probe was used at temperatures between 100 -
250°C. The spectra of the salts were recorded after addition of
$NaHSO_4$ (12).

Field desorption spectrum of cephalosporin C was obtained with
a MAT 731 mass spectrometer at ion source temperature of 60°C, app.
5 mA emitter current and 11 kV.

The cephalosporin C was spectrophotometrically determined by
nicotinamide method (13). The microbiological assay was performed
in the standard way, with Alcaligenes faecalis ATCC 8750.

RESULTS AND DISCUSSION

Studies on methionine. It is well known that methionine is the
most important precursor in the biosynthesis of cephalosporin by
Cephalosporium acremonium. An extensive description of known

metabolites of methionine during the cephalosporin synthesis has been given by Caltrider and Niss (14). In addition, two very wide-spread reactions are known to occur generally in lower plants (15): deamination and N-acylation. These reactions lead to 2-keto-4--methylthiobutyric acid, 2-hydroxy-4-methylthiobutyric acid, n--formyl methionine and n-acetyl methionine. It could be assumed that such substances could also be formed during the fermentation process in cephalosporin synthesis.

For the determination of methionine and its degradation products in fermentation broth, TLC rather than HPLC was used. Namely, methionine has its absorption maximum near 210 nm where the separation by HPLC under the described conditions was not satisfactory. Spraying with chloroplatinic acid was used to detect thioether amino acids and other organo sulphur compounds. It revealed the presence of two substances (R_f 0.52 and 0.69), which could not be detected by the other two methods. We tried to extract these unknowns selectively from the broth into a suitable organic solvent. Ethyl acetate proved to be a good solvent under the described conditions. The TLC of dried extracts showed that the metabolites are occasionally not completely separated, very probably owing to the poor separation of the two phases.

For identification of unknowns, their mass spectra were compared with the mass spectra of the above mentioned four standards. The recorded EI spectra of the standards were compared with data from the literature. According to our knowledge, only the spectra of methionine (16) (Fig. 1) and 2-keto-4-methylthiobutyric acid (16) (Fig. 2) have been reported.

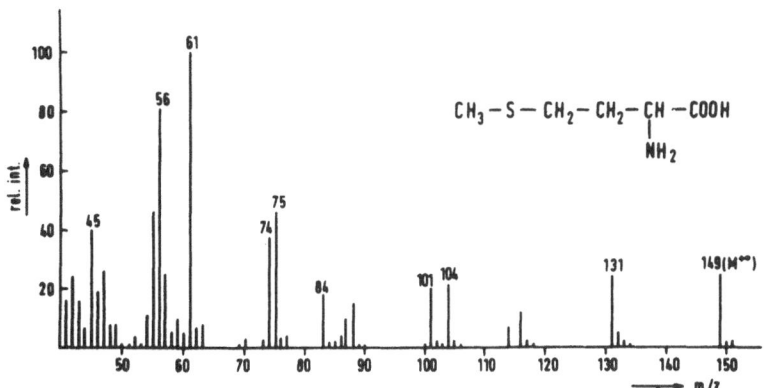

FIG. 1. EI mass spectrum of methionine.

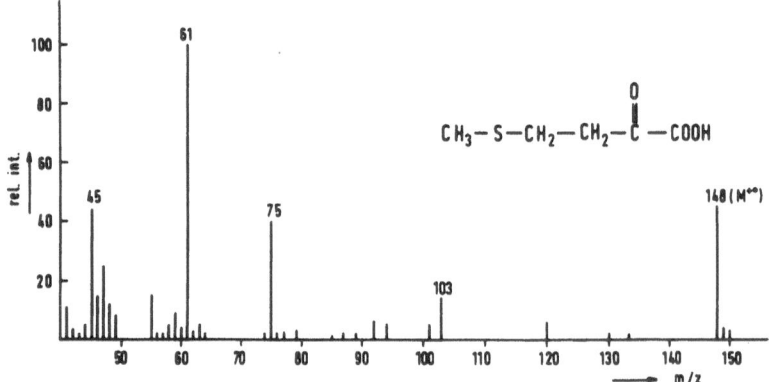

FIG. 2. EI mass spectrum of 2-keto-4-methylthiobutyric
acid (Ca-salt, NaHSO₄ added).

The differences in the intensities of some peaks in the mass spectra
compared can be explained by different experimental conditions (12).
The substances gave readily distinguishable EI spectra. Comparison
of the standard spectra with the EI mass spectrum of one of the
unknowns (R_f 0.69), shows that it matches with the spectrum of N-
-acetyl methionine (Fig. 3), though some additional peaks show that
the metabolite includes some impurities.

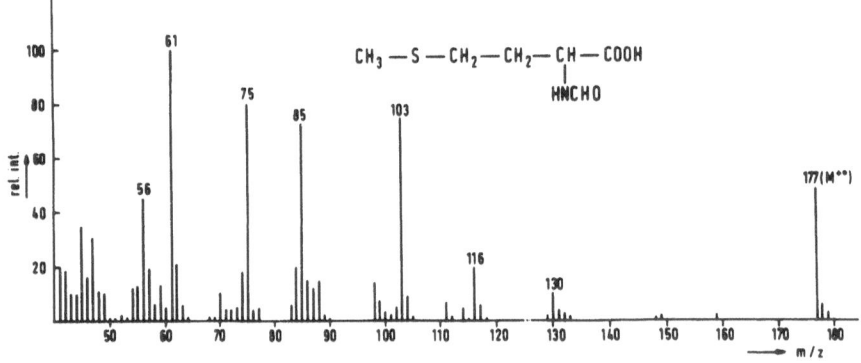

FIG. 3. EI mass spectrum of N-acetyl methionine.

   To the best of our knowledge this is the first time that N-
-acetyl methionine is detected as methionine metabolite in
Cephalosporium acremonium, although some studies indicate that this
metabolic pathway might be general for all fungi and for some higher
plants (15). Occurrence of this metabolite indicates a new metabolic
pathway of methionine in Cephalosporium acremonium.

   The mass spectra of 2-hydroxy-4-methylthiobutyric acid and of
N-formyl methionine are presented in Fig. 4 and Fig. 5, respectively.

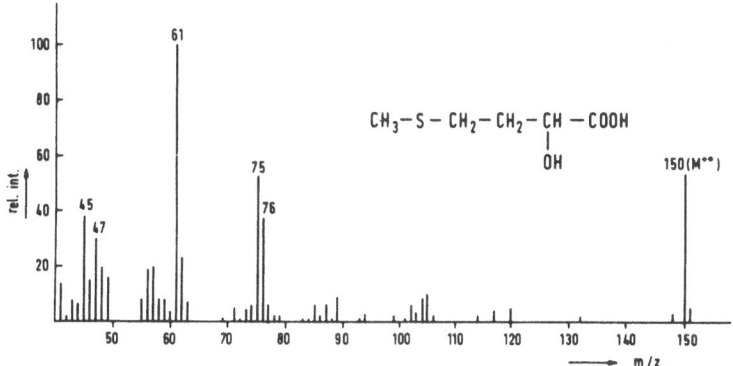

FIG. 4. EI mass spectrum of 2-hydroxy-4-methylthiobutyric
        acid (Na-salt, NaHSO$_4$ added).

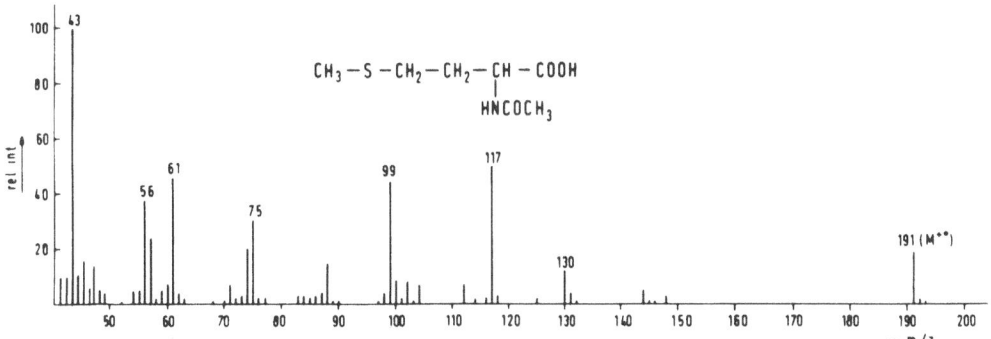

FIG. 5. EI mass spectrum of N-formyl methionine.

The mass spectrum of the second unknown (R_f = 0.52) reveals non characteristic peaks, indicating an impurity, also observed partially in mass spectrum of the first unknown (i.e. N-acetyl methionine).

Analyses of cephalosporin C. Working on the analysis of fermentation broth, our chief aim was to develop an HPLC method for determining various components during the fermentation process, especially cephalosporin C, its degradation products and its precursors. For mass spectrometric identification of separated compounds we had to avoid buffered solvents. An additional difficulty was the thermal lability of cephalosporin C and its derivatives. Highly volatile organic solvents could be used, but even small amounts of water require relatively high temperatures for evaporating the solvent. Another possibility was to use pure water as mobile phase, so that samples for mass spectrometric analyses could be directly dried by freeze-drying.

While trying to get a mass spectrum of cephalosporin C, we had to overcome some problems originating from low volatility and the ionic structure of the molecule. The EI spectrum shows at an ion source temperature of 150°C the highest mass at m/e 309 and some fragment peaks with low intensities in the lower mass range. The FD spectrum of cephalosporin C exhibits a peak at m/e 166 (corresponding to the ion m/e 165 in the EI spectrum), and two weak signals at m/e 283 and m/e 316/317. The attempt to cationize the compound with triethylamine was not successful. Finally, an FD spectrum was taken with an approximately tenfold excess of tartaric acid, which proved to be a suitable electrolyte. An association product of cephalosporin C and tartaric acid at m/e 566, as well as a protonated molecular ion of cephalosporin C at m/e 416 were observed. The appearance of the peak at m/e 373 can be explained in two different ways: a ketene can be released from the non--protonated molecular ion, or a CH_3CO fragment is eliminated from the protonated molecular ion.

The spectrum discussed represents that of cephalosporin C -free acid (provided by dr. A. Castelanni, Milano). Additionally, we prepared a standard by the well known method from the cephalosporin C-Zn salt. The zinc ion was precipitated with oxalate, the precipitate filtrated and purified on an adsorption chromatographic column. After these operations, a solution of cephalosporin C - free acid in water - acetone was obtained. The acetone could be readily evaporated under vacuum, and water was removed by lyophilisation.

This standard was analysed by three methods: spectrophotometrically, with a microbiological assay and by HPLC.

Significant differences in stability were observed between the two standards of cephalosporin C-free acid. After approximately 4 months, our standard showed, though kept air-tight in the refrigerator, a greater amount of decay products compared to the other one, which decomposed to a smaller extent when kept under the same conditions for a longer period.

Standard spectrophotometric assay (nicotinamide method (13)) showed a decrease in the cephalosporin C content from 89.2% to 80.5%. Similarly, the microbiological test with Alcaligenes faecalis ATCC 8750 showed lower activity (80.2% at the end of the 4 months storage compared to the 89.5% at the beginning). The previously described HPLC analysis of the standard at the beginning of the storage gives two peaks (cephalosporin C has the retention time of 3.9 min and the impurity 2.7 min). After decomposition has occurred the peak with t_R 2.7 min shows approximately a threefold increase and an additional peak appears at t_R 11 min.

One can suspect that degradation of cephalosporin C - free acid occurred in solid state, even at low temperatures. To explain this phenomenon we are trying to follow the decomposition of this compound under well defined conditions and sample the degradation products for mass spectrometric analysis. The peak with the retention time of 2.7 min could be contributed to the degradation product of cephalosporin C known to be formed in alkaline solution.

The chromatogram of the fermentation broth shows the two peaks with t_R 2.7 min and t_R 3.9 min as well. During the aging of the broth in refrigerator the peak intensity of cephalosporin C decreases and correspondingly that of the degradation product increases.

In conclusion we can say that the control of the fermentation process represents a complex analytical problem, owing to the various products formed under different conditions.

The application of FD mass spectroscopy seems to be a technique, which could contribute important additional information to that from other analytical methods.

ACKNOWLEDGEMENTS

The work was financially supported by the Research Community of Slovenia. We are indebted to Professor H. Budzikiewicz for scanning and interpretation of FD spectra. The authors wish to thank B. Kralj, M.S. from our Department for running EI spectra.

REFERENCES

1) T.A. Pursiano et al., Antimicrob. Ag. Chemother., 1973, 3, 33.

2) C.A. Claridge et al., Antimicrob. Ag. Chemother., 1969, 131.

3) M. Pokorny, N. Vitezic, M. Japelj and U. Valcavi, Il Farmaco, 1976, 31, 23.

4) D.L. Mays, F.K. Bangert, W.C. Cantrell and W.G. Evans, Anal. Chem., 1975, 47, 2229.

5) E.B. Lindström and K. Nordström, Antimicrob. Ag. Chemother., 1972, 1, 100.

6) R.D. Miller and N. Neuss, International Symposium on Micro-chemical Techniques, Davos, Switzerland, 1977.

7) J. Konecny, E. Felber and J. Grunder, J. Antibiotics, 1973, 26, 135.

8) A. Dinner, J. Med. Chem., 1977, 20, 963.

9) R.P. Buhs, T.E. Maxim, N. Allen, T.A. Jacob and F.J. Wolf, J. Chromatogr., 1974, 99, 609.

10) J.S. Wold, Antimicrob. Ag. Chemother., 1977, 11, 105.

11) M. Pokorny, N. Vitezic and M. Japelj, J. Chrom., 1973, 77, 458.

12) B. Kralj et al., Biomed. Mass Spectrom., 1975, 2, 215.

13) E.H. Flynn, Cephalosporins and Penicillins, Academic Press, 1972, p. 679.

14) P.G. Caltrider and H.F. Niss, App. Microbiol., 1966, 14, 746.

15) M. Pokorny, E. Mačenko, D. Keglević, Phytochemistry, 1970, 9, 2175.

16) M.S.D.C. Series Mass Spectral Data, AWRE Aldermaston, Berkshire.

GC-MS ANALYSIS BY ELECTRON IMPACT IONIZATION AND CHEMICAL IONIZATION

OF CARCINOGENIC 2-ACETYLAMINOFLUORENE AND ITS METABOLITES

C.Lallemant,G.Wood,M.F.Exilie,M.Chessebeuf,P.Gambert,

M. Hardy and P. Padieu, Laboratoire de Biochimie Médicale,

Université de Dijon, Dijon, France

INTRODUCTION

Among aromatic amines, fluorenyl amines are the most potent carcinogens which in addition exhibit a multi-tissular target effect. The development of fluorene chemistry 40-50 years ago led to the synthesis of 2-acetylaminofluorene (2-AAF) as a very efficient insecticide, the toxicity of which has been tested by Wilson et al. (1). Despite the lack of early toxic effects in the rat fed with 0.02% 2-AAF added to the diet, these authors pursued the appraisal and observed, after three months, the appearance of primitive cancerous tumors at various locations: liver, mammary gland, ear--duct and its sebaceous Zymbal gland, intestine, lung, kidney, bladder, salivary gland, adrenal, etc... Such substituted fluorenyl-amines also induce neoplastic growth in tobacco cell culture (2).

This princeps observation of Wilson et al. (1) constituted the first example of the effects on many tissues of a powerful carcinogen at doses devoid of any acute or long term toxic effect.

The carcinogenic effect of 2-acetylaminofluorene, the acetyl derivative of 2-aminofluorene (2-AF), is associated with metabolic activation into ultimate carcinogen (3). Two other fundamental concepts have been established during the study of the carcinogenic effects of such drugs: - the absence of pathological lesions at the place of entry; - the direct relation between carcinogen structure and activity (3, 4).

The liver, by means of its detoxification functions, represents the tissue which is mainly responsible for carcinogen activation

163

from the proximate to the ultimate form due to the presence of
enzymes of the endoplasmic reticulum, cytochrome P-450 monooxygenases,
which are found with microsome fraction upon tissue homogeneisation
(5).

In spite of a considerable amount of research on the carcinogenic effect of this substituted arylamine (around 500 publications),
not many trials to use gas phase analysis for metabolic studies
and carcinogenic effects in the animal as well as in tissue culture
can be found.

In 1971, Irving (6) showed the production of the glucuronide
conjugate of N-OH-2-AAF by gas chromatography analysis (GC) of the
aglycone moiety as trimethylsilyl (TMS) derivative after bacterial
β-glucuronidase hydrolysis of the glucuronide synthesized by liver
microsomes. Lhoest et al. (7) used this method and also methylation
and trifluoroacetylation to study by gas chromatography-mass
spectrometry (GC-MS) the in vitro tautomeric transformation of the
free form of N-OH-2-AAF.

The present article will report the preliminary development of
methods to study by GC and GC-MS the production of 2-AAF metabolites
in the rat and in tissue culture. Two methods of derivatization
were tried: peralkylation according to de Leenheer and Gelijkens
(8) and trimethylsilylation by a modification of the method
described by Irving (6).

MATERIAL AND METHODS

Reagents. - Anhydrous sodium sulfate was from Prolabo (Paris,
France). - Organic solvents: diethyloxide, dichloromethane, acetone
and dimethylsulfoxide were from Merck (Darmstadt, Germany).
- Derivatization reagents: i) for alkylation: methyl-iodide, or
propyl-iodide and potassium tert-butylate were from Merck, ii) for
trimethylsilylation: N,O-bis(trimethylsilyl)-acetamide (BSA) was
purchased from Supelco (Bellefonte, USA). - Carcinogens: 7-hydroxy-
-2-acetylaminofluorene (7-OH-2-AAF), 2-nitrofluorene (2-NO$_2$F),
2-AAF and 2-AF were bought from Aldrich Jansen Pharmaceutics N.B.,
(Beerse, Belgium) and recrystallized according to Weisburger et al.
(9) in the case of 7-OH-2-AAF and with ethanol/H$_2$O in the case of
2-AAF.

Three metabolites were synthesized: N-OH-2-AAF according to
Poirier et al. (10), N-acetoxy-2-acetylaminofluorene (N-OAc-2-AAF)
according to Gutmann and Ericksson (11), and 3-hydroxy-2-acetylamino-
fluorene (2-OH-2-AAF) according to Weisburger and Weisburger (12).

All compounds were controlled by melting point, thin layer
chromatography and direct inlet mass spectrometry.

Materials - A Packard-Becker (Delft, Holland) gas-chromatograph, model 427 with a flame ionization detector was equipped with a packed glass column (2% 2 m x 3 mm) of OV-17 coated on Chromosorb W 100-120 mesh (Spiral, Dijon, France) or with SE-30 coated glass capillary column (LKB 2101-203 type) (Spiral) 25 m long by 0.22 mm i.d. with 120 000 theoretical plate efficiency. - The mass spectrometry studies were carried out on the RIBERMAG GC-MS R1010 (Ribermag, Rueil-Malmaison), a quadrupole equipped with a single ion source working under electron impact ionization (EI) or chemical ionization (CI) by differential vacuum pumping and with dual disc data acquisition and processing computer. The outlet of the packed or the capillary glass column was interfaced directly to the vicinity of the ion source without any separator. The mass spectrometer was operated under routine settings for EI and CI analyses.

Methods. - Thin-layer chromatography (TLC): Separation of 2-AAF and its metabolites including fluorene (F) and 2-AF was done on TLC plates (F254, Merck) with a solvent phase of dichloromethane/acetone 90:10 according to Bartsch et al. (13). - Derivatization: The peralkyl derivatives were realized with methyl or propyl iodide following the procedure described by De Leenheer and Gelijkens (8). Trimethylsilyl derivatives were done with BSA: 200 µl for 50 µg of compound during 30 min at 85°C. Evaporation was carried out at room temperature under a nitrogen stream. - Gas chromatography: The initial temperature was 240°C for alkyl derivatives or 200°C for the trimethylsilyl (TMS) derivatives programmed at 2°C/min. Injection port and detector were set at 290°C.

The detection sensitivity by means of total ion current detection of the spectrometer with packed column was 2 ng 2-AAF, mono-TMS ($M^+=295(47)$)" with a signal/noise ratio of 3 in IE mode.

RESULTS AND DISCUSSION

Peralkyl derivatives. Experimentation was carried out with 2--NO$_2$F, 2-AAF, N-OH-2-AAF, 3-OH-2-AAF and 7-OH-2-AAF which were eluted in this order as permethyl derivatives on 2% OV-17 packed column. The compound N-OAc-2-AAF could not be studied at that time.

Mass spectra under EI mode showed conspicuously that the two hydrogen atoms of the 9th carbon were totally replaced by two methyl groups in addition to the derivatization of functional groups such as the secondary amine, the ring hydroxyl functions and the N--hydroxamic acid. Therefore, 2-NO$_2$-F was bis-methylated, 2-AAF and N-OH-2-AAF were tris-methylated (Figs. 1 and 2), while 3-OH-AAF and 7-OH-2-AAF were tetra-methylated (Figs. 3 and 4).

"The relative intensity of a mass spectrum peak is indicated in brackets after the ion mass value.

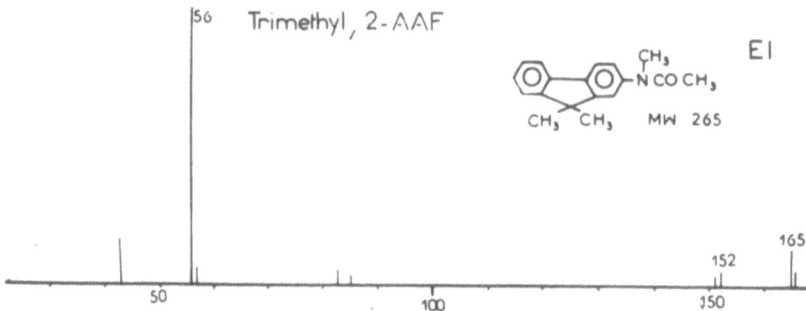

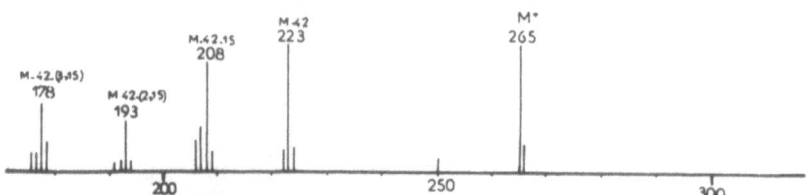

FIG. 1. EI mass spectrum of trimethyl, 2-acetylamino-
fluorene (trimethyl, 2-AAF).

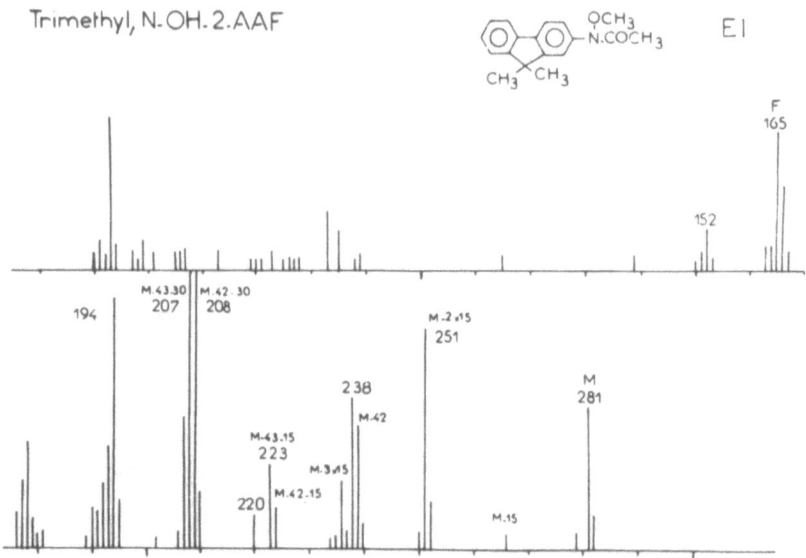

FIG. 2. EI mass spectrum of trimethyl,N-hydroxy-2-
-acetylaminofluorene (trimethyl,N-OH-2-AAF).

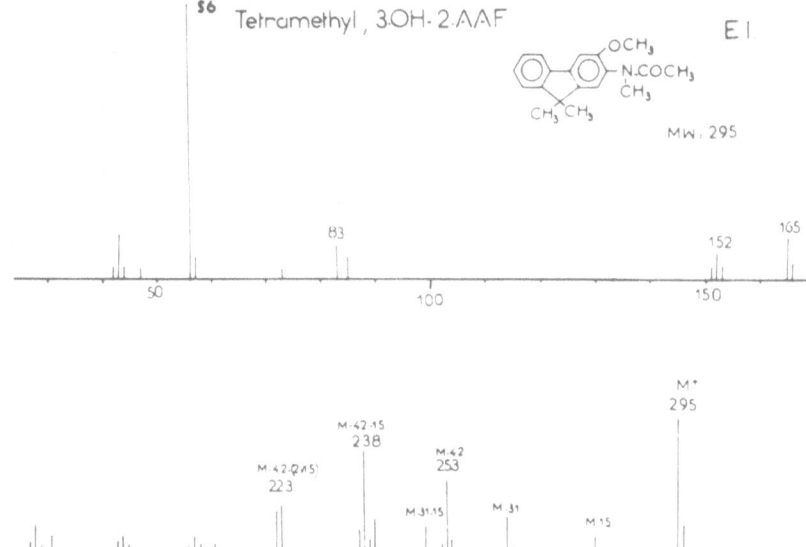

FIG. 3. EI mass spectrum of tetramethyl,3-hydroxy-
       -2-acetylaminofluorene (tetramethyl,3-OH-
       -2-AAF).

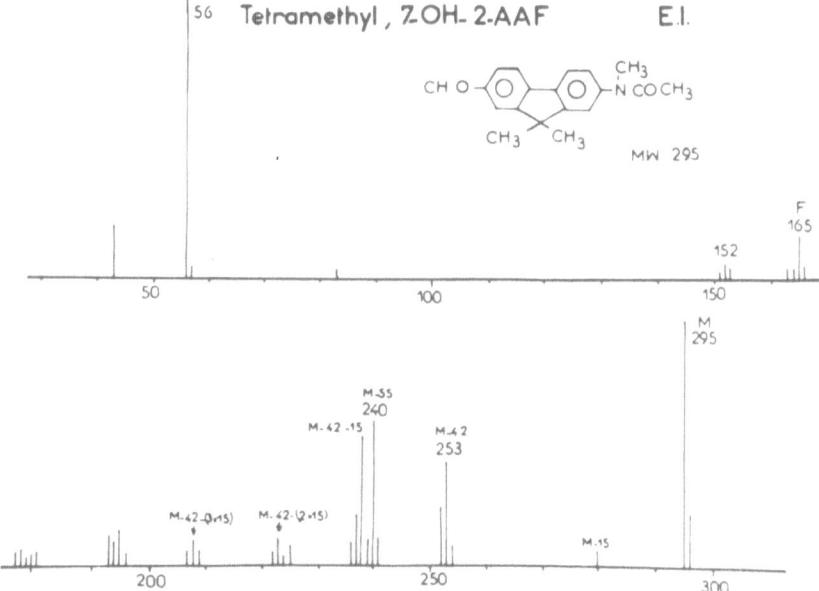

FIG. 4. EI mass spectrum of tetramethyl,7-hydroxy-2-acetyl-
       aminofluorene (tetramethyl,7-OH-2-AAF).

In the case of 2-AF the free amine should be acetylated with acetic anhydride before permethylation. Therefore a separate chromatographic run must be done to distinguish it from 2-AAF.

Among several other types of peralkyl derivatives, only per-propylation with propyl iodide was satisfactory, but permethylation with CH_3I was found easier and has the advantage of being done at room temperature.

Another very interesting aspect for isotopic dilution mass fragmentography (ID-MF) (14) is the possibility to perform the derivatization with perdeuterated methyl iodide (CD_3I). One dis-advantage of permethylation is the low increase of molecular weight upon derivatization, a situation which is not favorable for mass fragmentography (MF) since many biological compounds may interfere within this 250-300 mass range. In the case of perpropylation, reaction was slower and the temperature appeared critical since derivatization did not occur at the same rate for each compound.

Checking the occurrence of two tautomeric derivatives of N-OH--2-AAF during the methylation of the hydroxamic function (7), it was found that the mass spectrum of N-OH-2-AAF,trimethyl (Fig. 2) was only relevant to the presence of two methyl groups on the 9th carbon and one hydroxymethyl on the nitrogen atom.

The perpropylation of N-OH-2-AAF led to a final product whose molecular weight corresponded to a bis-propyl derivative with loss of an oxygen atom which could be interpreted as the loss of a propyloxy radical upon chromatography during the stages following the derivatization process.

Trimethylsilyl derivatives. Gas phase analysis of individual derivatives by GC and GC-MS demonstrated that the six N-fluorenyl compounds gave TMS derivatives with an increased retention time in the following order (Fig. 5): 2-AF and 2-NO$_2$F, 3-OH-2-AAF, N-OH--2-AAF, 7-OH-2-AAF and N-OAc-2-AAF, on an OV-17 packed column. But on an OV-17 glass capillary column 2-NO$_2$F was eluted before 2-AF. This compound which is not derivatized by BSA seems to be very interesting as a primary standard. Table 1 shows the principal fragmentations of each derivative under EI or CI modes. Molecular ions were found in EI and were asserted in methane CI (CH$_4$-CI) mode by the presence of the two adducts: (M+H)$^+$ and (M+C$_2$H$_5$)$^+$ for all compounds and also the (M+C$_3$H$_7$)$^+$ adduct except for 2-AF and 7-OH--2-AAF. In CH$_4$-CI, the (M+H)$^+$ quasi molecular ion was the base peak except for 7-OH-2-AAF,bis-TMS whose intensity was 40%.

Fragmentation pathways were consistent with the derivatization of: -NH-, -N(OH)- and aryl-OH groups. The 2-AAF carcinogen which could be eluted with a high retention time without being derivatized, was totally derivatized as an N-mono-TMS (M$^+$ = 295) (Figs. 6 and 7)

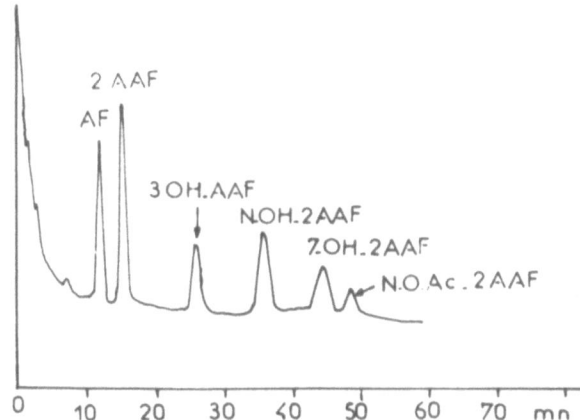

FIG. 5. Gas chromatographic electron pattern as the
total ionization current of trimethylsilyl
derivatives of 2-aminofluorene (AF), 2-acetyl-
aminofluorene (2-AAF), 3-hydroxy-2-acetyl-
aminofluorene (3-OH-2-AAF), N-hydroxy-2-
-acetylaminofluorene (N-OH-2-AAF), 7-hydroxy-
-2-acetylaminofluorene (7-OH-2-AAF) and N-
-acetoxy-2-acetyl-aminofluorene (N-OAc-2-AAF)
on a packed OV-17 column interfaced with the
GC-MS R10-10 quadrupole mass spectrometer
(Ribermag) (see text for parameter settings).

TABLE 1.

List of molecular ions and fragments in electron impact (EI) and methane chemical (CH_4-CI) ionizations of trimethylsilyl derivatives of fluorenyl compounds: 2-nitrofluorene (2-NO_2F), 2-aminofluorene (2-AF), 2-acetylaminofluorene (2-AAF), 3-hydroxy-2-acetylamino-fluorene (3-OH-2-AAF), N-hydroxy-2-acetylaminofluorene (N-OH-2-AAF) tautomers (1) and (2) and N-acetoxy-2-acetylaminofluorene (N-OAc-2-AAF) compounds (1) and (2) from computerized data of a quadrupole mass spectrometer coupled to gas liquid chromatography. Left digits are the mass number, right digits are the percent of peak height in relation to base peak (bp). (1) corresponds to a compound of m/e = 152 arising from the loss of 9th ring carbon to give dibenzocyclobutadione.

	2-NO_2F EI	2-NO_2F CI	2-AF EI	2-AF CI	2-AAF EI	2-AAF CI	3-OH-2-AAF EI	3-OH-2-AAF CI	N-OH-2-AAF tautomer 1 EI	tautomer 1 CI	N-OH-2-AAF tautomer 2 EI	tautomer 2 CI	7-OH-2-AAF EI	7-OH-2-AAF CI	N-OAc-2-AAF compound 1 EI	compound 1 CI	N-OAc-2-AAF compound 2 EI	compound 2 CI
M	211/20		253/bp	253/75	295/47	295/70	383/20	383/98	311/13		311/8	311/12	383/30	383/85	353/7	353/36	311/40	311/18
M+1			212/bp	254/bp	296/bp		384/bp			312/40		312/12		384/bp		354/bp		312/22
M+29			240/10	282/10	324/8		412/5			340/7								340/5
M+41			252/4		336/2		424/1			352/2								
M-13																340/12		
M-15			238/15	238/25	280/40	280/70	368/9	368/90	296/3	296/20	296/3	296/8	368/9	368/95		338/20	296/7	296/5
M-16	195/8																	
M-17	194/8																	
M-29	182/5																	
M-30	181/8																	
M-31			222/25						280/2								280/11	
M-41																312/95		270/5
M-42										269/-						311/36		269/20
M-43						252/5		340/5	268/8	268/-	268/-	268/-		340/8				
M-46	165/bp	165/8	207/20															
M-57	(1)														296/15	296/25		
M-58			195/18		237/8				253/1		253/-						253/60	
M-59	152/20	152/1								252/20		252/20						252/18
M-61					234/15		322/8		312/60		250/4							
M-71					224/30													
M-72	139/30				223/5								312/75				282/6	
M-73			180/60		222/8		310/5		238/1	238/1	238/-	238/5	311/5				238/10	238/4
M-84															269/18	269/6		
M-87									224/bp		224/bp		296/46				224/bp	
M-88			165/35				295/4		223/8		223/45	223/38						223/65
M-89					206/40	206/65	294/45					222/35	294/4	294/53				
M-90									221/4			221/-						
M-96	115/20																	
M-100															253/50	253/8		
M-101			152/25													252/36		
M-103								280/6				208/15		280/10				
M-105									206/52	206/8		206/25						206/5
M-113																240/20		
M-114					181/25			269/7					269/3	269/10		238/12		
M-115																		196/4
M-129									182/44									
M-130					165/bp	165/8	253/7	253/10	181/45	181/75	181/15		253/15	253/10	223/25			181/7
M-131										180/40		180/85				222/30		180/75
M-132											179/2					221/12		
M-133									178/25			178/12						
M-134			119/90															
M-143					152/8													
M-145														238/12			166/-	
M-146									165/30	165/6	165/45	165/5					165/22	165/-
M-159									152/bp	152/15	152/bp	152/5	224/15				152/bp	152/4
M-160							223/3						223/-					
M-161							222/7											
M-178									133/40									
M-180			73/70															
M-194			59/25						117/92									
M-201															152/22	152/4		
M-203													180/4	180/-				
M-220					75/30						91/45	91/8						
M-222					73/85						89/20	89/35						
M-231							152/2								152/1		152/-	
M-236									75/50		75/-							
M-238									73/35		73/-							
M-280															73/bp		73/bp	
M-292													91/65					
M-310							73/bp						73/bp					

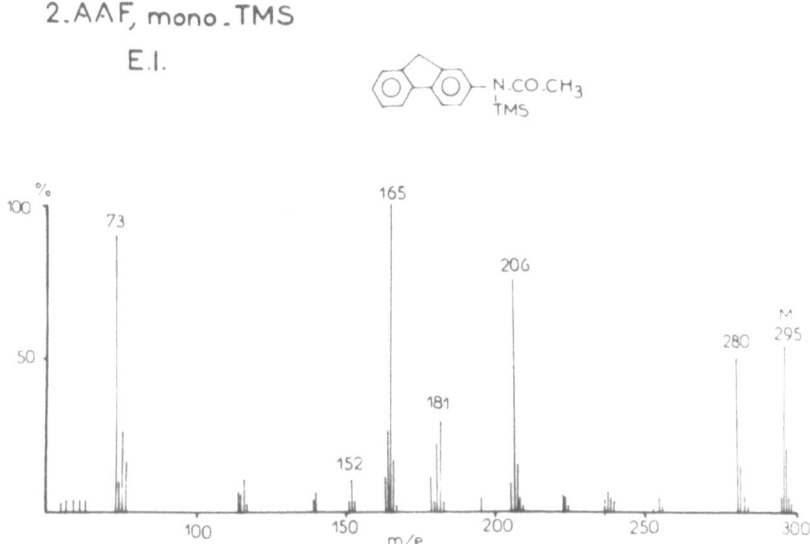

FIG. 6. EI mass spectrum of 2-acetylaminofluorene,
        mono-trimethylsilyl derivative (2-AAF,mono-TMS).

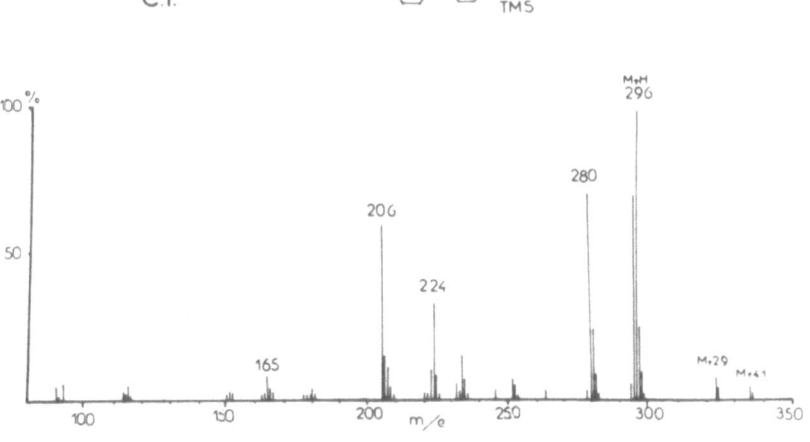

FIG. 7. CH₄-CI mass spectrum of same compounds as
        in Fig. 6.

as well as 2-AF (M$^+$ = 253) (Table 1). The ring hydroxylated compounds
were bis-trimethylsilyl, 3-OH-2-AAF (M$^+$ = 383) and 7-OH-2-AAF (M$^+$ =
383) (Table 1). The two activated carcinogens, N-OH-2-AAF and N-OAc-
-2-AAF appeared to undergo respectively a tautomerisation or a
structure modification. As a general comment on the main fragmentation of these TMS,fluorenyl compounds,ions m/e = 2-AF = 181, 2-AF - H
= 180 and F - H = 166 are major fragmentation products for 2-AF,
2-AAF, N-OH-2-AAF (tautomers 1 and 2). Dibenzocyclobutadiene (m/e =
152) (Table 1) arising from the loss of -CH$_2$- of the 9th position
of the ring was the end fragment of TMS derivatives of 2-AF, N-OH-
-2-AAF (tautomers 1 and 2) and N-OAc-2-AAF. In CH$_4$-CI mode this
fragment is produced by N-OH-2-AAF tautomers.

The chromatographic peak of N-OH-2-AAF,mono-TMS was associated
to a broadening of its base. Mass spectra from EI and CI modes of
tautomer 1 (Figs. 8 and 9) were fully compatible with the F-N(OTMS)-
-COCH$_3$ derivative as shown by the fragment m/e = M - COCH$_3$ = 311 -
- 43 = 268 which was found by mass spectrometric scanning in the
mass spectrum of tautomer 1 only.

In the EI spectra of the two forms the loss of TMS (M - 73 = 238)
was not abundant in tautomer 1 (1%) (Fig. 8) and nearly absent in
tautomer 2 (Fig. 10).

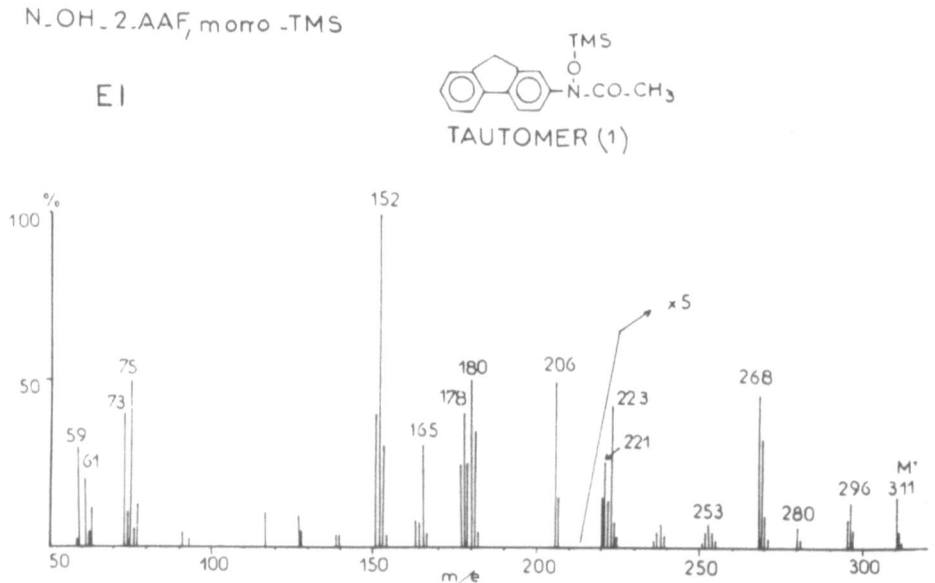

FIG. 8. EI mass spectrum of N-O-trimethylsilyl-
        -2-acetylaminofluorene (mono-trimethyl-
        silyl,N-OH-I-AAF derivative) or tautomer
        (1).

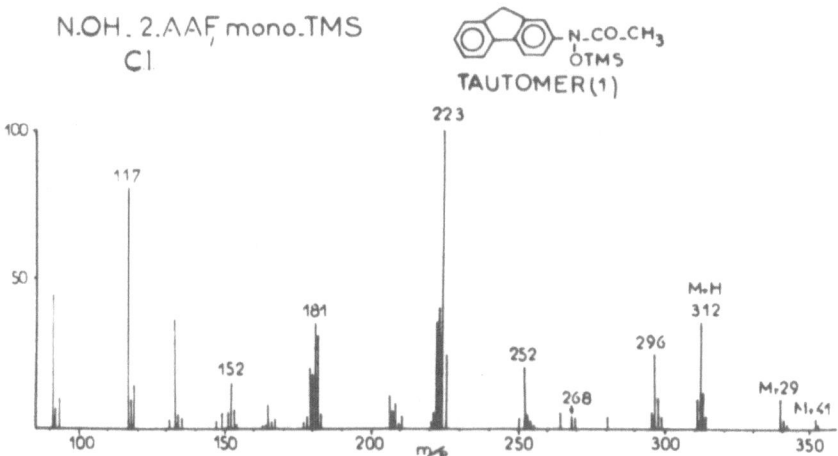

FIG. 9. CH$_4$-CI mass spectrum of same compound as in
        Fig. 8.

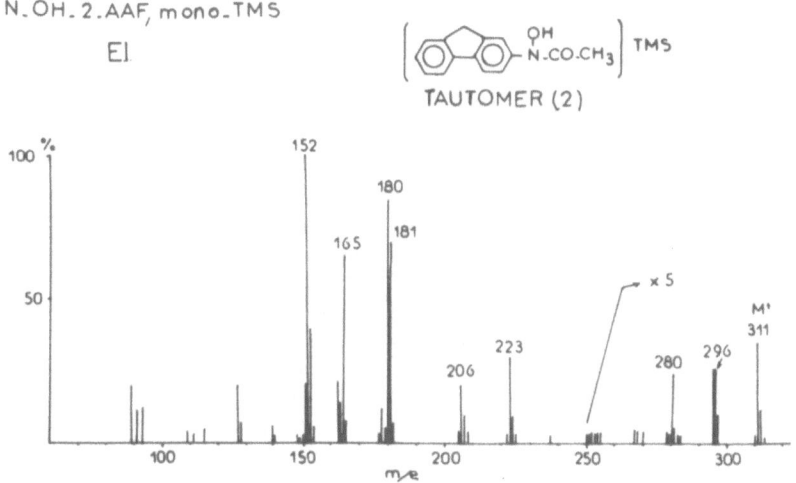

FIG. 10. EI mass spectrum of tautomer (2)
         corresponding to a mono-trimethylsilyl
         derivative of N-hydroxyl-2-acetylamino-
         fluorene.

In addition ion m/e = M-116 = 195 only described for tautomer
(1) (Fig. 8) by Lhoest (7) as F-NO or F-NH-CH$_3$ was absent in the two
tautomer mass spectra. The loss of the TMS radical as a trimethyl-
siloxy (TMSO) was weak for tautomer (1) m/e = M - 88 = 223(8) and
m/e = M - 90 = 221(4) (Fig. 8) and important in tautomer (2) m/e =
M - 88 = 223(38) (Fig. 10). In CI, the ion m/e = 206 = M 105 = M
TMSOH - CH$_3$ was present in the two tautomers, 52% in (1) (Fig. 8)
and 25% in (2) (Fig. 10), while it was expected to be absent in the
tautomer F-N(OTMS)COCH$_3$ or (1) in the above reference (7). The
production of a 2-NH$_2$-F ion (m/e = 178, 180 and 181) were abundant
for the two tautomers: m/e = 181(73), 180(85) and 178(12) for
tautomer (1) in EI and CH$_4$-CI and m/e = 180(40) and 178(25) for
tautomer (2) in EI only. In EI mass spectra, these ions were found
in the same proportions as by Lhoest et al. (7). The fragmentation
pattern of these two tautomers favors the statement of the following
structures: F-N(OTMS)COCH$_3$ for tautomer (1) and F-N(OH)C(OTMS)=CH$_2$
for tautomer (2). The occurrence of this last compound may be explained
by the effect of reactants during derivatization.

Tautomer (2) had the shape of a broad flat peak starting before
the base of tautomer (1) peak and ending with its descending side.
The trailing seems to be easily explained by the free -N(OH)-
function. In the present state of the work, tautomer (2) accounts
for 25% of the total peaks. No degradation into 2-AAF,mono-TMS was
found. Such a tautomerism did not occur during permethyl derivatization
(Fig. 2).

A similar event was observed with N-OAc-2-AAF whose molecular
peak M$^+$ = 353 was only relevant to a TMS derivatization (Figs. 12
and 13). The fate of N-OAc-2-AAF,monoTMS was found to be different
when chromatographed alone or in the mixture of all compounds. When
chromatographed alone the descending part of a sharp peak was
associated with a smaller but larger peak. But, surprisingly, only
this last peak was found in the chromatogram of the mixture. In the
first case this last peak accounted for 20% of the sharp peak. The
mass spectra taken during EI and CH$_4$-CI GC-MS made it possible to
assign to the sharp peak the formula m/e = M$^+$ = 353 = F-N(OCOCH$_3$)
COCH$_3$,TMS (Fig. 12) and to the slower peak the formula m/e = 311 -
- 42 = M$^+$ - COCH$_2$ (Fig. 13). Loss of ketene gave rise to a (FN(OH)
COCH$_3$,TMS)$^+$ molecular ion whose mass spectra in EI and CI modes
were absolutely identical to the mass spectra of the tautomer (2)
of N-OH-2-AAF,TMS (Figs. 10 and 11). Therefore, the TMS derivative
of N-OAc-AAF is plausibly F-(OCOCH$_3$)C(OTMS = CH$_2$. An extensive loss
of ketene radical seems therefore to arise from the acetyl ester
bond in association with the lability of CH$_3$ and TMS groups. The
base peak, not seen on Fig. 12, was m/e = 73 = Si(CH$_3$)$_3$.

For compound (2) of N-OAc-2-AAF,TMS, the only peak of the CH$_4$CI
mass spectrum which could be attributed to the molecular ion was the

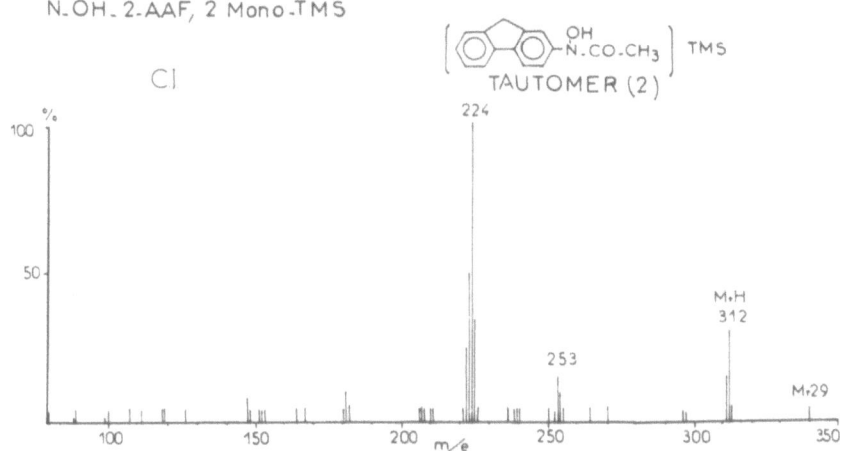

FIG. 11. CH₄-CI mass spectrum of same compound as in
         Fig. 10.

fragment m/e = $(M + H)^+$ = 312 of the former mono-TMS derivative of
N-OAc-2-AAF (compound 1) while no trace of m/e = 353 = (N-OAc-2-
-AAF,TMS)$^+$ was found since ions $(M + 29)^+$ and $(M + 41)^+$ were the
adducts of M$^+$ = 311.

CONCLUSION

This study shows that the presently known metabolites of 2-
-acetylaminofluorene (2-AAF) and a parent compound such as 2-NO$_2$F
are amenable to GC and GC-MS analyses.

The choice of derivatives has to be made in relation to the
eventual need of a preliminary separation of the compounds in
biological sample. This study also shows that N-OAc-2-AAF is stable
enough to undergo trimethylsilylation by reagents at moderate
temperature.

The procarcinogen 2-AAF is metabolized by liver epithelial cell
culture and the active carcinogen N-OAc-2-AAF induces caryological
alterations and increases the rate of sister chromatid exchanges in
these cultures (15).

Further studies by isotope labelling should help to elucidate
the fragmentation schemes completely and side reactions during
derivatization in order to facilitate experimental carcinogenesis
studies.

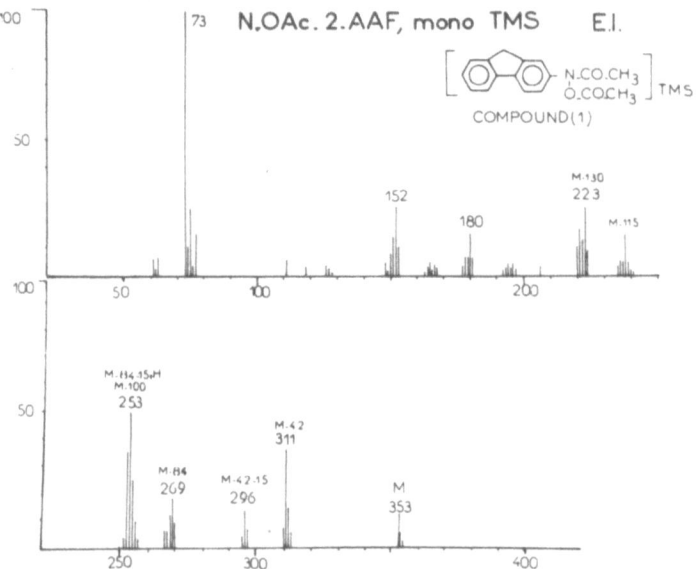

FIG. 12. EI mass spectrum of a mono-trimethylsilyl
         derivative of N-acetoxy-2-acetylamino-
         fluorene or compound (2).

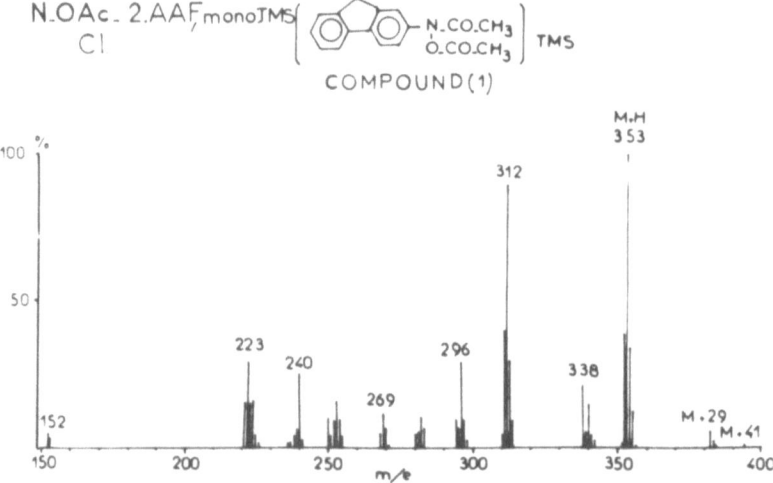

FIG. 13. CH₄-CI mass spectrum of same compound as in
         Fig. 12.

ACKNOWLEDGEMENTS

We thank Mrs Nicole Pitoizet and Mrs Anne Athias for their invaluable technical assistance and Miss Maroussia Bernot for manuscript realization.

Research was   supported by grants from Institut National de la Santé et de la Recherche Médicale (FRA n° 9, ATP 45-76-77-008 and a Visiting Research grant for one of us (G.W.) during 1976-1977), from Centre National de la Recherche Scientifique (ERA n° 267), Délégation Générale à la Recherche Scientifique et Technique (ACC n° 76.7.1674.01), Ligue Nationale Française contre le Cancer (Junior Research grant for one of us M.F.È.)and Université de Dijon.

REFERENCES

1) R.H. Wilson, F. DeEds and A.J. Cox, Jr., Cancer Res., 1941, 7, 444.
2) T.W. Bednar and E.M. Linsmaier-Bednar, Chem. Biol. Interactions, 1971, 4, 233.
3) E.K. Weisburger and J.H. Weisburger, Adv. Cancer Res., 1958, 5, 331.
4) J.W. Cramer, J.A. Miller and E.C. Miller, J. Biol. Chem., 1960, 285, 885.
5) J.A. Miller, Cancer Res., 1970, 30, 559.
6) C.C. Irving, Xenobiotica, 1971, 1, 387.
7) G. Lhoest, C. Razzouk and M. Mercier, Biomed. Mass Spectrom., 1976, 3, 21.
8) A.P. de Leenheer and C.F. Gelijkens, Anal. Chem., 48, 2203.
9) J.H. Weisburger, E.K. Weisburger and H.P. Morris, J. Nat. Cancer Inst., 1956, 17, 345.
10) L.A. Poirier, J.A. Miller and E.C. Miller, Cancer Res., 1963, 23, 790.
11) H.R. Gutmann and R.R. Erickson, J. Biol. Chem., 1969, 244, 1729.
12) J.H. Weisburger and E.K. Weisburger, J. Org. Chem., 1959, 24, 1165.
13) H. Bartsch, J.A. Miller and E.C. Miller, Biochim. Biophys. Acta, 1972, 273, 40.
14) P. Padieu and B.F. Maume, in "Quantitative Mass Spectrometry in Life Sciences", A.P. de Leenheer and R.R. Roncucci, Eds., Elsevier, Amsterdam, 1977, p. 49.
15) C. Lallemant, M. Chessebeuf, M.F. Exilie, G. Wood and P. Padieu, Abstracts of the IVth Meeting of the European Association for Cancer Research, Lyon (France), 13 September 1977, C.6, p. 79.

THE PROXIMO-DISTAL CARCINOGEN OF THE K REGION OF BENZO(a)PYRENE

G. Lhoëst

Université Catholique de Louvain

Brussels, Belgium

INTRODUCTION

It is well known that polycyclic aromatic hydrocarbons are metabolized by the microsomal NADPH-dependent P448/P450 containing mixed function oxygenases. It also seems to be well established that although serving a detoxification function, this system has a role in the activation of polycyclic aromatic hydrocarbons to toxic or carcinogenic forms such as oxide intermediates.

For the polyaromatic hydrocarbons and in the particular case of benzo(a)pyrene, in addition to theoretical calculations (1), carcinogenesis experiments suggested that the 4-5 K region should be the active region for carcinogenesis.

Non K region epoxides are becoming increasingly important as mutagens and are being hypothesized as the ultimate carcinogens (2, 3). Although current work in polycyclic hydrocarbon has steered markedly away from the K region epoxide, the fact remains that K region derivatives are being metabolically produced and that when tested for carcinogenicity they are positive (4).

So we are faced with the problem of looking more carefully at the overall chemistry of activation and detoxification of the K and non K region.

This work deals with the K region and with the discussion of some special intermediates and their biological significance.

EXPERIMENTAL

Materials. All solvents used were analytical grade solvents
(Merck). Benzo(a)pyrene and osmium tetroxide were respectively
purchased from Aldrich and EMscope. NADH, NADP, glucose-6-phosphate
and glucose-6-phosphate dehydrogenase (350 U/ml) were purchased
from Boehringer. Silicagel G plates (Macherey-Nagel) were used for
t.l.c. purposes.

Methods. Synthesis. A 3.9-nmol sample of benzo(a)pyrene to be
oxidized was dissolved in 15 ml of pyridine and stirred with 1.0 g
(3.94 nmol.) of osmium tetroxide for five days. To this mixture
a solution of 1.8 g of sodium bisulfite, 30 ml of water and 20 ml
of pyridine was added by stirring. The ratio of sodium bisulfite,
water and pyridine in the final mixture should be about 2:30:35.
After 30' the solution was extracted thoroughly with chloroform,
filtered, dried over sodium sulfate, filtered and evaporated to
dryness in vacuo. The crude extract contained cis 4,5-dihydrodiol
of benzo(a)pyrene, 4,5 benzo(a)pyrene-dione, small amounts of
benzo(a)pyrene and of a compound 1 (Rf: 0.65, $CHCl_3$).

Incubation. Male Wistar rats, R strain, (± 250 g) were given
free access to commercial food pellets and water. The animals were
exsanguinated by decapitation and the livers of the phenobarbital
treated rats (75 mg/kg aqueous solution 48 h and 24 h before
sacrifice) were excised, weighed, washed and fractionated according
to a method of De Duve as described by Amar-Costesec et al. (5)
to give a microsomal fraction P_1 and P_2 containing respectively
1.82 and 1.03 nmol. cyt. P450 per mg protein.

To the NADPH generating medium (3.3 ml) containing 2.8 mM
NADH, 1.9 mM NADP, 46.0 mM $MgCl_2$, 87.6 mM glucose-6-phosphate in
a histidine buffer 0.1 M (pH = 7.6) 20 μl of glucose-6-dehydrogenase
and 25 μl of an acetonitrile solution of benzo(a)pyrene (500 μg)
were added.

This solution was preincubated in a Gallenckamp shaking incubator
for 15' at 37°C in glass stoppered tubes to which, after incubation,
0.5 ml of P_1 and P_2 diluted four times was added. The tubes were
saturated with oxygen, stoppered and incubated for 30' at 37°C.

The incubation medium was vortexed three times for 2' with 3
ml of chloroform. The combined chloroformic solutions were
evaporated and the residue dried in vacuo.

Instrumentation. Mass spectrometric analysis was carried out
with an LKB 9000 S instrument. All mass spectra were recorded at
70 and 20 eV electron energy with 3500 V accelerating voltage,
trap current 60 μA and ion source temperature 270°C.

RESULTS AND DISCUSSION

When benzo(a)pyrene is treated with osmium tetroxide, reaction occurs at the K region to give complexes that can be decomposed to yield cis-4,5-dihydrodiols. In addition to this product, small amounts of a compound 1, the structure of which will be discussed hereafter, was isolated.

The mass spectrum of compound 1 (Fig. 1) shows a molecular ion of m/e 284 and a series of relevant fragmentation ions of m/e 282, 268, 254, 239, 226.

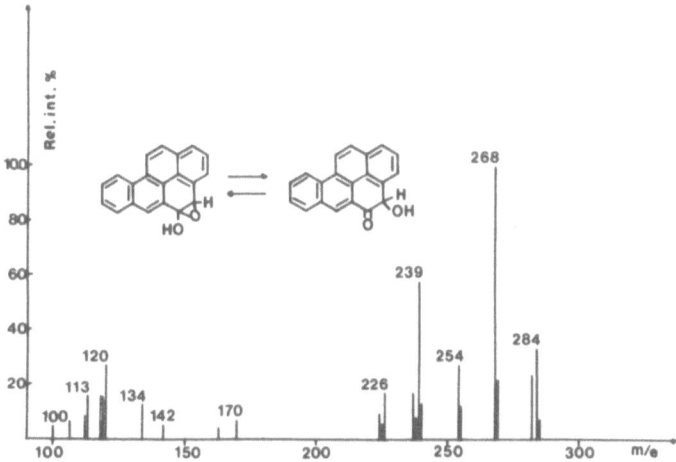

FIG. 1. Mass spectrum of compound 1.

Losses of M-H$_2$, M-O, M-(CO-H$_2$) may be observed in the mass spectrum of this derivative. If the mass spectra (Fig. 2) of 3-hydroxy and the 4,5 epoxide of benzo(a)pyrene are compared to that of compound 1, it may be observed respectively that the 3-hydroxy derivative loses directly M-29 mass units and the epoxide 16 mass units from the molecular ion, giving rise to a fragmentation ion of m/e 252 (M-O). A loss of 16 mass units is neither recorded for the 3--hydroxy derivative nor for cis-4,5 dihydrodihydroxybenzo(a)pyrene for which a loss of H$_2$O (m/e 268) and of H$_2$O$_2$ (m/e 252) is observed.

This observation strongly suggests that compound 1 contains a hemiketal function compatible with a structure 1$_a$ illustrated in Scheme 1 together with other tautomeric forms.

If a potential epoxide form is present in the structure of compound 1a, signals towards upwards fields, not observed in the N.M.R. spectrum of benzo(a)pyrene and its hydroxylated derivatives

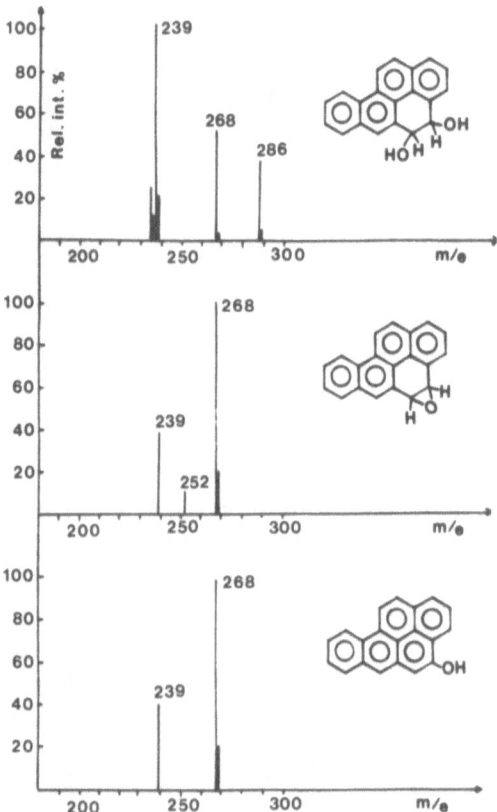

FIG. 2. Mass spectra of 3-hydroxy, cis-4,5-
-dihydrodiol and 4,5-epoxide of
benzo(a)pyrene.

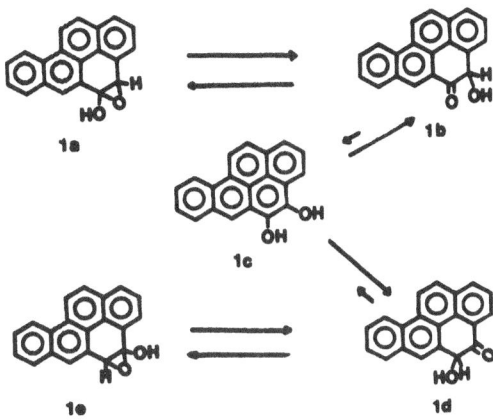

SCHEME 1. Structure of compound 1a and its
various tautomeric forms.

(6) have to be recorded.

The N.M.R. spectrum (270 MHz, CDCl$_3$) of compound 1 shows the
presence of signals between 7 and 8 ppm corresponding to 10
aromatic protons. It also reveals the presence of small broad
peaks at 5.45, 5.35, 4.65 ppm and of a complex signal at 4.2 and
3.8 ppm attributed respectively to the proton of the OH group
belonging to the different tautomeric forms 1a, 1b, 1c, 1d, 1e
and to the proton of the CH groups in position 4 or 5. The integration
curve between 5.5 and 3.8 ppm gives two protons confirming the
presence of protons in different chemical environment and the
existence of different tautomeric forms. Moreover it must be
observed that tautomer 1c cannot be a preponderant form taking into
account that no signals would be observed for 4,5-dihydroxy-benzo
(a)pyrene in between 5.5 and 3.8 ppm (6). The preponderant forms
consequently have to be the hemiketal forms 1a, 1e or the ketol
forms 1b, 1d.

That these last forms are found to be more stable than 4,5-
-dihydroxy-benzo(a)pyrene seems in this case to be more or less
in agreement with Pullman's theoretical calculation (7) predicting
a greater stability by 14 Kcal/mole of the ketonic forms of 4 or 5
hydroxybenzo(a)pyrene.

To know if significant amounts of compound 1 could be found as
metabolite, benzo(a)pyrene was incubated with liver microsomal
fractions of phenobarbital induced male Wistar rats. After extraction

TABLE 1. Abundances of the molecular ion and relevant fragmentation ions relative to m/e 254.

70 eV				20 eV			
m/e	Rel.int. %	m/e	Rel.int. %	m/e	Rel.int. %	m/e	Rel.int. %
284	8.6	284		284	42.8	284	6.7
282	6.7	282		282	28.6	282	
268	21.9	268		268	95.2	268	10.0
266	17.1	266	21.8	266	52.4	266	100
254	100	254	100	254	100	254	22.5
239	14.3	239	4.0	239		239	
237	19.0	237	15.3	237		237	

of the benzo(a)pyrene metabolites from the incubation medium, the crude extract was subjected to mass spectrometric analysis using the direct inlet system of the mass spectrometer.

The results obtained from two incubation experiments are reported in Table 1.

The mass spectra were recorded successively at 70 and 20 eV electron energy and all the ions mentioned in this table were not found in the mass spectrum of the crude microsomal extract not containing benzo(a)pyrene.

Surprisingly enough compound 1 was found as a metabolite of benzo(a)pyrene but moreover an ion of m/e 266 not present in the mass spectrum of the synthetic product was observed, resulting from a loss of water from the molecular ion.

In the first experiment, the observed intensity of the molecular ion in relation to m/e 254 is 8.6 and 42.8% successively at 70 and 20 eV electron energy. In the second experiment and for the same energy levels, the molecular ion is not observed (70 eV) and is weakly observed (6.7%) at the lowest energy level. The loss of water is also greatly enhanced in the second experiment.

The only way to explain the presence of ion m/e 266 (Scheme 2) is to consider that metabolite 1 exists under different tautomeric forms, hemiketal forms 1a, 1e and ketol forms 1b, 1d and that according to the presence of other compounds in the medium such as proteins or amino acids, the predominance of one of these tautomeric forms may be modified by association or reversible addition.

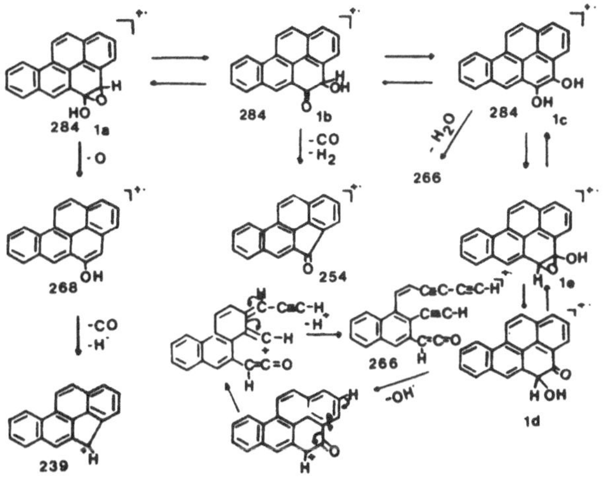

SCHEME 2. Fragmentation pathway of metabolite 1.

For the pure synthetic product, the predominant form is the
hemiketal form which loses oxygen from the molecular ion to produce
an ion of mass m/e 268. Carbon monoxide and a hydrogen radical are
lost from this last ion to give another one of m/e 239.

The intensity of ion m/e 254 is much smaller for the synthetic
product than for the same product isolated from incubation media
indicating most probably that the prevalence of the ketol form 1b
is unimportant in relation to the hemiketal form.

The situation is exactly inversed for the same compound 1
extracted from rat liver microsomes indicating that when interaction
with foreign compounds are possible, the predominance of the ketol
forms may be increased by association or by the formation of a
reversible addition product giving rise as a consequence of the
direct liberation of the ketol form in the ion source to a more
intense fragmentation ion of mass m/e 254. The observed loss of
water (m/e 266) may originate from a ketol form by a mechanism
similar to that proposed in Scheme 2 since catechols which may
rearrange to ketol forms are reported to lose water.

What has been revealed till now is that benzo(a)pyrene may
give rise to a very special derivative, metabolite , which may
preexist in biological fluids under different tautomeric forms,
mainly hemiketal and ketol forms having their own inherent reactivity.

Since the carcinogenic response of animals to polycyclic aromatic
hydrocarbons may be mediated by the metabolism of these compounds,
since K region epoxides formed in the microsomal metabolism of
hydrocarbons do not exhibit greater carcinogenic potencies than the
parent compound (8), since also 7-methylbenzo(a)anthracene 5,6 oxide,
a K region epoxide, does not react with DNA to yield products
identical to those formed when the parent hydrocarbon is bound to
DNA in cellular system (9), it must be admitted that this epoxide
may be not the true intermediate involved in the binding of certain
carcinogenic polyaromatic hydrocarbons to DNA in cellular systems.

These observations greatly enhance the importance of compound
1, metabolite of benzo(a)pyrene, which certainly plays an active
role in the molecular event initiating neoplasia.

The way by which this metabolite may react with macromolecular
cellular constituents is summarized in Scheme 3 and will be now
discussed.

It has been mentioned that metabolite 1 of benzo(a)pyrene may
exist under different tautomeric forms 1a and 1b and that one of
these forms 1a which contains a hemiketal function has to have
more chance to react with nucleophilic targets.

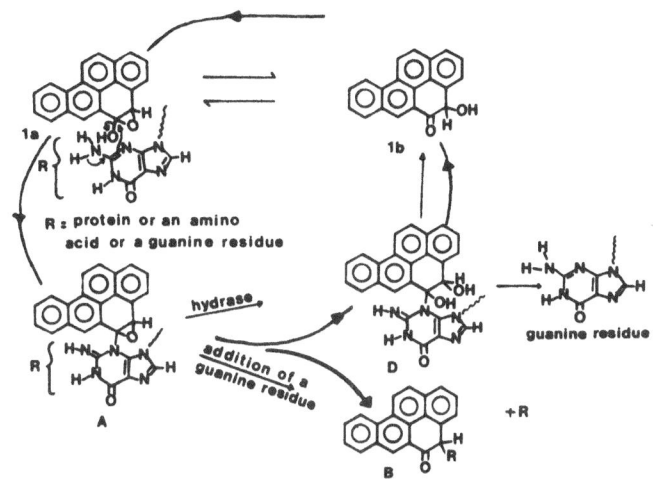

SCHEME 3. The proximo-distal carcinogenic forms
of benzo(a)pyrene and the continual
round way of detoxification and
toxification.

The effective concentration of the two tautomers 1a and 1b in
biological fluids must depend strongly on the local pH conditions,
the local concentration of the substrate, the intrinsic polarity
of the cellular membranes and stereochemical requirements.

Because the ketol form, which is the less reactive species,
contains its own inherent potent activation, this metabolite must
be considered as a proximo-distal carcinogen (10).

If this form 1a comes into interaction with aminoacids,
proteins or guanine residues (example illustrated in Scheme 3),
a covalent bond may be formed between position 4 or 5 of the benzo-
(a)pyrene ring and one of the nucleophilic centers of a guanine
residue.

This guanine addition product may be detoxified by the hydrase
enzymic activity leading to the splitting up, by a reversible
mechanism bound to the properties of the carbinolamine function, of
the guanine addition product into two moieties, the original guanine
residue and the original proximo-distal form 1b.

It is of course very interesting to note at this level that
hydrase activity may lead to repair mechanisms.

If metabolite 1a where an amino acid, a protein or another
residue has become attached reaches a region where the enzymic
hydrase activity is considerably reduced for one reason or another,

then this addition product may be forced to react with a nucleophilic center such as a guanine residue to give rise to an addition product B completely different from A and which is now unable to be detoxified.

Because of its way of working, this mechanism may be defined as the continual round way of detoxification and toxification well supported by the mass spectrometric evidence , as already mentioned, that the metabolite 1 extracted from rat liver microsomes gives a more intense ion of mass m/e 254 resulting from the loss of $M-CO-H_2$ from the ketol form which is released directly into the ion source by the splitting up into two moieties by temperature effects of the detoxified addition product D of metabolite 1 with a protein, an amino acid or some other products containing a nucleophilic center.

This continual round way mechanism of detoxification and toxification also supports very well Selkirk's observation (11) that 1,2-epoxy-3,3,3-trichloropropane stimulates benzo(a)pyrene binding to DNA, since being a potent epoxide hydrase inhibitor, it increases the chances of transformation of the addition product A to an adduct B which is unable to be detoxified.

Also because a great number of detoxifying cycles may be performed before a target is touched in an irreversible way, the mechanism may also be defined as the searching head mechanism proving once more the importance of cyclic phenomena in life and the importance of the hydrase activity. This continual round way mechanism of detoxification and retoxification also provides a fairly good explanation for the non carcinogenicity of benzo(a)-pyrene.

ACKNOWLEDGEMENTS

The author is grateful to Marie-Christine Fayt for skilful technical assistance. This work was supported by a grant from the "Banque Nationale de Belgique", Brussels, Belgium.

REFERENCES

1) P. Daudel and R. Daudel, in "Chemical carcinogenesis and molecular biology", Interscience, New York, 1966.
2) E. Huberman, L. Sachs, S.K. Yang and H.V. Gelboin, Proc. Nat. Acad. Sci., U.S.A., 1976, 73, 607.
3) P. Sims, P.L. Grover, A. Swaisland, K. Pal and A. Hewer, Nature, 1974, 252, 326.
4) E. Huberman, T. Kuroki, H. Marquargt, J.K. Selkirk, C. Heidelberger, P.L. Grover and P. Sims, Cancer. Res., 1972, 32, 1391.

5) A. Amar-Costesec, H. Beaufay, M. Wibo, D. Thinès-Sempoux,
   E. Faytmans, M. Robbi and J. Berthet, J. Cell. Biol., 1974,
   61, 201.
6) H. Yagi, G.M. Holder, P.M. Dansette, O. Hernandez, H.J.C. Yeh,
   R.A. LeMahieu and D.M. Jerina, J. Org. Chem., 1976, 41, 977.
7) B. Pullman, in "Physico-Chemical Mechanisms of Carcinogenesis",
   Academic Press, New York, 1969.
8) P. Sims, Int. J. Cancer, 1967, 2, 505.
9) W.M. Baird, A. Dipple, P.L. Grover, P. Sims and P. Brookes,
   Cancer Res., 1973, 33, 2386.
10) G. Lhoëst, M. Roberfroid and M. Mercier, Biomed. Mass Spectrom.,
    1978, 5, 38.
11) K.J. Selkirk, R.G. Croy, P.T. Roller and H.V. Gelboin, Cancer
    Res., 1974, 34, 3474.

QUANTITATIVE DETERMINATION OF VINCAMINE IN HUMAN PLASMA BY GAS

CHROMATOGRAPHY - MASS SPECTROMETRY

P. Devaux, E. Godbille and R. Viennet

Centre de Recherches Roussel Uclaf

Romainville, France

INTRODUCTION

Vincamine, a naturally occurring alkaloid, is a drug which is used for preventive or curative treatment in cerebral circulatory insufficiency. It was originally extracted from Periwinkles (Vincaminor) but is now produced by synthesis or hemisynthesis.

In order to facilitate clinical investigations of this drug, a method has to be designed for its quantitative determination in biological fluid which is not based on the use of radiolabelled compounds.

Two approaches have been described recently in the literature: one employs a fluorimetric assay after alkaline hydrolysis of Vincamine (1), the other one uses gas chromatography.

In the latter case, Vincamine was detected without prior derivatization (2) or after the formation of its trimethylsilyl ether derivative (3). However, none of these methods seems to present the specificity and the sensitivity required to carry out pharmacokinetic and bioavailability studies in man.

This paper describes a mass fragmentography assay for the quantitative determination of this drug in human plasma °).

°) During the time this paper was submitted for publication, Hoppen et al. (4) have described a similar assay but which does not use any internal standard or clean-up procedure.

EXPERIMENTAL

Materials. Human plasma was obtained from five volunteers
following an oral administration of three tablets of Vincadar
corresponding to 30mg of Vincamine. The samples were stored at
-20°C until analysis.

B.S.T.F.A. (N, O-Bis-(trimethylsilyl)-trifluoro-acetamide) and
d-9 TSIM (N trimethyl d-9 silyl imidazol) were purchased respectively
from Pierce Chemical and Merck, Sharp and Dohme, Canada.

Vincamine and its labelled analog (specific activity: 42.8
mci/m mole) were obtained from Roussel Uclaf.

Clean-up procedure was carried out by adsorption liquid
chromatography with 40 μm silica particle (Merck H60). The mobile
phase was a mixture of methylene chloride and ethanol (60/40; v/v).

Gas chromatographic analyses were performed on a 1.5m length
glass column packed with 1% SE30 on 100-120 mesh acid washed and
silanized Gas Chrom P.
Equipment. A Finnigan 3300 gas chromatograph mass spectrometer
was employed for mass fragmentographic measurements in the e.i. mode.
Two ions were monitored simultaneously: m/e = 367$^+$ (Vincamine) and
m/e = 376$^+$ (internal standard).

The instrument was operated under the following conditions:
injector: 250°C; column: 220°C; transfer line: 250°C; separator:
250°C.

The ionization potential and trap current were 70 eV and 0.5
ma. The electron multiplier voltage was set at 1627 volts.

A home-built liquid chromatograph (5) was used to clean up the
Vincamine extracted from the plasma. It consisted of a 15cm length,
18mm i.d. column packed with 20g of silica. The support was
longitudinally compressed under 7 bars with a pneumatic   jack.
The samples were injected using a 400 μl loop. Vincamine elution
was monitored with a U.V. detector (Spectrochrom from Gilson) set
at 280nm. The pressure and the flow rate were respectively 7 bars
and 4.8ml/min.

METHOD

Extraction. Prior to the extraction, a given amount of ^{14}C
Vincamine (30ng; 3.6nci) was added to 3ml of plasma. The sample
was then adjusted to pH 9 with 1.5ml borate buffer 0.1M. Vincamine
was then extracted three times by 5ml chloroform. The organic phases

were combined, washed with 5ml water and finally evaporated to
dryness under nitrogen.
     Clean-up. A liquid chromatographic procedure was set up to
achieve a better purification of the extracted material.

     First, the elution process was standardized by injecting a
detectable amount (4 μg) of pure Vincamine in order to determine
its retention and elution volumes. Then the residue from the
extraction was dissolved in 400 μl of the mobile phase and injected
onto the column.

     The fraction (20ml) corresponding to the elution of the drug
was collected. An aliquot (4ml) was taken for scintillation counting
to determine the extraction and clean-up recovery. The remaining
16ml were evaporated for further derivatization.
     Derivatization. The residue was taken up by 0.5ml of B.S.T.F.A.,
heated at 60°C overnight and evaporated to dryness. The deuterated
trimethylsilyl derivative was employed as internal standard: a
stock solution was prepared by heating 10mg of Vincamine at 60°C
overnight in 1ml of a mixture of pyridine and d-9 TSIM (90/10;
v/v).

     The biological residue from the trimethylsilylation was
dissolved by 40 μl of the stock solution. Aliquots (1-2 μl) were
injected onto the G.C. column.

RESULTS AND DISCUSSION

     Vincamine can be analysed by G.C. without derivatization (2).
However, two main difficulties may arise from the presence of a
free tertiary hydroxyl group in the molecule (Fig. 1): first, a
partial thermal degradation may take place leading to poor
quantitative results; second, at the a nanogram level severe
adsorption phenomenon may occur, decreasing the overall sensitivity
of the method.

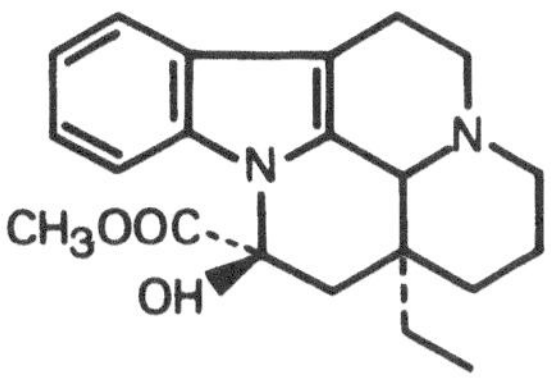

FIG. 1. Structure of Vincamine.

Therefore it seemed more reliable to achieve a derivatization
procedure by means of a trimethylsilylation.

The TMSi derivative of Vincamine exhibits good G.C. properties,
as shown in Fig. 2, and is stable for several days (Fig. 3).

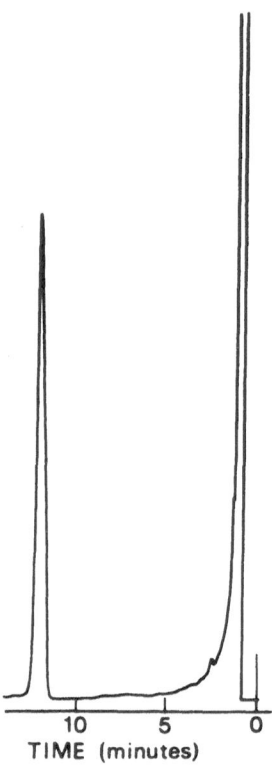

FIG. 2. Gas chromatogram of TMSi ether of Vincamine
on a 2m. 1% SE30 column at 240°C.

Moreover, as will be seen later, this step is very convenient
because it introduces a reproducible standardization of the mass
fragmentographic quantitation.

The mass spectrum of the TMSi derivative of Vincamine is shown
in Fig. 4. It exhibits a very intense ion at mass m/e = 367$^+$. This
fragment was selected for a sensitive and specific quantitation
of this compound by selected ion monitoring. It is formed by the
loss of the acetoxy moiety from the molecular ion (Fig. 5).

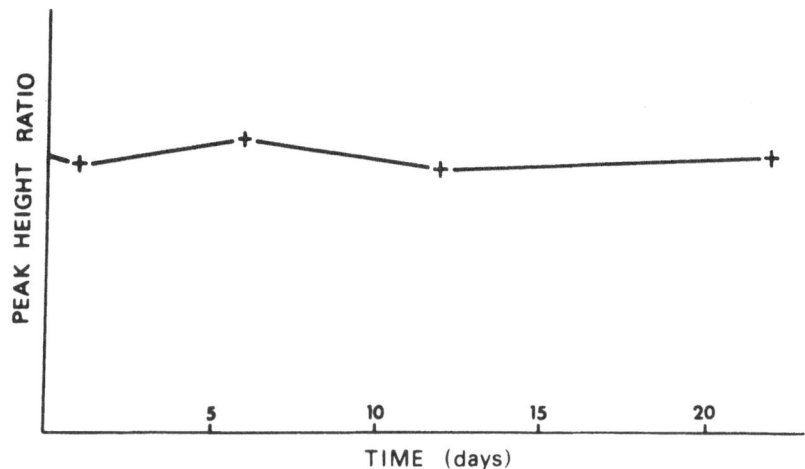

FIG. 3. Variations of the peak height ratio of the TMSi
derivative of Vincamine and internal standard versus
time in a solution of pyridine and bis trimethylsilyl-
acetamide (99/1; v/v).

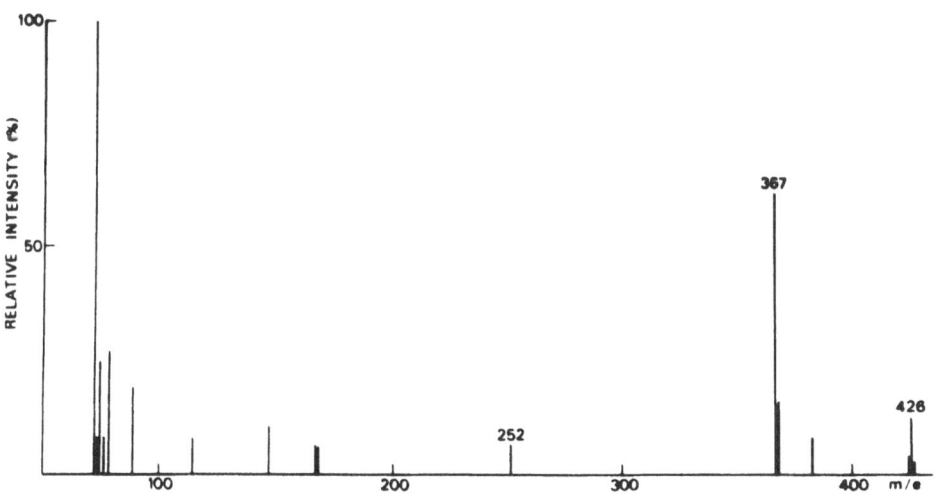

FIG. 4. The e.i. mass spectrum of TMSi - Vincamine.

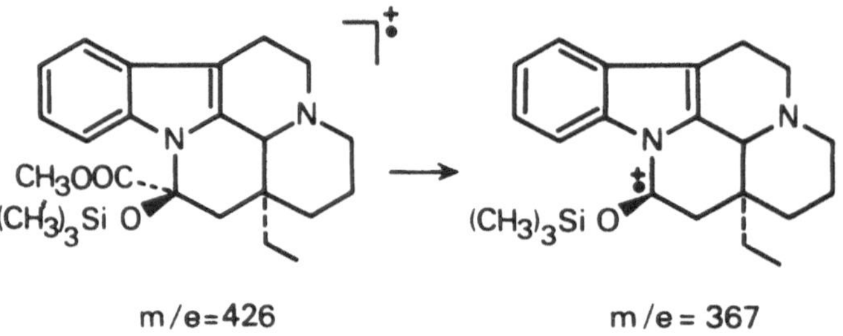

m/e=426   m/e = 367

FIG. 5. Fragmentation pathway leading to the
m/e = 367[+] ion.

Recovery determination. Prior to injecting the sample onto the
G.C. column, several stages are required such as extraction, clean
up and derivatization.

In order to account for recovery during these steps, an
internal standard is usually added to the biological material
before the extraction. It is detected simultaneously with the
compound under investigation. The ^{14}C labelled Vincamine, which was
synthetized for pharmacological tests on animals, was used for
this purpose. Our goal was to monitor the two fragment ions: m/e =
367[+] and m/e = 369[+], but this compound, having the ^{14}C atom on the
acetoxy group, did not look suitable as internal standard.

However, the clean-up procedure which was developed to purify
the extract was thought to introduce, besides the extraction,
another source of error in the methodology. For this reason, it
was decided to use the radioactive compound anyway to check for the
recovery. The estimation of Vincamine in plasma was then performed
according to the scheme given in Table 1.

The contribution of the internal standard to the intensity of
m/e = 367[+] ion reduced the sensitivity of the method. The limit of
detection is evaluated to 5ng of Vincamine per ml of plasma. This
sensitivity was however sufficient to obtain significant and useful
pharmacokinetic data concerning the elimination of this drug in
plasma after oral administration.

The injection was calibrated by using the deuterium labelled
derivative of Vincamine. The quantitation was then achieved by
measuring the peak height ratio of the m/e = 367[+] ions.

The intensity of the fragment 367[+] in the mass spectrum of the

TABLE 1
Analytical procedures for the quantitative
determination of Vincamine in plasma.

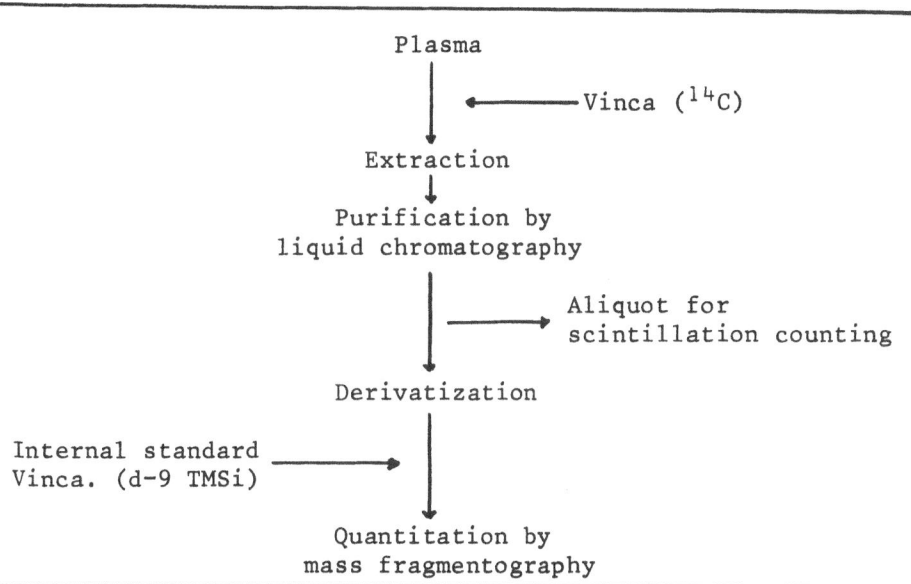

Plasma

Vinca (^{14}C)

Extraction

Purification by
liquid chromatography

Aliquot for
scintillation counting

Derivatization

Internal standard
Vinca. (d-9 TMSi)

Quantitation by
mass fragmentography

d-9 TMSi derivative was about 1% of the intensity of the fragment
376$^+$. Fig. 6 represents the variation of the peak height ratio of
the proteo and deuterio TMSi derivative versus the concentration
ratio of the two compounds. Within experimental errors, a straight
line is observed which is not different from the theoretical one.

   Purification. Attempts to inject directly the derivatized crude
plasma extract onto the gas chromatograph led to a rapid deterioration
of the column. Excessive peak tailing occurred after several
injections and high blank values were observed, as illustrated in
Fig. 7.

   It must be pointed out that peak A has the same retention time
as the TMSi ether of Vincamine. The presence of this endogenous
compound may drastically alter the reproducibility and the
sensitivity of the mass fragmentographic assay.

   To overcome this problem a clean-up method was set up using
H.P.L.C.. A similar procedure has already been described by De
Ridder et al. (6) to analyse in human plasma an antidepressant
drug by G.C. - M.S..

   Fig. 8 represents the liquid chromatogram of pure Vincamine

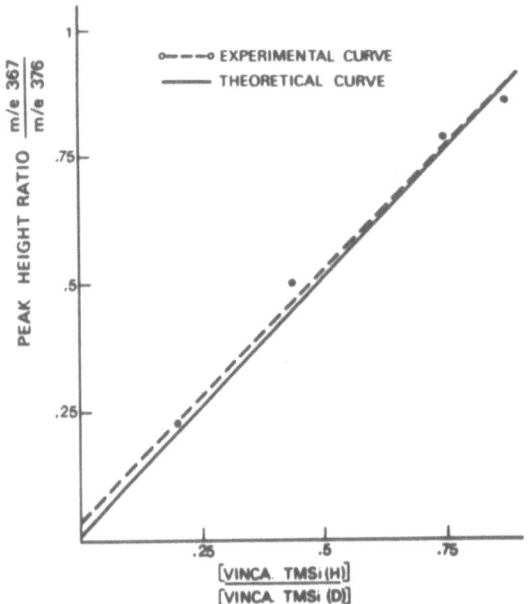

FIG. 6. Variations of
the peak height ratio
of 367$^+$ and 376$^+$ ions
versus the concentration
ratio of TMSi and d-9
ether of Vincamine.

FIG. 7. Multiple ion
recording of a blank
plasma sample derivatized
without prior
purification. Peak A
represents an
endogenous compound
which interferes with
TMSi ether of Vincamine.
Column: 1% SE30, 1.5m.
Temperature: 200°C.

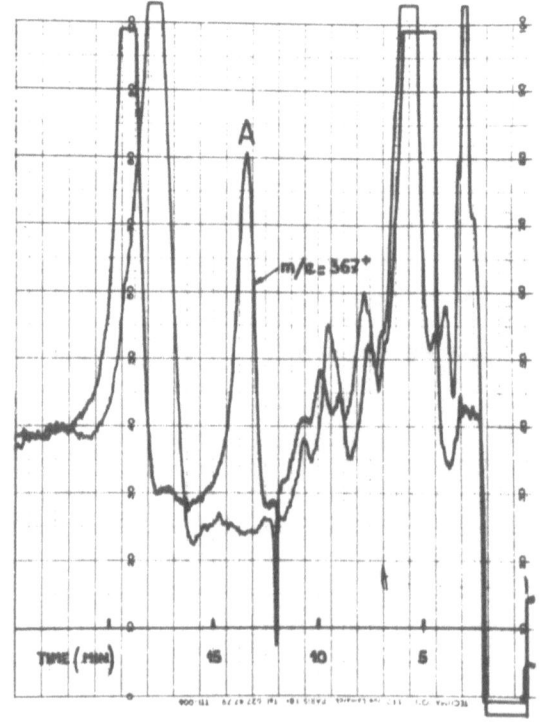

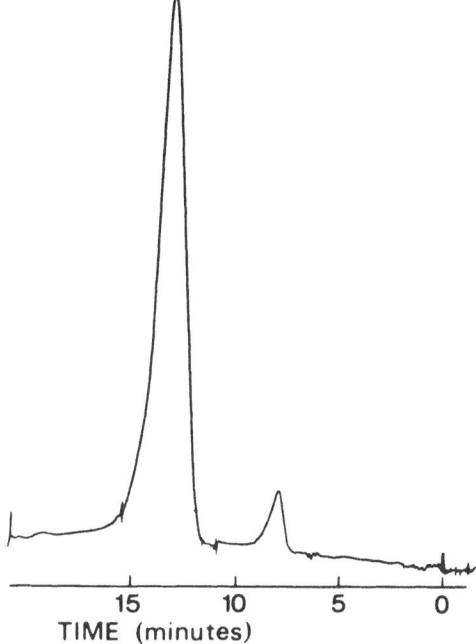

FIG. 8. Liquid chromatogram
of pure Vincamine.

FIG. 9. Liquid chromatogram of a
plasma extract containing
submicrogram amount of
Vincamine. Same experimental
conditions as for Fig. 8.

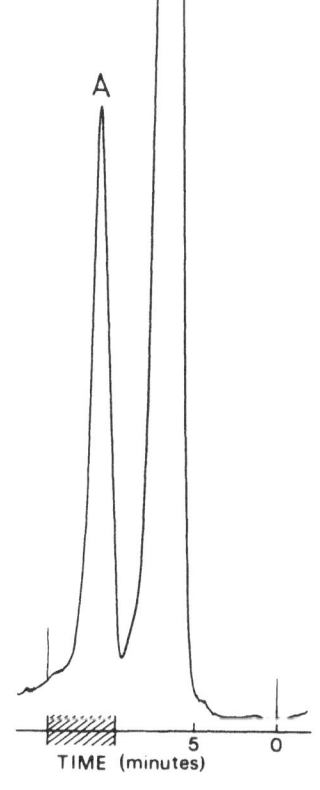

under the experimental conditions given above.

The injection, in the same experimental conditions, of a plasma extract containing submicrogram amount of Vincamine gave the elution profile shown in Fig. 9. At this concentration, the peak of Vincamine could not be detectable. The fraction of eluate corresponding to the elution volume of this compound is represented by the area noted in the figure. It was collected for further derivatization.

The product eluted with Vincamine and represented by peak A in Fig. 9 did not interfere with the mass fragmentographic detection, as illustrated in Fig. 10.

The chromatogram represents the multiple ion recording of plasma extract purified by liquid chromatography.

The comparison with Fig. 7 illustrates the usefulness of the liquid chromatography clean-up in terms of base line stability and peak shape. The chromatography and the evaporation of the eluate take about 30 mn.

Using this procedure, thirty samples per day can be analysed without significant deterioration of the column efficiency.

The mean overall recovery (extraction and clean-up) from 25 samples was estimated at 59% ± 2.3 ($\bar{y}$ ± $s_{\bar{y}}$).

<u>Vincamine quantitation</u>. A calibration curve was constructed by adding known amounts of Vincamine to blank plasma and by performing the entire procedure described above.

The recovery and the presence of the compound were taken into account for estimating the radiolabelled concentration of Vincamine in the plasma samples.

An example of calibration curve is given in Fig. 11 which represents the variations of the peak height ratio of m/e = 367^{+} ion and m/e = 376^{+} ion as a function of the total concentration of the drug in plasma (non labelled compound plus the internal standard). Each sample was injected twice.

The human plasmas were analysed according to the experimental procedure described above.

The result of the mass fragmentographic assay is shown in Fig. 12. Each point represents the mean of five different plasmas (± s_y).

In conclusion, this work has shown that G.C. - M.S. is very

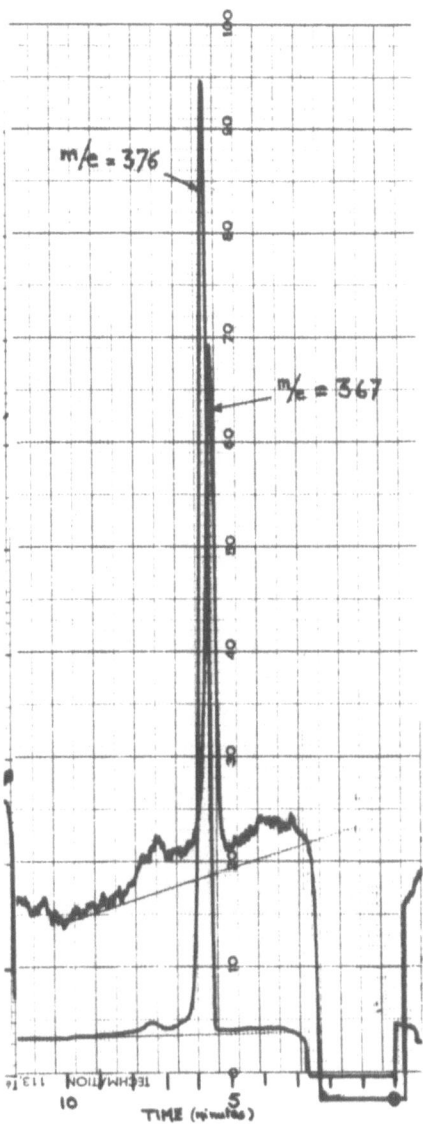

FIG. 10. Multiple ion recording of a plasma extract
        after purification by liquid chromatography
        and derivatization. Peak m/e = 367⁺: 2ng of
        TMSi ether of Vincamine. Peak m/e = 376⁺:
        20ng.Internal standard (d-9 TMSi ether of
        Vincamine). Column: 1% SE30, 1.5m.
        Temperature: 220°C.

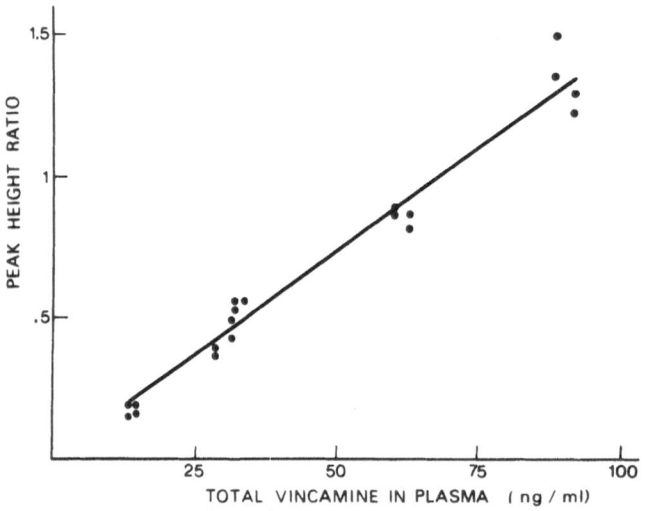

FIG. 11. Calibration
curve for the
quantitative
determination of
Vincamine in human
plasma.

FIG. 12. Variation of
Vincamine plasma concentration
versus time after oral
administration.

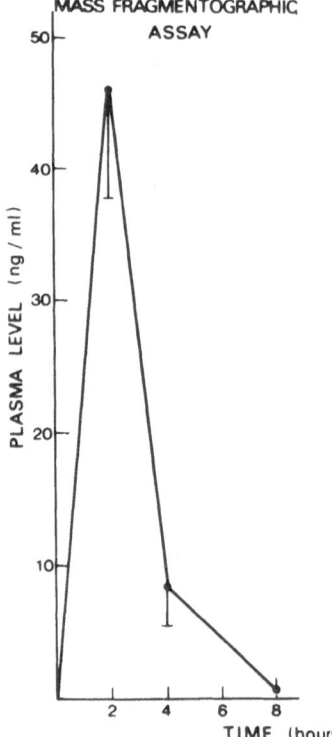

suitable to analyse Vincamine in biological fluid. However, unsatisfactory results can be obtained when the crude extract is injected onto the G.C. column.

In general, to be applicable on a routine basis, a purification step is required whose recovery must be evaluated by an internal standard.

ACKNOWLEDGEMENTS

The skilful technical assistance of A. Dubois and E. Lesage (Extraction and liquid chromatography) and R. Legrand (Mass spectrometry measurement) is gratefully acknowledged.

REFERENCES

1) V.H. Iven and C.P. Siegers, Arzneim.-Forsch., 1977, 27, 1248.
2) H. Kinsun and M.A. Moulin, J. Chromatogr., 1977, 144, 123.
3) V.H. Laufen, W. Juhran, W. Fleissig, R. Götz, F. Scharpf and G. Bartsch, Arzneim.-Forsch., 1977, 27, 1255.
4) H.O. Hoppen, R. Heuer and G. Seidel, Biomed. Mass Spectrom., 1978, 5, 133.
5) E. Godbille and P. Devaux, J. Chromatogr., 1976, 122, 317.
6) J.J. De Ridder, P.C. Koppens and H. Van Hal, J. Chromatogr., 1977, 143, 281.

MASS FRAGMENTOGRAPHIC QUANTIFICATION OF PHOSPHONOACETIC ACID IN THE

BLOOD OF MONKEYS INFECTED WITH HERPESVIRUS SAIMIRI

J. Roboz, R. Suzuki, R. Hunt and J.G. Bekesi

Department of Neoplastic Diseases, The Mount Sinai School

of Medicine, City University of New York, New York, U.S.A.

INTRODUCTION

Herpesvirus Saimiri° was isolated from a degenerating squirrel monkey kidney culture in 1968 (1). HVS induces leukemia and/or lymphoma when inoculated into a variety of nonhuman primates (2). The induced lymphoma appears to be analogous to those involved in EBV and Burkitt's lymphoma in humans, thus it is a good model for evaluating antiviral agents used alone or in combination with chemoimmunotherapy. In 1973 it was reported that phosphoacetic acid (PAA) and its disodium salt has an inhibitory effect on the in vitro and/or in vivo replication of several DNA-containing viruses which are assumed to be oncogenic, including HVS (3, 4). PAA inhibits the synthesis of late viral products by interfering with the activity of virus-associated DNA polymerase (5).

In cell culture studies PAA was found to be highly effective against HVS. Even when added 120 hr post infection with HVS, PAA was able to prevent further viral replication; moreover, cells appeared to be recovering after PAA treatment (6). PAA has little or no effect on RNA viruses; it is not toxic for normal monkey kidney cells. Based on these and some other related observations, a preclinical toxicological study of PAA was initiated using various normal and HVS-infected monkeys.

°Abbreviations: HVS=herpesvirus saimiri; EBV=Epstein-Barr virus; PAA=phosphonoacetic acid; PPA=phosphonopropionic acid; TMSi=trimethylsilyl; BSTFA=N,O-bistrimethylsilyltrifluoroacetamide; TMCS=trimethylchlorosilane; BUN=blood urea nitrogen.

In 1977 we reported a mass spectrometric technique for the
quantification of PAA in blood (7). PAA was quantified, after
removing proteins and lipids, as the silylated derivative (3 TMSi)
by monitoring the protonated molecular ions of PAA and phosphonopropionic acid (PPA, used as internal standard) obtained in
chemical ionization using methane as the reagent gas. Utilizing
this technique for initial toxicological studies two major
conclusions were made: (a) to achieve and maintain constant
therapeutic blood levels of PAA the drug must be administered in a
continuous infusion rather than in single doses; (b) dose levels
of 200-250 mg/kg/day by continuous infusion had therapeutic efficacy
and presented no toxicity. The corresponding blood levels were in
the 50-100 µg/ml range and remained constant during infusion. Upon
termination of treatment blood levels of PAA declined to 2 µg/ml
in 24 hr.

The objectives of the present work were to investigate how the
animals would tolerate multiple infusions and how these would effect
blood levels of PAA, and also what effect the condition of the
kidney of infected animals has on the blood levels of PAA.

EXPERIMENTAL

Infusion of PAA. In the early phase of this work, serious
vascular complications were encountered when PAA was administered
via repeated venipunctures and/or subcutaneous injections. A special
jacket was designed for the monkeys which includes a wire-coil
protected cannula, a Harvard liquid delivery pump, and a cannular
feed-through swivel. The apparatus now permits the animals to move
around freely in their cages while receiving continuous administration
of fluids; the monkeys accepted this apparatus without apparent
discomfort. Both intravenous and subcutaneous infusion can now be
accomplished continuously. Subcutaneous infusion of unbuffered
lactate solution of PAA resulted in cellulitis. Buffering to pH=7.3
and using lactate infusion during rest periods in multiple infusions
of PAA resulted in the animals surviving the drug regimen without
local or systemic evidence of untoward reactions.

Sample preparation. Blood obtained from the monkeys was
permitted to clot at room temperature for 15 min, centrifuged at
500 g for 10 min and the serum was removed and stored frozen until
analyzed. The first step in sample preparation was the addition of
PPA internal standard. The concentration of the PPA stock solution
(PPA dissolved in water) was selected in such a manner that 20 µl
of it, when added to 0.2 ml sample serum, would result in a PPA
concentration in the middle of the expected concentration range of
PAA. After adding the internal standard, the sample was diluted with
0.1 ml deionized-distilled water, placed in an ice-bath, and 10 µl
of $HClO_4$ (70%) added. Vortexing for 30 sec was followed by standing

at room temperature for 5 min and heating in a boiling water bath
for 1 1/2 min. Next, 0.8 ml water was added (for a total volume of
1.1 ml) and the tube kept in boiling water for an additional 3 min
(but not more than 5 min). After cooling to room temperature the
tubes were centrifuged at 48,000 g at 5°C for 20 min. This procedure
removed all proteins. The liquid phase could be decanted easily.

The next step was the removal of lipids by the addition of 2.0
ml of dichloromethane to the decanted liquid (about 1 ml) and mixing
by repeated squeezing with a Pasteur pipette. After clearing the
emulsion by centrifuging at 770 g for 10 min at room temperature,
the upper, clear aqueous layer was removed, placed in a small vial,
and evaporated to dryness with dry nitrogen in a water bath at
45–50°C.

The residue was silylated, prior to mass spectrometric analysis,
by adding 200 µl of a mixture of BSTFA+1% TMCS and pyridine (3:1
ratio) and the mixture heated at 100°C in a dry metal heating block
for 4 min.

Gas chromatography-mass spectrometry. Combined gas chromatography-
-low resolution mass spectrometry analyses were performed using a
Finnigan Model 3300 quadrupole type mass spectrometer connected to
a Finnigan Model 6000 interactive data system. The mass spectrometer
was equipped with a chemical ionization source, and the computer had
capability for selected ion monitoring.

Gas chromatographic conditions: 1 m long, 2 mm i.d. glass
column, filled with 3% SE-30 on Chromosorb W (HP), 80/100 mesh,
operated isothermally at 165°C. (For periodic cleaning the column
was heated to 250°C until no more eluent could be detected). Inlet
temperature: 250°C. High purity methane (Matheson Co.) was used
both as the gas chromatographic carrier gas and as the reagent gas
in the chemical ionization source. There was no separator between
the gas chromatograph and the mass spectrometer; the connecting
tubing was kept at 240°C.

Mass spectrometric conditions: methane carrier gas pressure
approx. 1000 µ (as read, uncorrected); electron emission current
0.3-0.6 mA, ionization energy approx. 100 eV, acceleration voltage
25-50 eV. Operational parameters were adjusted daily for optimum
sensitivity using pure PAA and PPA. For quantification of PAA the
instrument was operated in the "selected ion monitoring" mode,
tuned to monitor ions at m/e 371, 357, 355, and 341. Under these
conditions 100 scans took approx. 1 min.

After introducing a 4 µl sample into the gas chromatograph,
the effluent was vented for 30 sec to avoid contamination of the
ion source with the silylation reagents and solvents. The gas
chromatographic retention times of PAA and PPA were 1 min and 2 min,

respectively. Using the conditions given there were no interfering
constituents from monkey serum.

    Quantification of PAA. Normal pooled serum samples were spiked
with known amounts of PAA and PPA and handled in the same manner
as biological samples. Calibration curves were obtained this way
for all expected concentration ranges of PAA. To compensate for
possible errors in the course of a given set of experiments, a full
set of calibration sample was analyzed with every set of samples
from the animals. Peak areas were determined using the computer
system after background correction. After determining the area ratio
of the unknown PAA and known PPA peaks, the quantity of PAA in the
serum was determined from the experimentally obtained calibration
curve for the given set of analyses.

RESULTS AND DISCUSSION

    Analytical technique. Figs. 1 and 2 show the mass spectra of
PAA and PPA, respectively. These were obtained by operating the mass
spectrometer in the "full scanning" mode. The base peak in both
cases is the fully silylated compound (3 silyl groups taken up)
protonated in the chemical ionization mode. The only other peak
present corresponds to the loss of a methyl group. Thus, the
technique is well suited for selected ion monitoring to achieve
maximum sensitivity and to reduce the needs for elaborate sample
preparation.

    The limit of detection of PAA is 50 pg (0.5×10^{-12} mol) material
injected (limit refers to underivatized quantity). Using 0.2 ml
serum, the limit of detection in biological materials is 20 ng/ml.
For details of reproducibility studies the original reference
should be consulted (7).

    Two monitograms are shown in Figs. 3 and 4. These show two
data points from Fig. 5 (see later). Fig. 3 was obtained 16 hr
after infusion started. The area of PAA is considerably larger than
that of the PPA internal standard. The calculated amount of PAA in
this sample was 24.2 µg/ml. The $(M+1)^+$ ions were monitored to
further confirm the presence of the compounds. The areas of those
peaks were not utilized in the calculations. Fig. 4 was obtained
24 hr post infusion and represents a low PAA value. Here the
internal standard peak is considerably larger in appearance than
that of PAA. The actual amount of PPA added to both samples shown in
the two figures were the same (5 µg/ml). The PAA in the second
sample was 0.76 µg/ml.

    Multiple infusions of PAA. We have established during the early
phases of this work (7) that large doses of PAA in single injections
kill the animals. To achieve and maintain therapeutic levels it was
essential to administer the drug in slow continuous infusion so
that kidney clearance could come to equilibrium. The technical

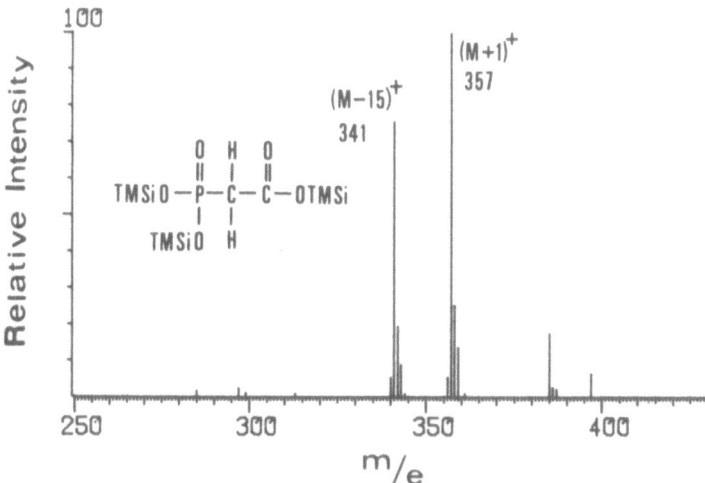

FIG. 1. Chemical ionization (methane) mass spectrum of phosphonoacetic
      acid.

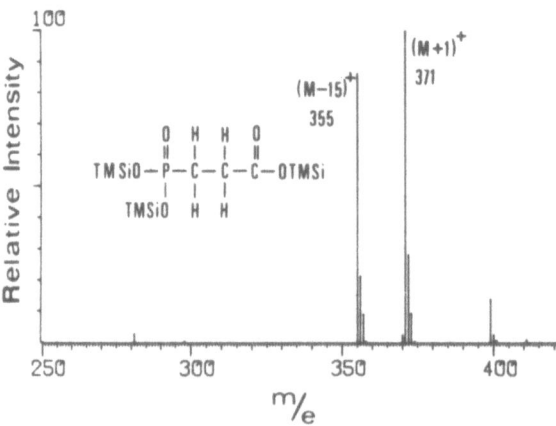

FIG. 2. Chemical ionization (methane) mass spectrum
of phosphonopropionic acid used as internal
standard.

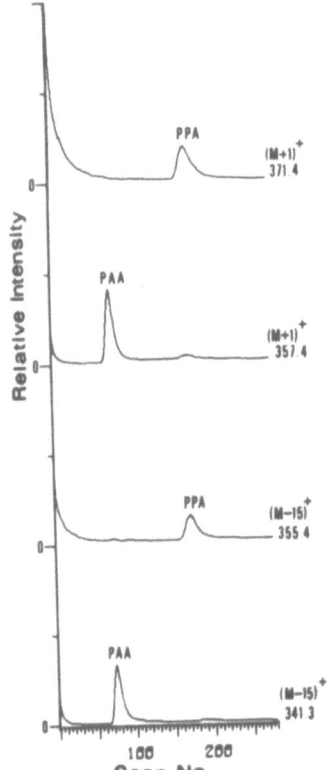

Fig. 3. Selected ion monitoring of
PAA and PPA. Sample corresponds
to 16 hr infusion point in
Fig. 5.

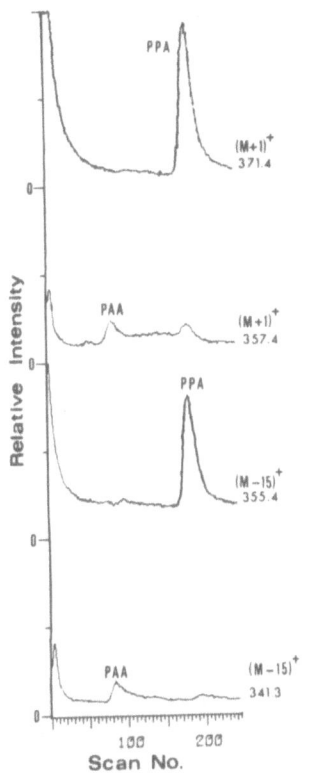

FIG. 4. Selected ion monitoring of PAA
and PPA. Sample corresponds to
24 hr post infusion point in
Fig. 5.

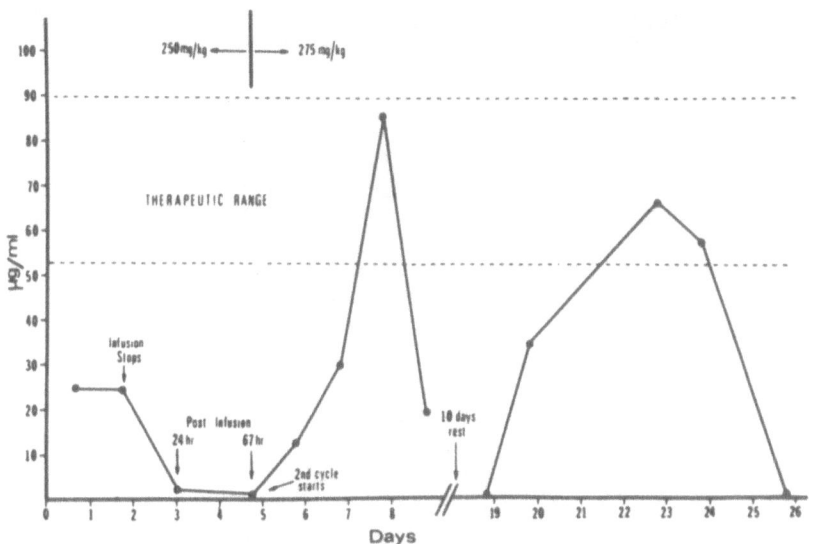

FIG. 5. Three cycles of continuous infusion of PPA in
monkey.

problems of continuous infusion were solved by the design of the special infusion apparatus mentioned earlier.

Fig. 5 shows results of a typical multicycle infusion experiment. Starting with an infusion of 250 mg/kg PAA, the 16 hr post infusion sample gave a rather steady blood level of about 25 µg/ml. After the infusion was stopped the blood level of PAA dropped. When the second cycle of PAA started the dose level was raised to 275 mg/kg in an attempt to achieve the therapeutic range for this particular animal. It is clearly seen that this was achieved by the 7th day. After the second cycle the animal was given a rest period of 10 days. In the third cycle therapeutic level was once again reached and maintained. The entire treatment was well tolerated by the animal.

It is concluded that the proposed therapeutic range for PAA can be reached and maintained by continuous infusion and several cycles of treatment may be made without ill effects. Since the rate of drug elimination varies from animal to animal, therapeutic monitoring of PAA is essential to establish if the therapeutic range was reached for a given dose. When indicated, such as in the case shown in Fig. 5, a change in the dose should be made for the second and subsequent cycles.

Effect of kidney condition upon PAA blood levels. When an animal with diagnosed glomerulonephritis was given a dose of 300 mg/kg/day, the PAA level in the blood increased rapidly. When the PAA level increased over 100 µg/ml the animal became sick. Blood urea nitrogen increased to 42 mg/100 ml, and shortly after the infusion was terminated the animal died. In the control animal, also receiving the same dose of PAA, the blood level of PAA reached that considered to be therapeutical (Fig. 5) and blood urea nitrogen remained normal during infusion. In all cases where the cause of death was determined as toxic nephrosis, the blood levels of PAA (and also that of blood urea nitrogen) were excessively high. The observed PAA blood levels appeared proportional to the severity of toxicity. Since PAA is cleared mainly through the kidneys, an equilibrium between excretion and infusion must be established to avoid toxicity. Monitoring the PAA content of the blood is an important means of following the maintenance of this equilibrium during multiple infusion since toxicity does not occur as long as the PAA level is below 100 µg/ml.

REFERENCES

1) L. Melendez, M.D. Daniel, R.D. Hunt and F.G. Garcia, Lab. Animal Care, 1968, 18, 374.
2) F. Deinhardt, in "The Herpesviruses", A.S. Kaplan, Ed., Academic Press, New York, 1973, p. 595.
3) N.L. Shipkowitz, R.R. Bower, R.N. Appell, C.W. Nordeen, L.R.

   Overby, W.R. Roderick, J.B. Schleicher and A.M. Von Esch, Appl.
   Microbiol., 1973, 27, 264.
4) L.R. Overby, E.E. Robishaw and J.G. Schleicher, Antimicrob.
   Agents Chemother., 1974, 6, 360.
5) J.C. Mao, E.E. Robishaw and L.R. Overby, J. Virology, 1975, 15,
   1281.
6) N.W. King, H. Barahona, M.D. Daniel, J.G. Bekesi and T.C. Jones,
   Lab. Invest ., 1978, 38, 181.
7) J. Roboz, R. Suzuki, R. Hunt and J.G. Bekesi, Biomed. Mass
   Spectr., 1977, 4, 291.

THE DETERMINATION OF BLOOD CYANIDE AND PLASMA THIOCYANATE BY GAS CHROMATOGRAPHY-MASS SPECTROMETRY

I. Thomson, R.A. Anderson and W.A. Harland

Department of Forensic Medicine, The University of

Glasgow, Glasgow, Scotland

INTRODUCTION

Cyanide is a normal constituent of blood, present at very low levels, usually less than 20 μmoles/l blood (1, 2). It is involved in a number of metabolic reactions of which the conversion to thiocyanate is quantitatively the most important (3). Thiocyanate may therefore be used as an index of transformed cyanide (4). Methods for the analysis of these two species in body fluids are necessarily extremely sensitive and the existing techniques include colourimetry (4, 5), fluorimetry (6, 7) and gas chromatography (8-10).

The toxicity of cyanide is well known and although it was formerly used as a sedative (9), cyanide has traditionally been of interest in cases of poisoning arising as a result of homicide, suicide or industrial accidents. Currently cyanide is of much interest in the field of fire research. The combustion of many natural and synthetic nitrogenous polymers produces hydrogen cyanide in a mixture with a complex series of combustion products (11-13). Over the last two decades there has been a generally widespread introduction of modern polymers as materials of domestic furnishing and construction. During the same period the number of fire deaths due to inhalation of smoke and fire gases has increased by a factor of three (14), and concern has been expressed that this may be due to an increase in the toxicity of the fire atmosphere caused by the combustion of these new polymeric materials.

This paper describes the development of a GC-MS method for the determination of cyanide and thiocyanate in which the two species were converted to cyanogen chloride by reaction with

215

chloramine-T (10). Cyanide labelled with nitrogen-15 was used as
the internal standard for analysis by specific ion monitoring
(S.I.M.). The method was applied to a study of fire casualties in
the Strathclyde Region of Scotland.

EXPERIMENTAL

Reagents and standards. All reagents were of analytical grade.
n-Heptane (B.D.H., Poole, Dorset) was redistilled before use through
a 1 m fractionating column packed with glass spirals. Chloramine-T
(sodium-p-toluene sulphonchloramide, Hopkins and Williams, Chadwell
Heath, Essex) was recrystallised from methanol-water. A buffered
solution of chloramine-T was prepared immediately before use by
mixing 3 parts 1M NaH_2PO_4 and 1 part 0.25% w/v chloramine-T in
distilled water. Potassium cyanide containing 95% ^{15}N was obtained
from Prochem, B.O.C. Ltd., London. Standard cyanide solutions were
prepared in 0.1N NaOH solution to minimise loss of HCN and were
determined by titration against $AgNO_3$, using p-dimethylaminobenzyl-
idene-rhodanine as indicator.

Methods. Blood samples were stored at 4°C until analysed,
normally within 3 days. If analyses could not be performed within
this period, samples were processed through the first stage of
Cavett flask diffusion (CN) or protein precipitation (SCN) and the
resulting NaOH solutions or deproteinised plasma were stored at
-28°C. (a) Cyanide. Whole blood (2ml) was added to a Cavett flask,
the diffusion cup of which contained 1 ml of a standard solution
of $KC^{15}N$ in 0.1N NaOH (10-50 μM in $KC^{15}N$). To the flask was added
2 ml 6N H_2SO_4. The flask was shaken gently and the diffusion cup
replaced immediately. After 2 hr the NaOH solution was transferred
to a 5 ml Reactivial (Pierce) containing 0.2 ml heptane and 1 ml
of freshly prepared chloramine-T solution. After mixing of the
contents, the vial was placed in an ice bath for 10 minutes. The
vial was then removed and reshaken. When the layers had separated
1-10 μl of the heptane layer containing the dissolved CNCl were
used for GC-MS. Standard solutions of $KC^{14}N$ (50 μM in distilled
water) were similarly treated. (b) Thiocyanate. Plasma was prepared
by centrifugation of whole blood at 3000 rpm for 5 minutes. Proteins
were removed by the addition of 10% w/v trichloroacetic acid (TCA)
solution (3 ml) to 1 ml plasma and repeating the centrifugation.
A 1 ml aliquot of the supernatant was transferred to a Reactivial
containing 1 ml of a $KC^{15}N$ standard solution (10-50 μM in 0.1N
NaOH), 1 ml of freshly prepared chloramine-T solution, 0.1 ml of
0.25% w/v $FeCl_3$ solution and 0.2 ml heptane. The method then
proceeded as for cyanide analyses. Standard solutions of KSCN (50
μM in water) were treated similarly.

Interference of cyanide in thiocyanate analyses. $KC^{15}N$ was
added to a blood sample to produce an approximate concentration
of 50 μmole/l $C^{15}N^-$. After an equilibration period of 1 hr, the

blood was mechanically haemolysed and then centrifuged (3000 r.p.m., 5 min). To 1 ml of the supernatant was added 3 ml 10% w/v TCA and after centrifugation 2 ml supernatant were removed for cyanide analysis.

Carbon monoxide in blood. The level of carboxyhaemoglobin in blood was estimated spectrophotometrically using the CO oximeter (15) (Instrumentation Lab., Altrincham, Cheshire) or by gas chromatography (16). Clotted blood from post-mortem examination was treated initially by the method of Freireich and Landau (17).

Gas chromatography-mass spectrometry. A Pye 104 gas chromatograph interfaced to a V.G. 16F mass spectrometer was used for the analyses. GLC was carried out on a glass column (9' x 1/4" i.d.) packed with 7% Hallcomid M-18 on 80/100 mesh Chromosorb W AW DMCS at 80°. The helium carrier gas flow rate was 30 ml/min. Under these conditions, CNCl and solvent had retention times of 1.8 min and 3 min respectively (Fig. 1).

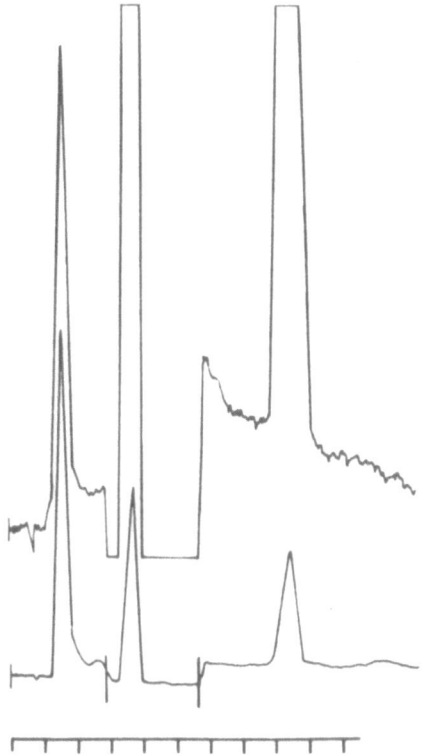

FIG. 1. A typical selected ion profile of blood
cyanide as $C^{14}NCl$ (m/e 61) using $KC^{15}N$
as internal standard (m/e 62).

The column was clear for injection of the next sample after 10 min. The spectrometer was operated in the S.I.M. mode at 70 eV ionising energy, filament current of 200 μA and multiplier voltage of 3.0 kV. The source and interface temperatures were 220°C and 100°C respectively. A four-channel peak selection unit, equipped with a masslock accessory, was used to monitor the ions at m/e 61 and m/e 62, corresponding to the molecular ions of $C^{14}NCl$ and $C^{15}NCl$ respectively. The background ion at m/e 44 was used as a reference ion. With a recorder voltage of 100 mV, a 2 ml blood sample containing 10 μmole/l CN^- would typically give a deflection of 50% of full scale on injecting 5 μl of heptane.

Subject groups and statistics. Cyanide and thiocyanate levels were measured by GC-MS in three subject groups: 8 firemen on active service with the Dundee Fire Brigade, 3 non-fatal casualties taken to hospital suffering from the effects of smoke inhalation in domestic fires, 5 fire fatalities who died in dwelling house fires, and 2 controls, not previously exposed to smoke who were attending a hospital outpatient clinic. Where possible the smoking history of each subject was recorded.

Statistics were carried out using the Statistical Package for the Social Sciences through the Northumbrian Universities Multiple Access Computer (NUMAC) Service.

In vivo metabolism of $KC^{15}N$. The in vivo conversion of $C^{15}N$ to $SC^{15}N$ was studied using two albino rabbits, body weight 2.5 and 3.0 kg respectively as experimental animals. Following withdrawal of control blood samples, 1 ml of a solution containing 50 μM $KC^{15}N$ in isotonic saline was injected intravenously. Further blood samples were removed from a vein in the ear at intervals over a 90-minute period following dosage. The samples were separated into red blood cell and serum fractions for cyanide and thiocyanate analyses respectively.

RESULTS

The GC-MS method was evaluated using aqueous standards of cyanide and thiocyanate and found to be linear over a range of concentrations from 1 - 100 μmole/l (Fig. 2). Four replicate analyses were carried out for each of 5 standard solutions with concentrations in the range 1 - 100 μmole/l. The standard deviation was less than 3.5% for cyanide analyses or 9% for thiocyanate analyses. The minimum detectable quantity of cyanogen chloride was 100 pg (signal:noise ratio 3:1) which permitted cyanide to be measured in blood at levels as low as 0.1 μmole/l and thiocyanate in plasma at 0.5 μmole/l. For comparison purposes, 8 replicate analyses of a single blood sample were carried out by colourimetry, GLC and GC-MS. The results are summarised in Table 1. The GC-MS method showed the lowest standard deviation for CN^- analysis.

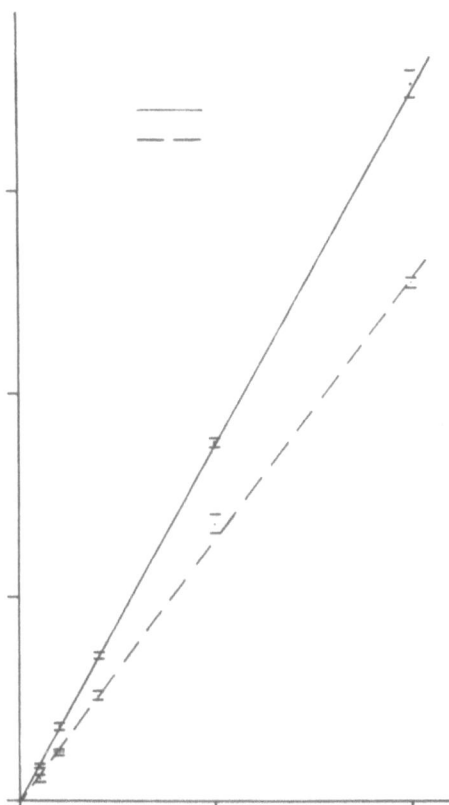

FIG. 2. Calibration curves determined with aqueous
solutions of cyanide and thiocyanate in the
range 5-100 µmoles/l, using a 50 µmole/l
solution of $KC^{15}N$ as internal standard.
The ratio of the peak heights of CN-Cl at
m/e 61 and m/e 62 is plotted against (CN⁻)
and (SCN⁻).

TABLE 1
Comparative analysis of blood cyanide and plasma
thiocyanate by colourimetry, GLC and GC-MS^.

Method	No. of samples	Mean (μmoles/l)	Range (μmoles/l)	Std. Deviation (μmoles/l)
Cyanide				
Colourimetry	8	114.5	102-128	± 10.3
GLC	8	101.0	78-121	± 5.4
GC-MS	8	121.7	118-131	± 4.8
Thiocyanate				
Colourimetry	8	92.5	88.3-97.7	± 3.3
GLC	8	95.1	88.4-103.2	± 5.6
GC-MS	8	99.5	91.1-110.1	± 6.0

^ Analyses performed on a blood standard (9).

An experiment was carried out to measure the degree of
interference by cyanide in the analysis of thiocyanate. $KC^{15}N$ was
added to a sample of blood to produce a concentration of 50 μmole/l
and after an equilibration period, the plasma was deproteinised
and analysed for the presence of ^{15}N-labelled cyanide. No detectable
quantities were found.

The results of analyses of blood samples from firemen, fatal
and non-fatal fire casualties and controls are given in Table 2.
To permit a meaningful interpretation of these results, mean levels
of cyanide and thiocyanate measured in similar but more extensive
groups by GLC and colourimetry (2) are also included in Table 2.
In the present series of firemen, smokers showed higher mean cyanide
and thiocyanate levels than non-smokers, and this difference was
significant when examined using Student's t-test (for CN^- t=6.83,
P<0.01, for SCN^- t=7.16, P<0.01). There was also a significant
increase in the mean cyanide level in firemen compared to controls
(t= 2.49, P<0.05).

The metabolic interconversion of ^{15}N-cyanide to ^{15}N-thiocyanate
was studied in the rabbit. Following injection of a 50 nM dose of
$KC^{15}N$, peak blood ^{15}N-cyanide levels were achieved on average within
2 minutes and of thiocyanate approximately 15 minutes later. The
CN/SCN interconversion was approximated to an open two-pool system
described by the equation:

$$\frac{q_{CN}}{q_{CN_o}} = H_1 e^{-g_1 t} + H_2 e^{-g_2 t} \qquad (1)$$

TABLE 2
Cyanide and thiocyanate levels in firemen, fire
casualties, fire fatalities and controls.

Subject group	Blood CN⁻ (μmole/l)	Plasma SCN⁻ (μmole/l)	HbCO%
Controls			
Mean levels for non-smokers			
(n=30)	2.9 ± 2.4^	30.7 ± 28.8^	1.8**
Mean levels for smokers (n=25)	6.8 ± 4.2^	59.8 ± 26.1^	5.3**
Firemen non-smokers			
Case 1	4.5	14.6	1.0
2	1.0	3.4	0.2
3	14.7	77.7	3.0
4	2.4	26.0	1.3
5	7.4	43.8	2.0
Mean level for group ± S.D.	6.0 ± 5.43	33.1 ± 29.1	1.5 ± 1.1
Firemen smokers			
Case 6	21.1	123.5	5.0
7	23.7	192.0	5.8
8	26.2	131.4	6.0
9	20.0	58.2	4.0
Mean level for group ± S.D.	22.6 ± 2.7	126.3 ± 54.7	5.2 ± 0.9
Non-fatal fire casualties			
Case 10	22.4	140.0	16.0
11	7.0	41.0	8.5
12	5.0	30.2	5.0
Fire fatalities			
Case 13	22.0	61.0	46.0
14	18.1	46.6	9.5
15	96.5	82.2	75.2
16	6.0	15.0	6.0
17	34.0	34.0	32.5

^ Determined by colourimetry and GLC.
** Figures from population surveys.

where q_{CN} is the quantity of tracer in the cyanide pool at time t
and q_{CN_0} is the quantity present at zero time.

Values of H_1, H_2, g_1 and g_2 were obtained using a digital
computer to optimise data fitting to the curve defined by $Ae^{-\alpha t} +$
$Be^{-\beta t}$. Kinetic parameters for the model in Table 3 were then
calculated from these values (18).

TABLE 3
Kinetic parameters for the metabolic interconversion
of cyanide and thiocyanate in the rabbit^.

Experiment No.	Dose of KC^{15}N (μmole)	Coefficients and exponents				Rate constants (min^{-1})			Pool sizes (μmole)	
		H_1	H_2	g_1	g_2	k_{na}	k_{ab}	k_{ob}	Q_a	P_b
1	7.7	0.78	0.22	0.26	0.014	0.20	0.043	0.019	2.7	8.8
2	4.2	0.81	0.19	0.12	0.003	0.098	0.021	0.004	0.57	2.2

^ Kinetic parameters defined by Fig. 4 and equation (1).

DISCUSSION

Conventional colourimetric and gas-liquid chromatographic
methods have been used for some time in the Department of Forensic
Medicine of the University of Glasgow for the analysis of small
quantities of cyanide and thiocyanate in body fluids. Both
techniques involve separation of cyanide from blood by gas-phase
diffusion with subsequent trapping in alkali and both depend on the
absence of significant quantities of cyanide from plasma to permit
isolation of thiocyanate (the ratio SCN:CN in human plasma is
normally 80:1) (4). Colourimetric procedures have several dis-
advantages: the sensitivity is relatively low and internal standards
can not be used. In addition the pyridine dyestuffs formed are unstable over a period of more than a few minutes and their preparation
involves the use of carcinogenic reagents. In contrast, chromatographic procedures offer considerable advantages in terms of
sensitivity and specificity. Internal standards may be used but may
be added only at the final stage, immediately before analysis.

The GC-MS method described in this report was developed from
a GLC procedure reported by Valentour (10) in which cyanide and
thiocyanate were converted to cyanogen chloride, a volatile gas
readily detectable by electron-capture GLC. Modifications were
introduced to permit the use of ^{15}N-labelled cyanide as the internal
standard which is added early in the sample preparation sequence, thereby permitting corrections to be made for losses or incomplete
recovery. Analyses were carried out by SIM using the molecular ions
of CNCl and C^{15}NCl at m/e 61 and m/e 62. The stationary phase
selected for the GLC column (Hallcomid M-18) gave a very low
background at these ions and, because the solvent was eluted after
the cyanogen chloride, large sample volumes (10 μl) could be
injected without interference in the analysis.

The method has been found to be as sensitive as the GLC method

and to have a lower standard deviation than the GLC method for
replicate analyses of blood samples.Most samples encountered in the
course of our work have cyanide and thiocyanate levels in the range
0-100 μmoles/l (Table 2), within which the method was found to give
a linear response (Fig. 2). In practice, 24-30 samples per day may
be processed using this method by a single worker. A wide variety
of blood samples have been analysed using this method including
fresh blood samples from humans and experimental animals and old
blood samples, some putrefied, from post-mortem dissections. No
problems have been encountered as a result of interference by
other blood constituents.

The method is now used routinely in our laboratory in a study
of fire deaths in the Strathclyde Region of Scotland. Previous
studies of fire deaths have shown that blood cyanide levels may be
elevated within the toxic range (1, 19). Non-fatal fire casualties
suffering from the effects of smoke inhalation have also been found
to have significantly elevated cyanide and thiocyanate levels when
compared to controls but this was not shown in the case of firemen
(2). The absence of high levels in firemen was considered to be
due to the routine use of self-contained breathing apparatus and
to the length of time which had elapsed on average between
exposure to smoke and the drawing of blood samples.

The present series of firemen were exposed for a 15-minute
period to a fire involving plastic waste in a disused building.
Because the fire was small, breathing apparatus was not used, and
without exception the firemen suffered shortness of breath and
chest pains after the exposure. The cyanide and thiocyanate levels
in this group (Table 2) show significant elevation compared to
control figures (2) and indicate that a potential risk exists to
firemen from cyanide poisoning.

Cyanide in blood is normally contained principally within the
red blood cells, bound to haemoglobin. In many of our fire
fatalities considerable haemolysis has occurred which might have
caused interference by cyanide in the thiocyanate determination.
This cross reaction was examined by the addition of ^{15}N-labelled
cyanide to blood and, after an equilibration period, the blood was
haemolysed. The resultant plasma was stained deep red by the
liberated haemoglobin but following treatment with T.C.A. a
colourless supernatant was obtained. No detectable amounts of $C^{15}N$
were found on analysing the supernatant by GC-MS.

Cyanide levels in blood have been found to change rapidly in
cases of accidental poisoning (5). Little data is available however
and further work is needed to permit an accurate assessment of the
rates of metabolic conversion of cyanide in vivo. In view of the
extreme toxicity of cyanide, obvious problems arise in using human

subjects for such measurements. Tracer experiments are therefore necessary although the use of radioactive species in human subjects again poses problems. Species labelled with stable isotopes have been used to investigate hormone metabolism in pregnancy (20), a condition in which radiolabelled tracers are not acceptable. The application of stable isotope-labelled cyanide to the study of cyanide metabolism has been assessed in the rabbit, a convenient animal model.

A dose of cyanide was administered approximately equivalent to 10% of the LD-50 for intravenous injection (LD-50=820 µg/kg) (21). No immediate or delayed ill-effects were observed in any of the animals. A semilog plot of fraction of dose/µmole against time after injection was drawn (Fig. 3).

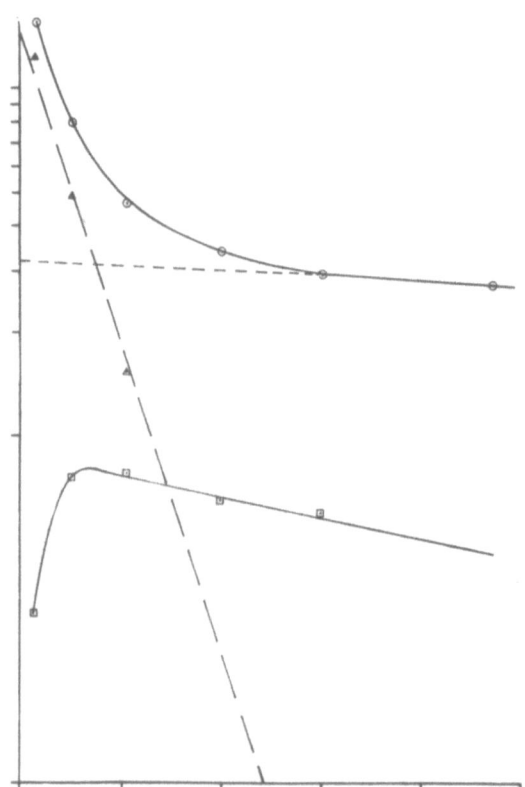

FIG. 3. A semilog plot of fraction of dose/µmole, of blood $C^{15}N$ and plasma $SC^{15}N$, against time following injection of 4.2 µmoles of $KC^{15}N$ into a rabbit (Experiment 2 Table 3).

Cyanide is principally converted to thiocyanate in the blood by the enzyme rhodanase (glutathione S-transferase). Thiocyanate lebelled with ^{15}N was observed to increase and reach a maximum 15 min after dosing. As the cyanide curve could be fitted to a double exponent equation and the curves for cyanide and thiocyanate declined in parallel the system could be approximated to a two-compartment model (Fig. 4).

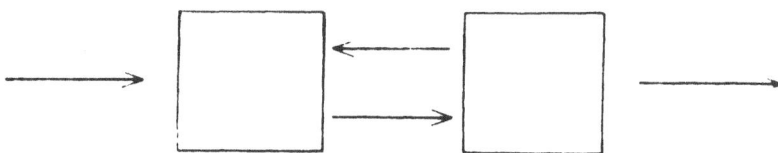

FIG. 4. An open two-compartment model describing the metabolism of CN$^-$ in the rabbit. k_{ab}, k_{ba} and k_{ob} are rate constants and F_{ao} is the rate of flow into the cyanide pool.

However as the two curves do not intersect it seems very likely that the thiocyanate pool is subject to a large inflow of thiocyanate from another compartment. Therefore, in the rabbit at least, cyanide is not the main precursor for thiocyanate.

The dose of cyanide administered in these experiments was sufficient to produce a maximum ratio of labelled:unlabelled cyanide of 6.5:1 which places the experiment more in the field of a pharmacological rather than a tracer study. However the sensitivity of the analytical technique is sufficient to permit the administration of lower doses and the method should be readily applicable to the study of cyanide metabolism _in vivo_ in man.

ACKNOWLEDGEMENTS

The authors acknowledge the assistance of Dr. I.S. Symington and the Dundee Fire Brigade in the provision of blood samples, the technical assistance of Mr. T. Randall and Mr. T. Duffy in the animal experiments and the assistance of Dr. D. Sumner and Dr. J. Lawrence with computer optimised curve-fitting. The work was supported by a contract (No. FRO/28/059) awarded to the Forensic Medicine Department, Glasgow University by the Building Research Establishment Fire Research Station, Borehamwood.

REFERENCES

1) Y.H. Caplan, "Relationship of Cyanide to Deaths Caused by Fire", Johns Hopkins University, Applied Physics Laboratory, FPP TR31, June, 1977.
2) I.S. Symington, R.A. Anderson, I. Thomson, J.S. Oliver, J.W. Kerr and W.A. Harland, Lancet, in press.
3) M. Ansell and F.A.S. Lewis, J. Forensic Med., 1970, 17, 148.
4) A.R. Pettigrew and G.S. Fell, Clin. Chem., 1972, 18, 996.
5) A.R. Pettigrew and G.S. Fell, Clin. Chem., 1973, 19, 466.
6) G.G. Guilbault and D.N. Kramer, Anal. Chem., 1966, 38, 834.
7) J.S. Hanker, R.M. Gamson and H. Klapper, Anal. Chem., 1957, 29, 879.
8) R.E. Isbell, Anal. Chem., 1963, 35, 255.
9) R. Altman, "The Microdetermination of Cyanide in Fire Fatalities", University of Maryland, Ph.D. Dissertation, 1976.
10) J.C. Valentour, V. Aggarwal and I. Sunshine, Anal. Chem., 1974, 46, 924.
11) W.D. Woolley,Br. Polymer J., 1972, 4, 27.
12) D.N. Napier, Med. Sci. Law, 1977, 17, 83.
13) K. Sumi and Y. Tsuchiya, J. Fire Flammability, 1973, 4, 15.
14) P.C. Bowes, Med. Sci. Law, 1976, 16, 104.
15) K.M. Dubowski and J.L. Luke, Ann. Clin. Lab. Sci., 1973, 3, 53.
16) D.J. Blackmore, Analyst, 1970, 95, 439.
17) A.W. Freireich and D. Landau, J. Forensic Sci., 1971, 16, 112.
18) R.A. Shipley and R.E. Clark, "Tracer Methods for In Vivo Kinetics. Theory and Applications", Academic Press, New York, 1972.
19) H.R. Wetherall, J. Forensic Sci., 1966, 2, 167.
20) T.A. Baillie, R.A. Anderson, M. Axelson, K. Sjövall and J. Sjövall, "Proceedings of International Symposium on Stable Isotopes: Applications in Pharmacology, Toxicology and Clinical Research", Royal Postgraduate Medical School, London, January,1977, The Macmillan Press Ltd.,
21) "Toxic and Hazardous, Industrial Chemicals Safety Manual", The International Technical Information Institute, Tokyo, 1975, 1976, p. 273.

PRESENT STATE AND FUTURE TRENDS OF NEGATIVE ION MASS SPECTROMETRY

IN TOXICOLOGY, FORENSIC AND ENVIRONMENTAL CHEMISTRY

H. Brandenberger

Department of Forensic Chemistry, University of Zürich

Zürich, Switzerland

INTRODUCTION

Negative ion mass spectrometry (or anion MS, as we call it) is gaining ground. It is entering the field of biochemistry and medical analysis. Two years ago and earlier, no papers on anion MS were presented in the international meetings organized by Dr. Frigerio and his group in Italy. Last year at Riva del Garda, there was one (1), and this year there will be 4 or 5.

Since we have been dealing intensively with anion MS for about 15 months, we have been asked to present an introduction to the field. It should outline where we stand today and what anion MS may mean to us in the future.

The presentation will contain: 1. A very short appraisal of the role of anion MS by electron impact (EI). 2. A brief review of the publications on anion MS by methods related to chemical ionization (CI). 3. An outline of our own work with anion CI-MS at relatively low pressure. 4. An outlook into the future which of course will have to be somewhat hypothetical.
Most of the time will be devoted to part 3. In choosing the anion spectra to be presented, we have taken care a) to consider examples from several different fields (toxicology, forensic chemistry, environmental analysis); b) to present typical examples which can illustrate the various advantages and draw-backs inherent in anion MS by CI.

ANION MASS SPECTROMETRY BY ELECTRON IMPACT

The existence of stable gas phase anions has been known since
the earliest days of MS (2). The interaction of electrons with
atoms or molecules can lead to electron gain (and negative ions) or
to electron emission by the excited species (and positive ions).
It could be expected that similar amounts of research work would
have been carried out with anion and with cation MS. But this is
not the case. Way over 99% of the analytical applications in
chemistry have been based on positive ion formation. MS by EI is a
positive ion method.

The main reasons for this discrepancy are: 1. The low relative
abundance of negative ions which is usually 3 or 4 or more orders
of magnitude lower than for positive ions. 2. The fact that most
of the anions produced by EI belong to the low mass range and can
only have minor analytical significance. 3. The much greater
sensitivity of anion MS toward experimental variations (pressure,
temperature and geometry of the ion source) as compared to cation
MS.

Table 1 lists some examples of anions produced by conventional
EI (at 70 eV and near 10^{-6} torr).

TABLE 1
Negative ions by EI-MS at 70 eV and near 10^{-6} torr.

Aliphatics: H^-, $/\overline{C_2}/^{\overline{\cdot}}$, $/\overline{C_2H}/^{\overline{\cdot}}$, $/\overline{C_x}/^{\overline{\cdot}}$, $/\overline{C_xH}/^-$
Aromatics: $/\overline{C_2H}/^-$, $/\overline{C_4H}/^-$, $/\overline{C_6H}/^-$
Oxygen compounds: $O^{\overline{\cdot}}$, $/\overline{OH}/^-$, $/\overline{C_2HO}/^-$
    - alcohols little $/\overline{M-1}/^-$ and $/\overline{M-3}/^-$
    - acids some $/\overline{M-1}/^-$
Nitrogen compounds: $/\overline{CN}/^-$, $/\overline{C_xN}/^-$
Halogen compounds: F^-, $\overline{Cl}^-$, $\overline{Br}^-$, J^-
negative ion yield $\leqslant 10^{-4}$ of positive ion yield

The underlined species often figure as base ions in the anion spectra:
a. the acetylide anion with mass 25 for aliphatic and aromatic
hydrocarbons; b. O^- and OH^- for oxygen compounds; c. the cyanide
anion with mass 26 for nitrogen compounds; d. chloride, bromide
and iodide anions for halogenated molecules.

They all belong to the low mass range and are seldom of
analytical value, except perhaps for detection of halogens. There
are - however - some exceptions. Alcohols usually give 2 anions in
the molecular range, by loss of 1 and of 3 hydrogens. Acids yield

an M-1⁻ peak. Such anions could be useful for structural studies provided that their abundances are higher.

In contrast to pure analytical chemistry, gas phase anions have played an important role in physics and in the field of instrumentation. Two examples are: 1. The determination of the accurate isotope abundance in chlorine and bromine. 2. The development of a sniffing device for explosives such as trinitrotoluene (TNT) and other polynitro aromatics.

Fig. 1 shows two anion EI mass spectra of TNT, at 2 and 6 eV. It is taken from a paper by Yinon et al. (3). / NO₂ /⁻ with mass 46 is the base anion. At 6 eV, its intensity corresponds to almost 60% of the total ion current. The explosive sniffer is a mass spectrometric detector focussed on this anion. It is strongly selective for nitro-compounds and does not pick up other ions of similar mass, as a positive ion detector with equal sensitivity to all types of volatile organics would do.

In 1965, Aplin et al. (4) reviewed the anion mass spectra of some simple organic compounds. They conclude: "It appears that negative ion mass spectra measured with a commercial instrument are not suitable either for molecular weight or for structure determination. With regard to the fragmentation observed, it is clear that hydrogen stripping of the organic compounds to acetylide and polyacetylide anions occurs. If heteroatoms are present, abundant species corresponding to stable anions are formed. The overall similarity of the spectra of alkylbenzenes and aliphatic hydrocarbons, alcohols, and ketones, acids and esters, etc., shows that the type of cleavage observed does not reflect the type of functional group, but only the nature of the heteroatom present in the molecule".

REVIEW OF ANION MASS SPECTROMETRY BY METHODS RELATED TO CHEMICAL IONIZATION

Also in 1965, Munson and Field (5), in their description of Chemical Ionization MS, mentioned in a brief footnote that their new method would probably permit analytical work with anions, since in principle, their ionization approach was similar to the procedure of v.Ardenne.

v.Ardenne and his group in Dresden have been working with negative ion MS since 1958 (6). They have built an anion Duoplasmatron mass spectrograph. It uses an Argon gas discharge source at 10⁻² torr for the production of low energy electrons which interest with sample molecules by electron attachment. The anions formed belong predominantly to the molecular weight region. Only minor fragmentation reactions occur.

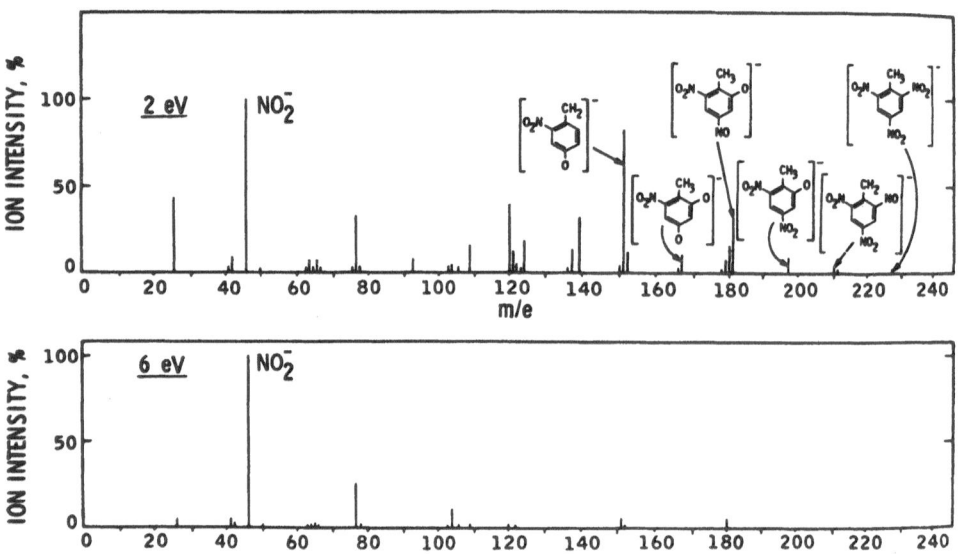

FIG. 1. Anion MS of trinitrotoluene by EI at 2 eV and 6 eV (from
        lit. ref. 3).

v.Ardenne and his collaborators have used their anion spectrograph mainly for the determination of molecular masses and called
their method at first "Molekül-Massen-spektrographie". Later, they
changed the name to "Elektronenanlagerungs-Massenspektrographie"
(7). Fig. 2 gives an example from the first application to natural
products, taken from a paper published in 1963 (8).

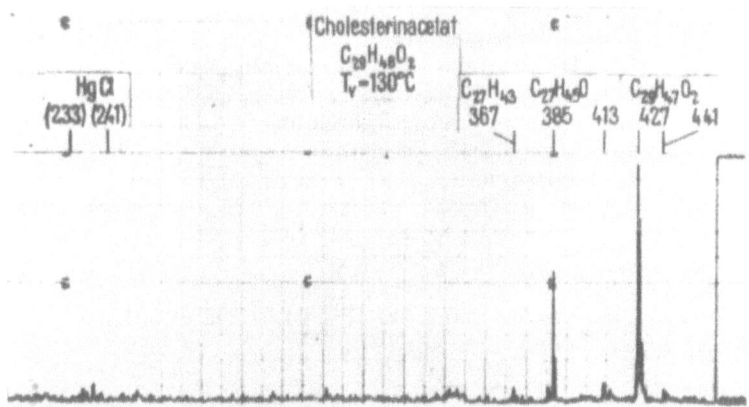

FIG. 2. Photometric registration of the electron
attachment MS of cholesterol acetate
according to the technique of v.Ardenne
(from lit.ref.8).

It shows the anion spectrum of cholesterol acetate with MW 428.
The different peaks represent: a. mass 427 the base anion $\underline{/}\,M-1\,\underline{/}^-$;
b. mass 413 the fragment $\underline{/}\,M-15\,\underline{/}^-$; c. mass 367 an anion resulting
probably from loss of acetic acid; d. mass 385 presumably the base
anion from non acetylated cholesterol.

Table 2 lists the approaches of other research groups which
have successfully applied techniques related to chemical ionization
to negative ion formation. They all work in the U.S.: Dougherty in
Tallahassee, the Hornings in Houston, Hunt at the University of
Virginia. In every case, reagent gases such as argon, nitrogen,
oxygen, methylene chloride and others are used to transmit the
negative charge to the substrate molecules. The chamber pressure is
usually kept high, at 0.1 to 1 torr (analogous to positive CI) by
Dougherty and by Hunt, at atmospheric pressure by Horning.

Table 3 shows an example from Dougherty's work with methylene
chloride as reagent (9), which yields – on electron impact –

TABLE 2

Negative ion formation by methods related to chemical
ionization prior to 1977.

Research group	Principle of anion formation	Literature
v.Ardenne	Electron attachment–MS in an Argon Duoplasmatron at 10^{-2} torr.	(7)
Dougherty	Negative CI–MS with CH_4 and CH_2Cl_2 at up to 1 torr. Negative ion cluster formation.	(23–25)
Horning	API in external reaction chamber with radio–Ni-foil.	(26–28)
Horning	API with a Corona ion source	(29–30)
Hunt	Negative CI–MS with Townsend discharge source at 1 torr.	(11)

TABLE 3

Formation of anion clusters by CI with CH_2Cl_2 as
reagent gas (from lit.ref.9).

Compound	Relative abundance, %		
	$(M + Cl)^-$	$(M - H)^-$	$(2M + Cl)^{-b}$
Aliphatic and aromatic carboxylic acids			
Benzoic acid	100	6	
Cyclohepten-5-carboxylic acid	100	...	
Cyclopropyl carboxylic acid	100	...	
p-Fluorobenzoic acid	100	12	
p-Methoxybenzoic acid	100	...	
p-Nitrobenzoic acid	100	14	
Amides			
Octadecanamide	100	...	
Urea	100		33
Amino acids			
2-Amino-2-methylbutyric acid	100	7.5	
Glycyl-L-tyrosine	100		
Aromatic amines and phenols			
Aniline	100	7.5	5.6
Phenol	100		53
Hydroquinone	100		4

[a] Ions from the methylene chloride spectrum are not included in
the listing; source pressure ~1 Torr, temperature ~200 °C, intensities are relative to the most intense ion in the entire spectrum,
namely, [M + Cl]−. [b] The relative intensities of these ions are
highly temperature and substrate-concentration dependent. They
may be eliminated in all cases by increasing the source temperature.

chloride anions. They combine with sample molecules to form negative
cluster ions. The base peaks of all the substances listed in this
table are / M+Cl /⁻ ions, if we disregard the anions from the
reagent. However, the relative intensities of the cluster ions are
highly dependent on temperature and substrate-concentration.

Fig. 3 shows an example from Horning's work (10).

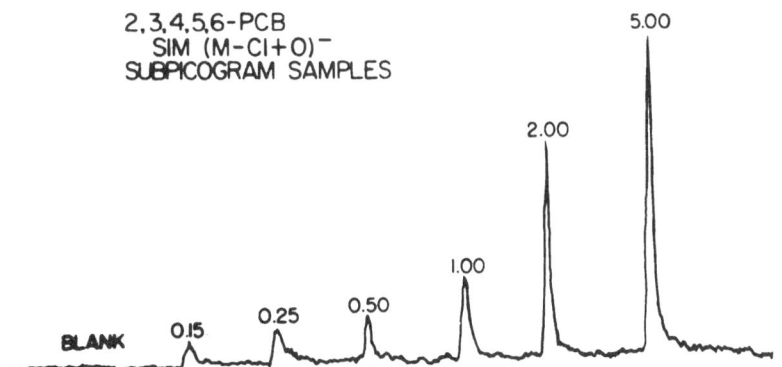

FIG. 3. Response from pg and sub-pg samples of
2,3,4,5,6-pentachlorobiphenyl by anion
MS at atmospheric pressure with air as
reagent. The tetrachlorophenoxide ion
(an oxygen substitution product) with
m/e 307 was monitored (from lit.ref.10).

He has developed two atmospheric pressure ionization methods. In
one, an external chamber with a radio-Ni-foil, in the other, a
Corona discharge source is used for ionization of the reagent.
This permits the use of air or oxygen as reagent gas, which would
not be possible in a conventional ion source with filament.
Superoxide anions are formed. They can substitute hydrogens or
halogens in sample molecules. In the present example, pentachloro-
-diphenyl is converted to the tetrachloro-phenoxide ion. Mass
specific detection of this anion permits the recording of subpicogram
quantities of the polychloro derivative.

Fig. 4 is a scheme of a pulsed positive and negative ion mass
spectrometer developed by Hunt et al. (11). With this quadrupol
instrument with two multipliers, cation and anion spectra can be
recorded simultaneously. A conventional CI source can be used, but
Hunt replaced it by a Townsend discharge source which permits a

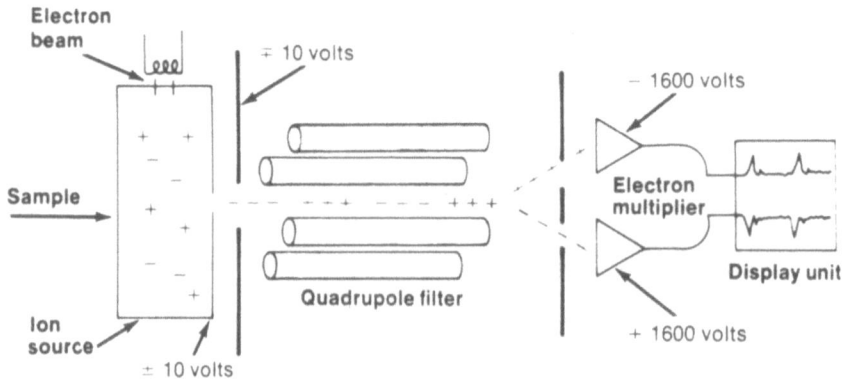

FIG. 4. Scheme of a pulsed positive and negative ion
        MS according to Hunt (11), from C and EN,
        Nov. 8, 1976.

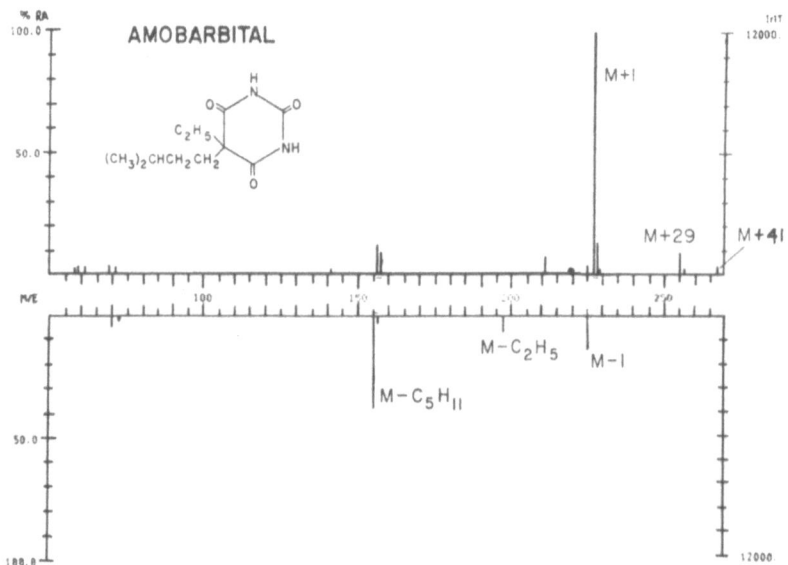

FIG. 5. Dual CI-MS (cation and anion) of Amobarbital
        with N_2 as reagent gas (from lit.ref.11).

wider choice of reagent gases and also accepts oxygen.

Fig. 5 reproduces just one of the applications desscibed by Hunt, the dual CI mass spectra of Amobarbital. We recognize: a. in the positive mode the molecular ion clusters $\underline{/}$ M+1 $\underline{/}^{+}$, $\underline{/}$M+29 $\underline{/}^{+}$ and $\underline{/}$ M+41 $\underline{/}^{+}$; b. in the negative mode $\underline{/}$ M-1 $\underline{/}^{-}$, $\underline{/}$ M-29 $\underline{/}^{-}$ and $\underline{/}$ M-71 $\underline{/}^{-}$, that is the anions produced by loss of H and of the two side chains.

All these methods report the formation of significant anions in good yields. But in most cases, the investments were high, both with respect to instrumentation and time.

OUR APPROACH TO NEGATIVE ION MASS SPECTROMETRY

When we decided to start with negative ion MS by CI, we chose a very simple approach. In collaboration with Dr. R. Ryhage from the Karolinska Institute for Mass Spectrometry in Stockholm, two units of commercially available magnetic sector instruments with combined EI/CI source (LKB-2091) were equipped with new power supplies which permit reversion of magnet current, accelerating voltage and repeller voltage by shifting a single switch. The modified instruments are capable of recording anion or cation spectra. The switch-over time is less than a minute (1, 12, 13).

A first report of the work with the modified instrument was presented at Riva del Garda 1 year ago (1). We have now dealt with anion MS by CI for 15 months. We have tried over 10 different reagent gases, optimized some of the conditions and made a few systematical studies (12-15). Even though the field is still new and a great deal of development remains to be done, anion CI-MS has already become – in our forensic laboratory in Zurich – one of the most powerful analytical tools in the daily toxicological and forensic service work (16-18).

In the following section we will illustrate 12 positive aspects of negative ion MS. We expected quite a bit before we started our work. The results have by far exceeded our expectations.

TWELVE POSITIVE ASPECTS OF NEGATIVE ION MASS SPECTROMETRY

Anion MS by CI is a soft ionization method. Fig. 6 contains the mass spectra, obtained by cation EI and anion CI of Etonitazene, one of the most potent analgesics ever synthesized. The cation spectrum (19, 20) yields mainly information about the side chains (m/e 86 and 58 for the N-chain, m/e 135 and 107 for the O-chain, m/e 30 for the nitrogroup), but the molecular ion could easily be overlooked. In the anion spectrum, the molecular ion with mass 396

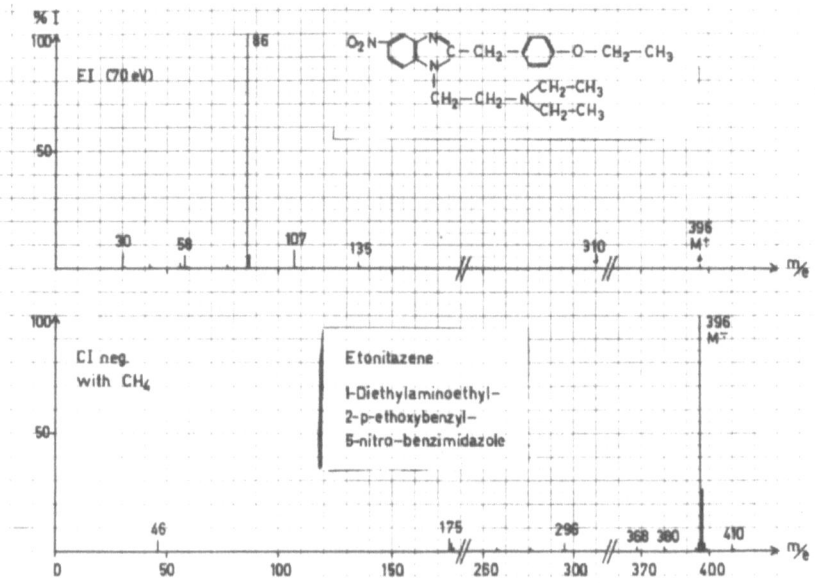

FIG. 6. Cation EI and anion CI-spectra of the
analgesic Etonitazene.

is the base peak. The anion 296 results from the loss of the
entire N-chain (which is typical for anion spectra by CI). The
nitro group is signalized by the anions m/e 380 (loss of oxygen)
and m/e 46 (17, 18).

Fig. 7 illustrates that the anion in the molecular range is
not always M⁻, but more often $/$ M-1 $/$⁻ and sometimes also $/$ M-2 $/$⁻.
It is interesting to note that in contrast to morphine and codeine,
dihydrocodeinone, which does not possess a hydroxy group, gives a
stable molecular anion (17, 18).

Anion spectra are often simple, interpretation fairly easy.
Fig. 8 shows 3 mass spectra of the hypnotic Novonal resp. Epinoval
(2,2-diethyl-2-allyl-acetamide): the positive EI-spectrum and the
negative CI-spectra with 2 different reagent gases. The contrast
between the 2 modes is striking. The anion spectra only contain 2
peaks, the base anion $/$ M-1 $/$⁻ and the fragment m/e 42, the cyanate
anion typical for the presence of amide groups.

Anion CI-spectra are complementary to cation EI-spectra.
Already from Figs. 6 and 7 we could see that cation EI- and anion
CI-spectra are complementary. For identification work, they should
be used side by side. Fig. 9 gives a further example (18). It contains 3 mass spectra for Butethal (5-butyl-5-ethyl-barbituric acid):
the cation spectra in EI and in CI, and the anion CI-spectrum. In
contrast to the EI-presentation, both CI-spectra permit the

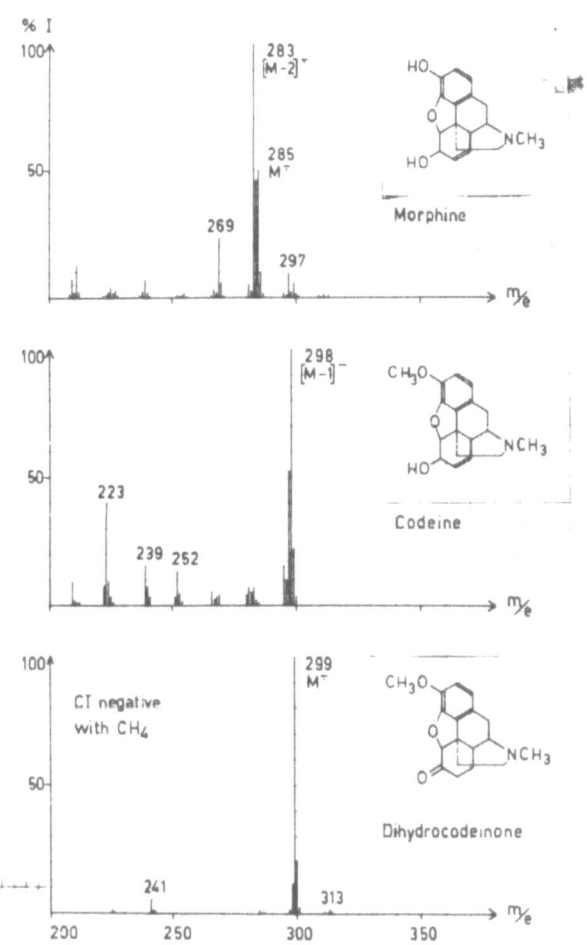

FIG. 7. Anion CI spectra of Morphine, Codeine
and Dihydrocodeinone.

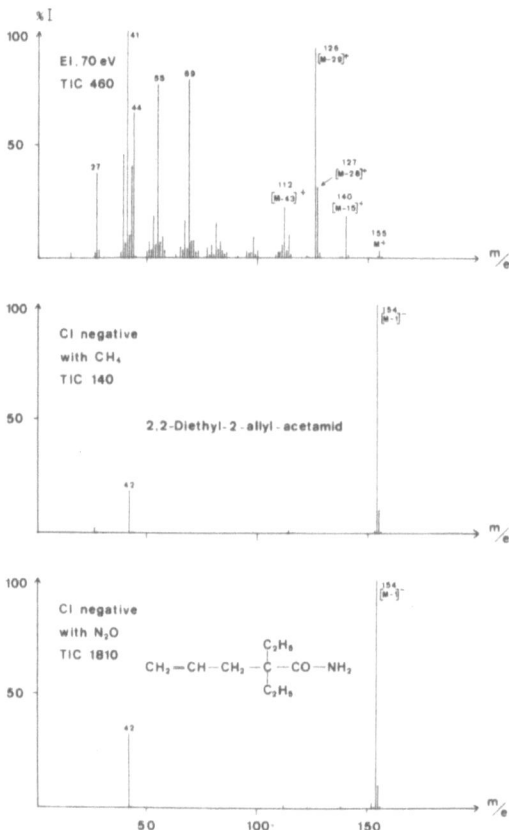

FIG. 8. Cation EI-MS and anion CI-MS (with CH₄
and with N₂O as reagents) of the
hypnotic Novonal (Epinoval).

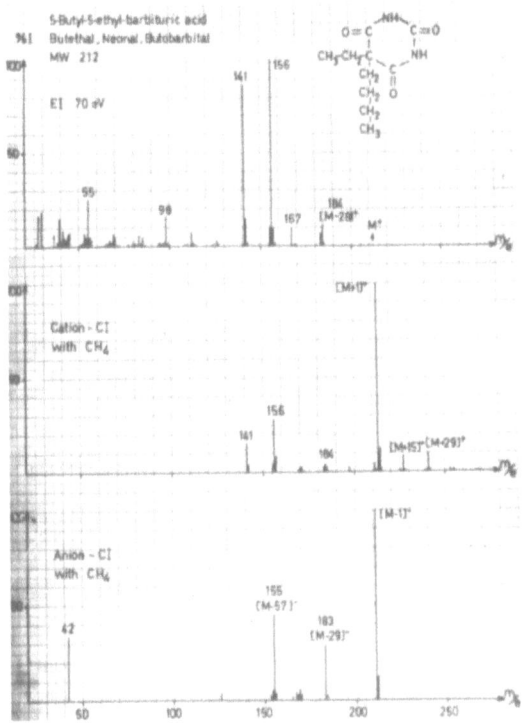

FIG. 9. Cation EI, cation CI and anion CI spectra
of a saturated barbiturate.

recognition of the molecular mass of the saturated barbiturate. In
addition, the cation CI-trace does not yield much information; it
rather duplicates the EI-data. The anion spectrum, however, does
not duplicate, it complements the information obtained by EI.
Our last year's paper contained the three corresponding spectra of
an unsaturated barbiturate (5-allyl-5-isobutyl-barbituric acid).
Allyl abstraction is favored over proton abstraction; / M-41_/⁻ is
the base anion (1). This is the case for all allyl-barbituric acids.
The complementary nature of cation EI and anion CI is illustrated
by the fact that the 2 respective base ions yield - when added -
the molecular mass of the compound investigated. A similar ion pair
formation often occurs with halogenated compounds: the halogen
becomes the base anion, the rest of the molecule the base cation
(in EI).

    Anion MS by CI is a flexible method. As in cation CI, it is
also possible in anion CI to guide the fragmentation process by
proper choice of reagent gases. Up to now we have worked with more

than ten different reagents. Fig. 10 gives the anion mass spectra
for eight of them (18).

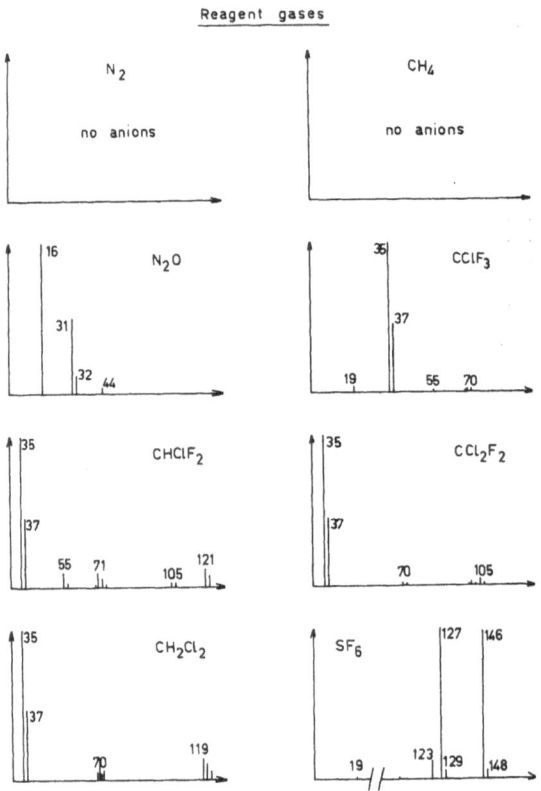

FIG. 10. Anion spectra of 8 reagent gases
used for anion MS by CI.

Noble gases, N_2 and essentially also CH_4 do not yield negative
ions on electron impact. They act primarily as moderators for the
production of low energy electrons which can convert substrate
molecules to anions by electron attachment and dissociative electron
attachment.

For the analysis of compounds with low electron affinity,
reagents which readily form negatively charged ions are often better
choices. The negative charge can be transmitted from the reagent
anion to the substrate molecule by charge exchange or by anion
cluster formation. Methylene chloride has been used for this
purpose (9). We prefér Freons. Since they are gases, dosage is easier.
Cl^- and F^- are good cluster forming agents.

Sulfur-hexafluoride was initially chosen as a charge exchange
agent, since it produces, on electron impact, SF_6^- and SF_5^- in high

yields. However, the small contribution of F$^-$ also leads to cluster formation.

Fig. 11 illustrates the influence of the reagent gas on the anion spectrum. With CH$_4$, tertiary alcohols such as tertiary butanol give mainly $/$ M-1 $/$$^-$ and only little loss of OH. With N$_2$O, however, $/$ M-17 $/$$^-$ becomes the base anion.

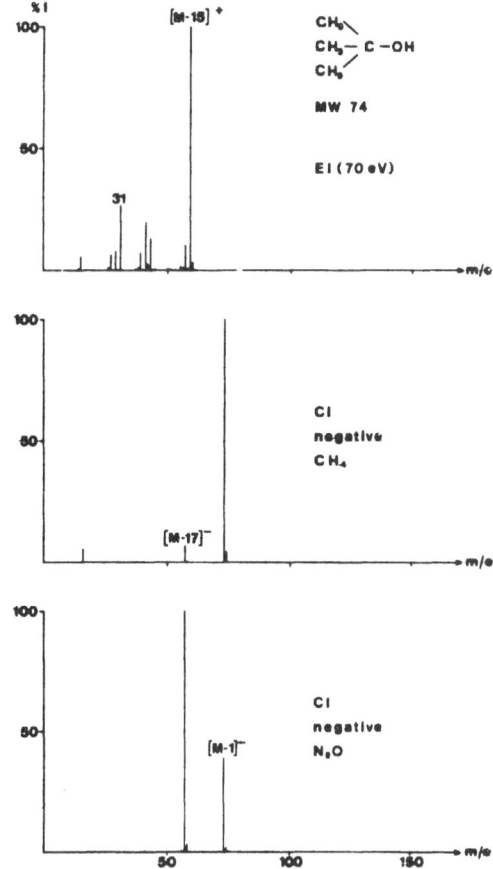

FIG. 11. Cation EI spectrum and anion CI spectra with CH$_4$ and N$_2$O as reagent gases for tertiary butanol.

Oxidations in a conventional CI-source are possible by N$_2$O. Nitrous oxide (N$_2$O) yields O$^-$ and $/$ NO $/$$^-$ on electron impact. The strong base O$^-$ is an excellent charge exchange reagent. But it also leads to oxidations in the ion source, since it can replace H or other atoms (1, 12, 13). Such oxidations may be a desirable future. Other authors have effected them by $/$ O$_2$ $/$$^-$, produced from air or oxygen. This necessitates a special ion source. With N_2O, oxidations in a conventional CI-source are possible, in presence of the heated filament.

Fig. 12, the anion CI-spectrum of Methaqualone (14), is an illustration for the action of the reagent N_2O.

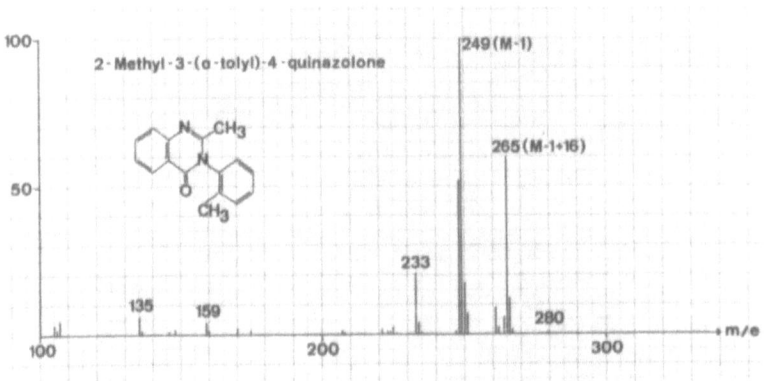

FIG. 12. Anion CI-spectrum of Methaqualone
with the reagent N_2O.

The 2 highest peaks represent the anions with masses 249 (hydrogen abstraction) and 265 (substitution of H by O^-). An anion with 15 mass units above the molecular mass is usually an oxidation product.

By using deuterium-labelled compounds, it was possible to prove this assumption (21). Fig. 13 shows the anion spectra of benzene and hexa-deuterium-benzene. Such aromatic compounds easily loose 2 H (resp. 2 D). This leads to the anions with masses 76 resp. 80. The substitution of 1 H resp. 1 D by O^- produces the anions with masses 93 (that is 78 + 16 - 1) resp. 98 (84 + 16 - 2). The substitution of 2 H resp. D by O^- is also visible. It is responsible for the anions with masses 108 (76 + 32 - 2) resp. 112 (84 + 32 - 4).

Anion MS is a sensitive technique. In Fig. 14, the total ion currents obtained from some barbiturates in identical GC-runs followed by different mass spectrometric recordings are compared (18). The relative values (rel.TIC) show that for saturated barbituric acids, a sensitivity equal at least to that in the positive modes can be obtained with negative MS. For allyl-barbituric acids, negative CI is over a factor of 10 more sensitive than positive EI and CI.

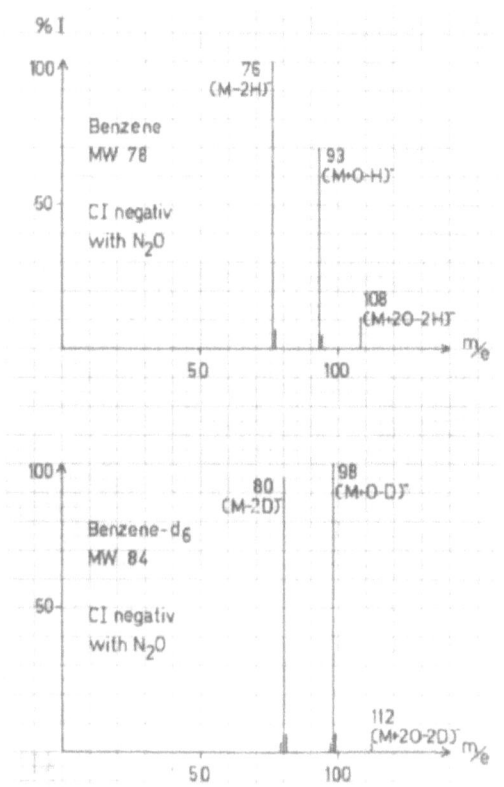

FIG. 13. Anion CI-spectra of benzene and hexadeuterium-benzene with the reagent N_2O.

FIG. 14. Intensity of different ionization techniques for the mass spectrometric characterization of some barbiturates (Rel.TIC = rel. total ion current).

R_1	R_2	mode	reagent	rel TIC	base peak	% of TIC
$CH_2 \cdot CH_3$	$CH_2 \cdot CH_3$	EI pos	-	500	$[M-28]^+$	21
		CI pos	CH_4	80	$[M+1]^+$	49
		CI neg	CH_4	45	$[M-1]^-$	33
		CI neg	$CH_4 \cdot SF_6$	1100	$[M-1]^-$	50
		CI neg	N_2O	460	$[M-1]^-$	61
$CH_2 \cdot CH_3$	— ⬡	EI pos	-	190	$[M-28]^+$	21
		CI pos	CH_4	70	$[M+1]^+$	52
		CI neg	CH_4	360	$[M-77]^-$	55
		CI neg	$CH_4 \cdot SF_6$	120	$[M-1]^-$	60
		CI neg	N_2O	360	$[M-77]^-$	30
CH_3	— ⬡ with $N \cdot CH_3$	EI pos	-	440	$[M-15]^+$	16
		CI pos	CH_4	530	$[M-79]^+$	32
		CI neg	CH_4	180	m/e 42	26
		CI neg	$CH_4 \cdot SF_6$	520	$[M-1]^-$	67
		CI neg	N_2O	390	$[M-1]^-$	51
$CH_2 \cdot CH \cdot CH_2$	$CH_2 \cdot CH \cdot CH_2$	EI pos	-	430	m/e 41	10
		CI pos	CH_4	310	$[M+1]^+$	19
		CI neg	CH_4	6700	$[M-41]^-$	89
		CI neg	$CH_4 \cdot SF_6$	2000	$[M-41]^-$	84
		CI neg	N_2O	1250	$[M-41]^-$	88

Fig. 14 also illustrates some other interesting aspects: a. the base anions are /_M-1_/⁻ for saturated barbiturates, /_M-1_/⁻ or /_M-77_/⁻ for phenobarbital, /_M-41_/⁻ for allyl-barbituric acids; b. the anion m/e 42 is usually quite abundant (amide structure); exceptionally, it may become the base anion; c. highest anion sensitivity is obtained with CH₄ as reagent (electron attachment) for allyl compounds, with charge exchange reagents for saturated barbiturates.

GC-combination with anion MS has advantages over GC with cation MS. This is a point we have emphasized earlier (1, 12, 13). Anion backgrounds are usually much simpler and easier to subtract than cation backgrounds; less ions from column bleed are present. The effect of GC temperature programming on the base line is often small or negligible.

Anion CI-MS is the electron capture detector of the future. For substances with high electron affinity (which are often analyzed by GC with ECD), the sensitivity of anion CI is far superior to the sensitivity of cation CI and EI (16-18).

Fig. 15 shows the cation EI- and the anion CI-spectra of the benzodiazepine Flunitrazepam, the active component of the hypnotic Rohypnol (18).

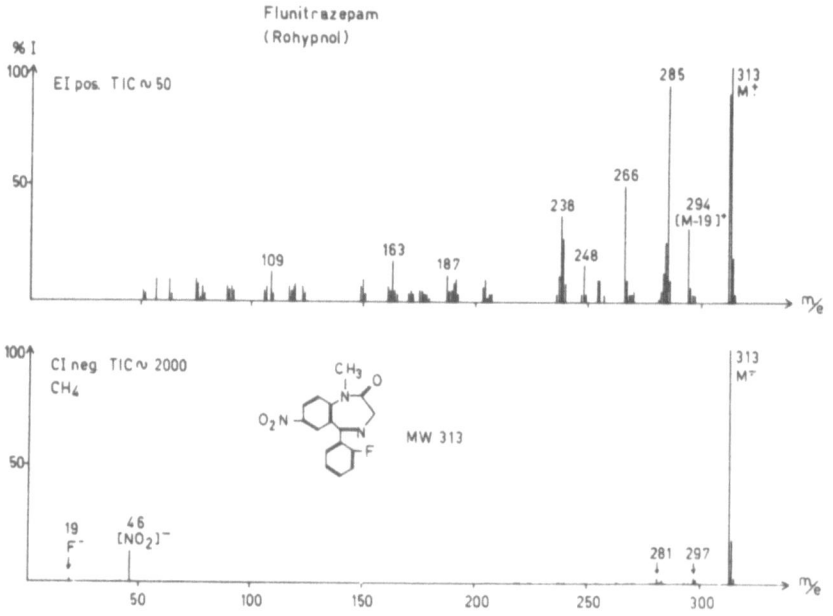

FIG. 15. Cation EI and anion CI-spectra of
Flunitrazepam, with relative TIC.

The anion spectrum is dominated by the molecular ion 313. In addition, the presence of the nitro group (m/e 46, M-16 and M-32) and the fluoride atom (m/e 19) is evident. The total ion current is 40 times higher in anion CI than in cation EI; by far the largest contribution comes from the molecular anion.

Fig. 16 contains the 2 corresponding spectra of the tranquilizer Diazepam, another compound with benzodiazepine structure (18).

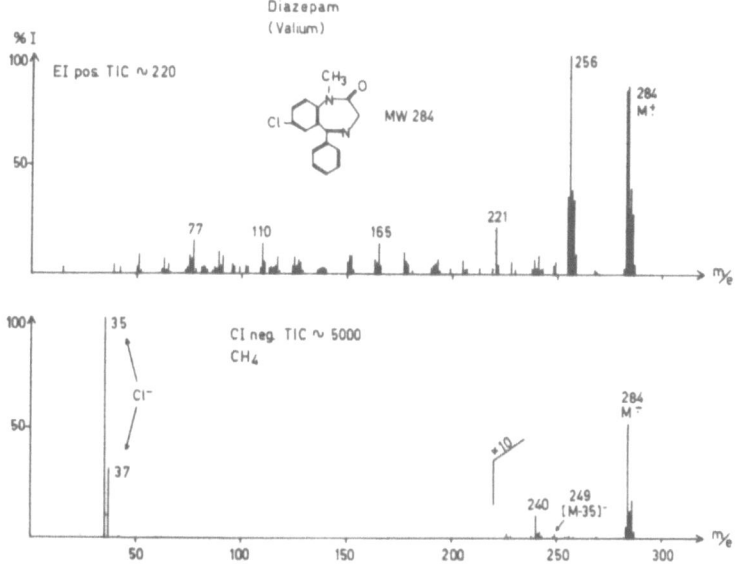

FIG. 16. Cation EI and anion CI-spectra of Diazepam, with relative TIC.

Again, the total ion current is (almost 50 times) higher in the anion mode. But in this case, the largest contribution comes from the chloride ions; the relative intensity of the molecular anion is only 50%.

These 2 examples (Figs. 15 and 16) are typical for anion CI-MS of compounds with high electron affinity. The total ion current is always very high, especially with CH_4 as reagent gas (electron attachment MS). Nitro- and fluoro-substituted compounds give molecular base anions. Chloro- and bromo-compounds are split and the halogens usually form the most abundant peaks. The following rules should therefore be respected when working with highly electron attracting compounds: a. for identification work, anion MS by CI should be used in conjunction with cation EI-MS; b. for trace detection of nitro- or fluoro-compounds, mass specific detection

of the molecular anion is usually by far the best and most sensitive
approach; c. for trace detection of chloro- and bromo-compounds, the
mass spectrometer should be focussed on the 2 pairs of chloride
or bromide ions for best sensitivity. Of course, only group specificity and not individual compound specificity results with such an
approach. The latter must be obtained by the GC-data or by additional
runs with other tracer ions.

An example for such an application is given in Fig. 17 (12, 13).

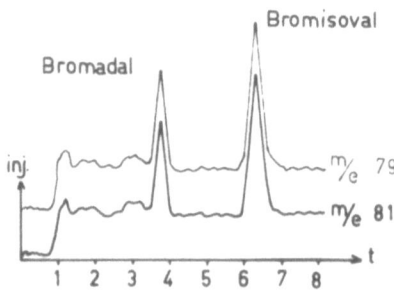

20 pg Bromadal + 25 pg Bromisoval

LKB - 2091 , CI neg. CH$_4$
3 % XF - 1150 , 150 °C
m/e 79 and m/e 81
multiplier 500
preamplifier 10
MID gain 500

FIG. 17. Trace detection of 2 bromo-sedatives
         by GC with simultaneous recording of
         the 2 bromide anions with masses 70
         and 81.

20 resp. 25 pg of the 2 bromo-sedatives Bromadal and Bromisoval
are separated on a packed column with XF-1150. The appearance of
the bromide isotopes (anions with masses 79 and 81) are recorded.
The base line is stable. A larger amplification factor could be
used and would permit sub-pg detection, as long as background
problems and column adsorption can be handled.

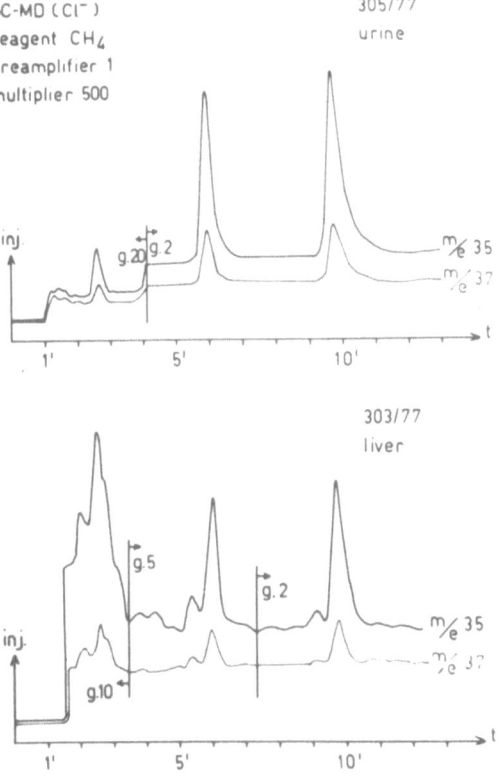

FIG. 18. Simultaneous detection of tri-, tetra- and
pentachlorophenols by GC with recording of
the 2 chloride anions m/e 35 and 37.

The chromatograms with specific ion recording reproduced in
Fig. 18 have been obtained in an environmental investigation we
were charged with (16). In a new wooden house, the inhabitants
became ill, the cat died. Since pentachlorophenol had been used for
wood treatment by the builders, we searched the urine of 2 children
and a small residue of cat liver for the presence of this pesticide.
GC with mass specific detection of the molecular ions 264, 266 and
268, applied to the corresponding extracts, established the presence
of pentachlorophenol. By means of GC with mass specific recording
of the 2 chloride anions (Fig. 18) it became evident that trichloro-
phenols (eluting after 2.5') and tetrachlorophenols (retention time
6') were also present beside pentachlorophenol (10').

The GC in Fig. 19 is taken from a forensic investigation (17).
A prisoner was admitted to the hospital in an unconscious state.
He could be saved. Some days later he claimed that colleagues had

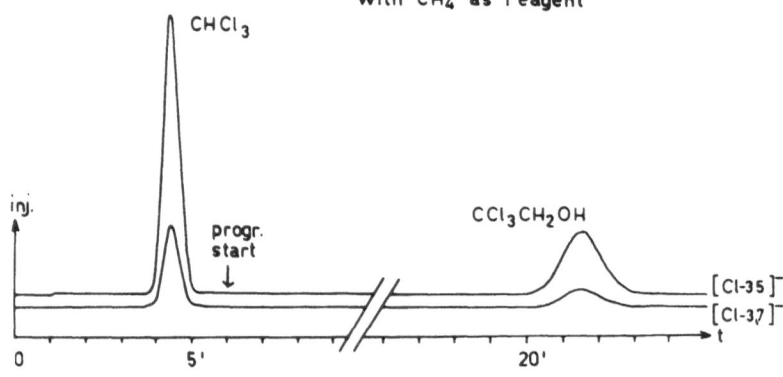

Case 73/78
blood , 1 μl
Porapak Q , 160 °C + 10 °C/min.
Chloride anion recording
with CH₄ as reagent

FIG. 19. Simultaneous detection of chloroform and
trichloroethanol, 2 metabolites of chloral
hydrate, by GC with chloride anion recording.

poisoned him. A toxicological investigation was requested.
Unfortunately, by that time, stomach wash and urine had already
been discarded at the hospital. From the clinical laboratory, we
could obtain some left-over serum vials, each containing only a
small fraction of a ml of serum from the second day after hospital-
isation. We used it for a search for sedatives and hypnotics. In
an attempt to find chloral, 1 μl of serum was injected onto a GC-
-column with Poropak Q. The appearance of the 2 isotopic chloride
ions was recorded by anion CI-MS using CH₄ as reagent. Chloral
would elute after about 8'. It could not be detected (Fig. 19).
However, two other chlorine-containing compounds were signalized,
one with the retention time of chloroform (4.5'), the other with
the retention of trichloroethanol (21.5'). Since chloroform is the
decarboxylation product of trichloro-acetic acid, the GC-run with
a single μl of serum had been able to detect - simultaneously -
both main metabolites of chloral hydrate, trichloroethanol and
trichloro-acetic acid.

Fig. 20 compares ECD with chloride anion recording for the
analysis of the 2 pesticides Lindane and DDT. Anion MS has a larger
linear range and considerably higher sensitivity than ECD. It is
of course also more specific. We are convinced that it is going to
replace conventional ECD in the future.

Anion MS may function as specific nitrogen detector. We detect
and measure cyanide in body fluids by conversion to hydrocyanic

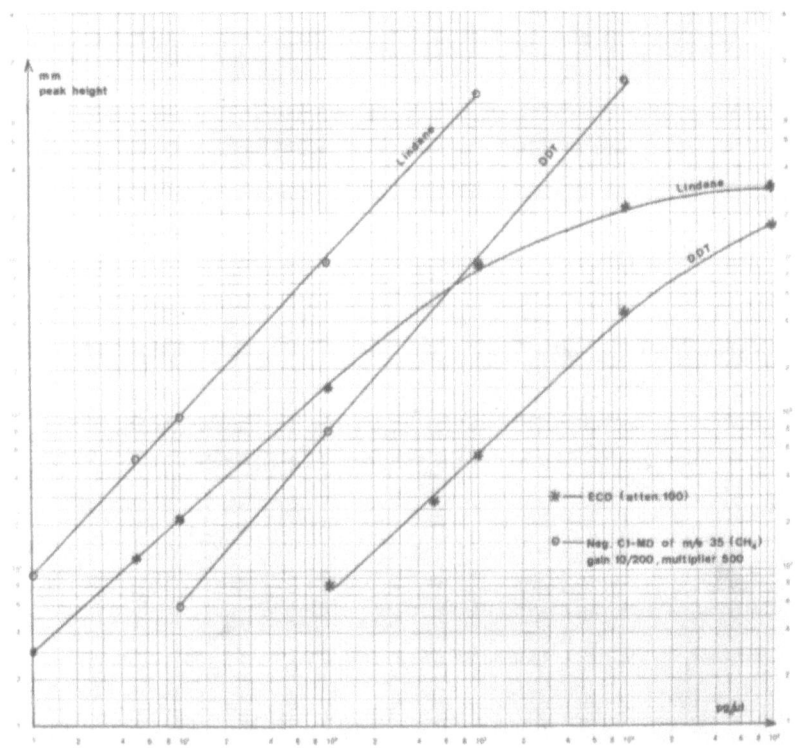

FIG. 20. Comparison of GC with ECD and GC with chloride
anion recording for the trace analyses of DDT
and Lindane.

acid and GC-analysis on Poropak Q or a similar column (21). The
procedure is illustrated in Fig. 21. At first, GC was carried out
with a normal FID. In order to improve the sensitivity, a thermionic
nitrogen detector was used later. Today, we employ mass specific
recording of the anion CN⁻ (22). This detection system is much more
reliable and as sensitive as a thermionic detector (Fig. 22).

Traces of other nitrogen-containing compounds may be detected
by a similar approach. But this is a field we have just entered.

Anion MS is often a good trace detector for compounds with a
common structural sub-unit. Trace detection of structurally related
compounds can usually be carried out more easily by GC with anion
than with cation MS. We have already given some examples. They
concerned the simultaneous detection of chloro- or bromo-compounds
(Figs. 16-18). Further examples have been described in other papers

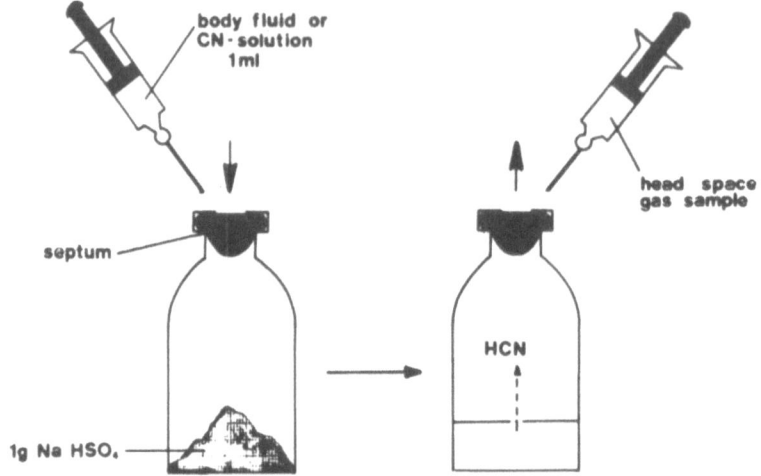

FIG. 21. Sample preparation for GC-analysis of cyanide
in body fluids and other liquids.

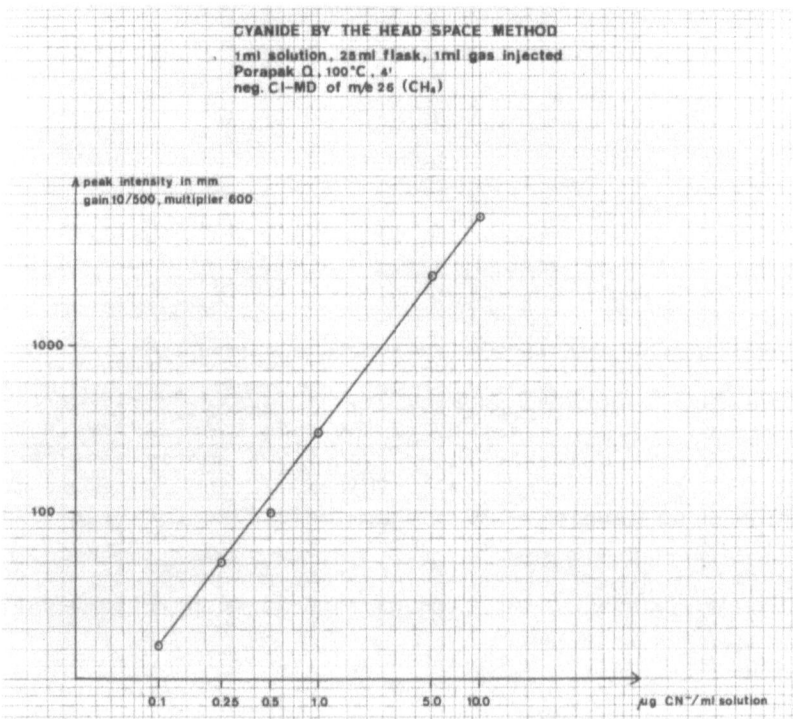

FIG. 22. Linearity of cyanide analysis by GC with mass
specific recording of the cyanide anion m/e 26.

(12-15). They dealt with the simultaneous trace detection of compounds belonging to a specific group of tricyclic antidepressants. In anion MS with N_2O as reagent gas, complete side chain abstraction often occurs and the tricyclic ring becomes the base anion. It is therefore possible to reveal, in a single GC-run, the presence of most phenothiazines by focussing the mass spectrometer on the ring anion with mass 198, the presence of imino-dibenzyls by recording the appearance of the corresponding ring anion m/e 194 and the presence of imino-stilbenes by incorporating the ion 192 into the detection scheme (Fig. 23).

Phenothiazines

    m/e   198 (200)

Azaphenothiazines

    m/e   199 (201)

Iminodibenzyls

    m/e   194

Iminostilbenes

    m/e   192

Dibenzocycloheptatrienes

    m/e   191

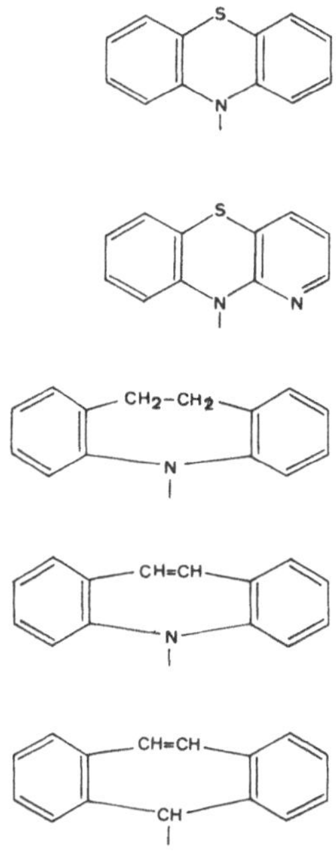

FIG. 23. Tracer anions for the detection of tricyclic psychotropic drugs by GC with mass specific anion recording (N_2O as reagent gas).

Anion MS by CI can be carried out at relatively low pressure. All the anion spectra from our own laboratory which we have reported in this review have been obtained at ion source pressures between 10^{-3} and 10^{-2} torr. This is low when compared with the pressures reported by the US authors. They have used the same conditions as for cation CI-MS, that is 10^{-1} to 1 torr or even atmospheric pressure. We believe that low pressure work has considerable advantages;

some of them are: a. the ion source stays cleaner; b. there is less
danger of spark formation; c. electron attachment and charge
exchange are favored over cluster ion formation.

Our anion CI-approach has only moderate pressure dependence.
Fig. 24 is a tabulation of the anions obtained by chemical ionization
of Methaqualone with N_2O as reagent (18).

Methaqualone

2-Methyl-3-(o-tolyl)-
4-quinazolone

Anion-MS with N_2O

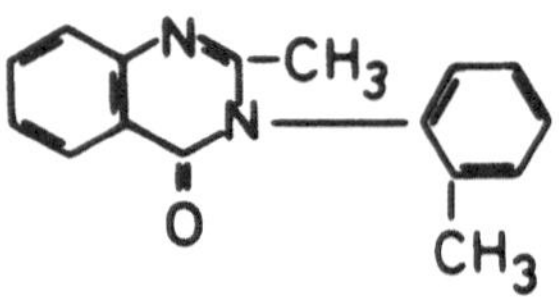

m Torr	135	159 M-91	233 M-17	247 M-3	248 M-2	249 M-1	261 M+11	265 M+15	280 M+30
1	6.6	3.6	26.1	17.7	51.1	100	0.9	63.6	-
3	7.9	4.4	25.4	19.7	55.7	100	4.4	67.2	0.7
5	8.4	4.3	25.1	18.9	56.8	100	6.2	67.6	0.8
7	8.1	2.7	25.7	21.6	59.5	100	9.5	67.6	1.4
9	7.7	4.6	26.2	20.0	58.5	100	13.8	67.7	3.1

FIG. 24. Anion spectra of Methaqualone at different
ion source pressures with reagent N_2O
(% rel.I are listed).

The relative intensities of the anions are listed for different
source pressures from 10^{-3} to 10^{-2} torr. For most anions, the
pressure influence is negligible. The fragmentation picture is not
much affected by the tenfold increase of reagent pressure, and
even the intensity of the O-substitution product $/\overline{M+15}/^-$ is
fairly constant. There are just 2 exceptions. One is the ion
$/\overline{M+11}/^-$ of unknown origin, the second the ion $/\overline{M+30}/^-$ which is
probably a cluster formation $(M + NO^-)$. In our recent work, we are
trying to avoid cluster ion formation. It seems to us that electron
attachment and charge exchange are more rewarding reaction mechanisms
for anion MS by CI.

SUMMARY AND OUTLOOK INTO THE FUTURE OF ANION MASS SPECTROMETRY

Due to the fact that EI-MS is essentially a cation method,
negative ion work has been neglected. Even after the development
of CI, it took over ten years until anion MS started to get a hold.
By now, it has become clear that EI, the hard ionization approach

which produces interesting fragment cations, and anion CI, the
soft ionization process which yields analytically useful anions
predominantly in the molecular range, are truly complementary
methods.

In conjunction with GC, anion MS by CI is an ideal detection
technique which can be used: a. for identification purposes with
a scanning technique; b. for trace analysis as a selected anion
detector.

Many toxicologically important compounds, be they pharmaceuticals
or environmental poisons, possess high electron affinity. They can
be detected with much better sensitivity by anion than by cation
recording.

Selected anion recording is superior to detection by electron
capture since it combines equal or better sensitivity with
substantially improved specificity. It will undoubtedly become the
ECD of the future.

Selected anion recording can also advantageously replace other
semi-specific GC-detectors, i.e. a thermionic detector.

Trace analysis of all substances belonging to a class with a
common structural sub-unit is very often easier by GC with anion
than with cation recording.

Anion MS will certainly not replace EI-MS as an identification
tool, but it may replace cation MS by CI. On the other hand, in
the field of selected ion monitoring, the applications of anion MS
may outgrow the applications of cation CI and EI even within the
next few years.

It is too early to tell which modification of anion MS will
prove to be more efficient: atmospheric pressure ionization or
ionization at relatively low pressure. The different possibilities
still have to be compared. Also we do not yet know how an anion MS
will be finally designed. For the time being, a commercial CI mass
spectrometer with modified power supply is all we need to explore
the wide open field.

REFERENCES

1) H. Brandenberger and R. Ryhage, in "Recent Developments in Mass
   Spectrometry in Biochemistry and Medicine", A. Frigerio, Ed.,
   Vol. 1, Plenum Press, New York, 1978, p. 327.
2) F.W. Aston, "Mass Spectra and Isotopes", E. Arnold, London,
   1933, p. 27.

3) J. Yinon, H.G. Boettger and W.P. Weber, Anal. Chem., 1972, 44, 2235.

4) R.T. Aplin, H. Budzikiewicz and C. Djerassi, J. Am. Chem. Soc., 1965, 87, 2926.

5) M.S.B. Munson and F.H. Field, J. Am. Chem. Soc., 1966, 88, 2621, see footnote 13 on p. 2630.

6) M.v.Ardenne, Kernenergie, 1958, 1, 1029.

7) M.v.Ardenne, K. Steinfelder and R. Tümmler, "Elektronenanlagerungs--Massenspektrographie organischer Substanzen", Springer, Berlin, 1971.

8) M.v.Ardenne, K. Steinfelder, R. Tümmler and K. Schreiber, Experientia, 1963, 19, 178.

9) H.P. Tannenbaum, J.D. Roberts and R.C. Dougherty, Anal. Chem., 1975, 47, 49, and Lit. Ref. in Table 2.

10) I. Dzidic, D.I. Carroll, R.N. Stillwell and E.C. Horning, Anal. Chem., 1975, 47, 1308, and Lit. Ref. in Table 2.

11) D.F. Hunt, G.C. Stafford, F.W. Crow and J.W. Russell, Anal. Chem., 1976, 48, 2098.

12) H. Brandenberger and R. Ryhage, in "Blood Drugs and Other Analytical Challenges", E. Reid, Ed., Ellis Horwood Publishers, Chichester, 1978, p. 173.

13) H. Branderberger and R. Ryhage, Mass Fragments, 1977, 1, 1.

14) D. Frangi-Schnyder and H. Brandenberger, Fresenius' Z. Anal. Chem., 1978, 290, 153.

15) R. Ryhage and H. Brandenberger, Biomed. Mass Spectrom., 5, in press.

16) H. Brandenberger, U. Amsler and D. Frangi-Schnyder, in "8th Annual Symposium on the Analytical Chemistry of Pollutants", Geneva, 1978.

17) H. Brandenberger, in "8th International Meeting of Forensic Sciences", May 1978, Wichita, Kansas.

18) H. Brandenberger, in "International Symposium of Instrumental Drug Application in Forensic Drug Chemistry", May 1978, Washington D.C., in press.

19) H. Brandenberger, in "Mass Spectrometry in Drug Metabolism", A. Frigerio and E.L. Ghisalberti, Eds., Plenum Press, New York, 1977, p. 379.

20) H. Brandenberger, Deutsche Lebensm. Rundschau, 1974, 70, 31.

21) H. Brandenberger, in "Clinical Biochemistry", Vol. II, H. Ch. Curtius and M. Roth, Eds., W. de Gruyter, Berlin, 1974, p. 1431.

22) U. Amsler and H. Brandenberger, in preparation.

23) R.C. Dougherty, J. Dalton and F.J. Biros, Org. Mass Spectrom., 1972, 6, 1171.

24) H.P. Tannenbaum, J.D. Roberts and R.C. Dougherty, Anal. Chem., 1975, 47, 49.

25) R.C. Dougherty, J.D. Roberts and F.J. Biros, Anal. Chem., 1975, 47, 54.

26) E.C. Horning, M.G. Horning, D.I. Carroll, I. Dzidic and R.N. Stillwell, Anal. Chem., 1973, 45, 936.

27) D.I. Carroll, I. Dzidic, R.N. Stillwell, M.G. Horning and E.C. Horning, Anal. Chem., 1974, 46, 706.

28) E.C. Horning, D.I. Carroll, I. Dzidic, S.-N. Lin, R.N. Stillwell and J.-P. Thenot, J. Chromatogr., 1977, 142, 481.

29) E.C. Horning, D.I. Carroll, I. Dzidic, K.D. Haegele, M.G. Horning and R.N. Stillwell, 1974, 99, 13.

30) D.I. Carroll, I. Dzidic, R.N. Stillwell, K.D. Haegele and E.C. Horning, Anal. Chem., 1975, 47, 2369.

MASS FRAGMENTOGRAPHIC DETERMINATION OF POLYAMINES IN THE URINE

OF PREMATURE BABIES

H. Milon[°], I. Antener[°] and S. Nordio[^]; [°]Nestlé Products

Technical Assistance Company Ltd., Orbe, Switzerland;

[^]"Burlo Garofolo"Paediatric Hospital, Trieste, Italy

INTRODUCTION

Polyamines have been intensively investigated during the last decade for their implication in the control of cellular growth (1). Relations have been established between polyamine levels and different types of cancer (2, 3), and the hormonal regulation of their biosynthesis has also been assessed (4).

From a certain point of view, cancer and fetal cell growth can be compared (5), and polyamines could play an important role in fetal tissue growth. Fig. 1 shows a simplified representation of the pathway of polyamine biosynthesis. It has been shown that the pathway of methionine metabolism in human fetal liver and brain is affected by the lack of cystathionase (6, 7), thus enhancing RNA, proteins and DNA syntheses (8) (Fig. 2). Proteins are of prime importance in the feeding of babies. Urinary polyamine levels may be of interest in relation to the characteristics of various commercial cow milk formulas as compared with maternal milk (9).

For these reasons, determinations of polyamine levels in the urine of prematures have been included in the preliminary investigations of a large clinical study concerning some nutritional and metabolic aspects of maternal lactation (9). As the expected levels were very low and the amounts of sample limited, polyamines were measured by a mass fragmentographic method adapted from Denton et al. (10) and Walle (11). The results obtained will be discussed from a methodological point of view together with their physiological indications.

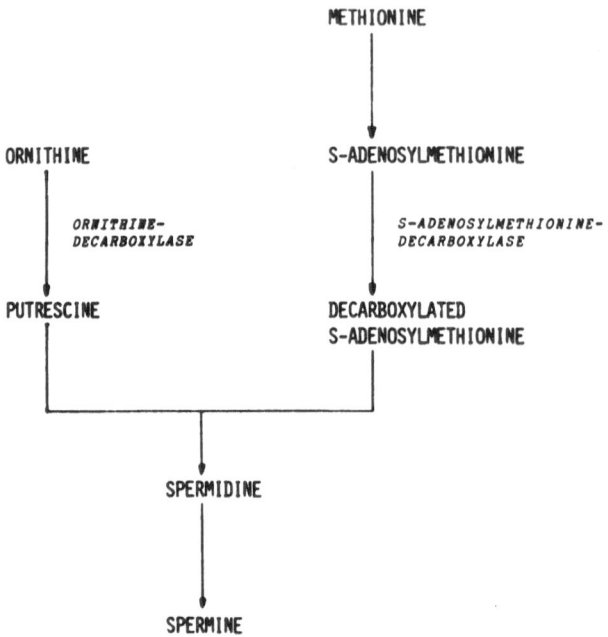

FIG. 1. Pathway of polyamines biosynthesis.

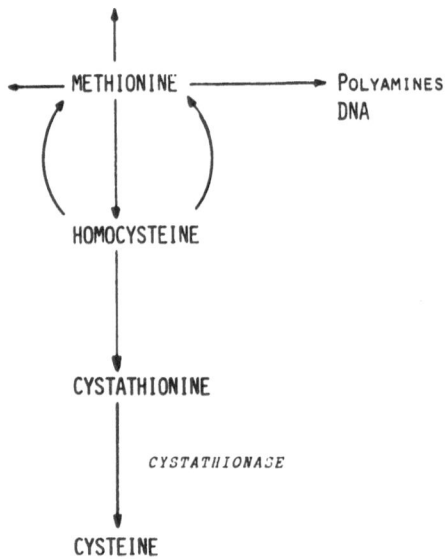

FIG. 2. Methionine-cysteine pathway.

EXPERIMENTAL

Subjects. Table 1 shows the characteristics of the subjects. They are divided into four groups according to the different treatments.

TABLE 1
Subjects characteristics.

Group	N	Weight		24 hr urine
		Birth kg	Sampling day kg	ml
1	5	2.25 ± 0.12	2.59 ± 0.05	135 ± 17
2	5	2.24 ± 0.13	2.56 ± 0.02	143 ± 15
3	5	2.38 ± 0.07	2.54 ± 0.03	135 ± 30
4	3	2.36 ± 0.17	2.68 ± 0.13	129 ± 12
Overall mean		2.30 ± 0.06	2.58 ± 0.13	136 ± 10

Values are expressed ± S.E.M.

Treatment. Table 2 gives the composition of the three commercial cow milks compared to that of maternal milk. It can be seen that Nan represents a cow milk which has been adapted to resemble maternal milk, whereas Nido and Eledon remain close to their original composition.

Table 3 shows a more detailed picture of the protein composition and its influence on the babies' daily intake.

Analyses. The analytical method is given in Table 4. Two mass fragments (m/e 126 and m/e 154) were followed for each amine, except for the internal standard (for which the more specific m/e 196 was followed) and spermine (m/e 126 which was too weak). The ratio of their intensities was used to control the specificity of the peak attributions.

Figs. 3 and 4 show examples of the M.I.D. response and Figs. 5, 6 and 7 show the polyamine calibration curves.

RESULTS

Table 5 gives the average amount of the urinary polyamines excreted in 24 hours. Cadaverine, although it does not seem to be

TABLE 2
Characteristics of the milks
% composition.

	Maternal	Nan	Nido	Eledon
Proteins	8	12.5	20.1	30.8
Lipids	36	26.0	12.0	14.0
Lactose	54	55.5	28.6	40.0
Saccharose	-	-	15.0	-
Honey	-	-	5.0	-
Starch	-	-	10.7	-
Lactic acid	-	-	-	4.7
Mineral salts	3	3.3	5.3	6.5
Water	-	2.7	3.3	4.0

TABLE 3
Protein contents of the milks.

	Maternal	Nan	Nido	Eledon
Total (milk %)	8	12.5	20.1	30.8
% Casein	65	40	82	82
% Whey proteins	35	60	18	18
Daily intake (g/kg)	2.5	2.5	5	7

TABLE 4
Polyamines analytical method.

Sample	10 ml urine
	+
	2 μg internal standard (1.7-diaminoheptane)
Extraction	pH 13
	3 x 10 ml 1-butanol
	dried over Na_2SO_4
	2 x 5 ml HCl 1N
	evaporated / N_2
	1 ml HCl 0.1N, -20°
Derivatization	evaporated / N_2
	25 μl TFAI + 75 μl CH_3CN, 100°, 15 min.
Separation	GC-MS-MID LKB 2091
and	glass column, 1 m, 2 mm I.D.,
detection	3% QF-1, helium, 180° / $15°.min^{-1}$ / 275°,
	separator: 260°, ionisation source: 275°
	electrons energy: 20 eV,
	slits: collector: 0.35 mm, source: 0.12 mm,
	m/e: 126, 154 and 196

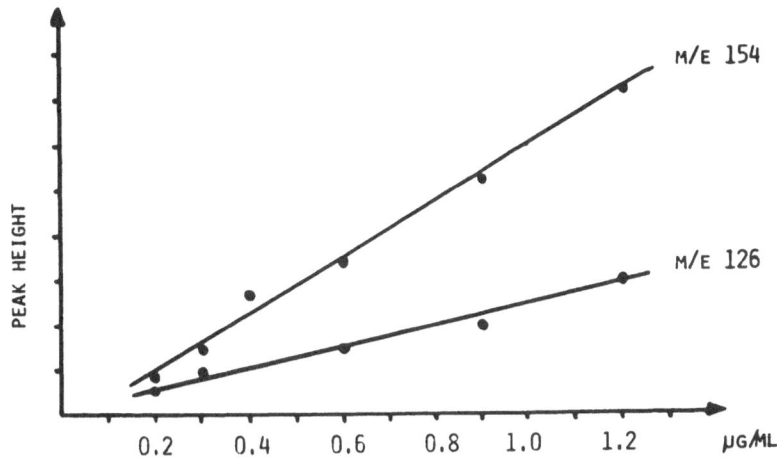

FIG. 3. M.I.D. response curve for putrescine.

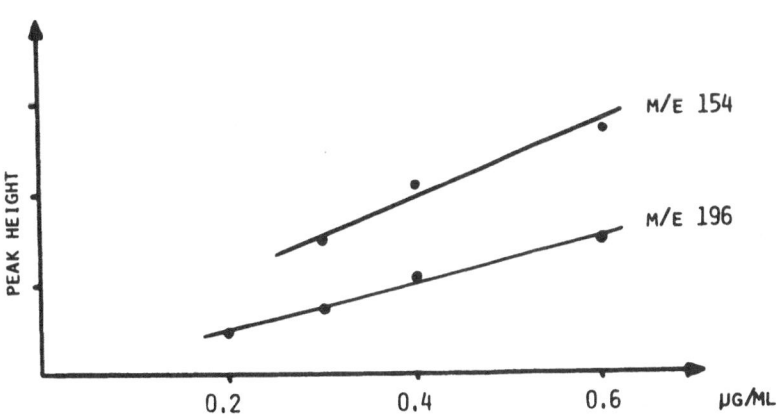

FIG. 4. M.I.D. response curves for spermidine and
1.7-diaminoheptane (internal standard).

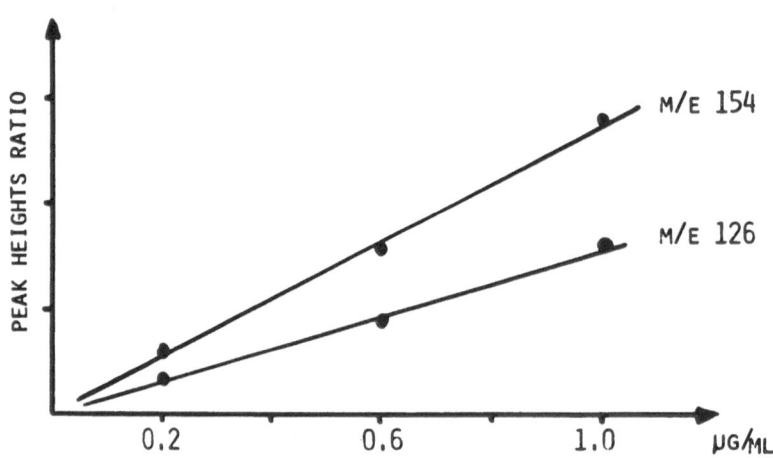

FIG. 5. Putrescine calibration curves.

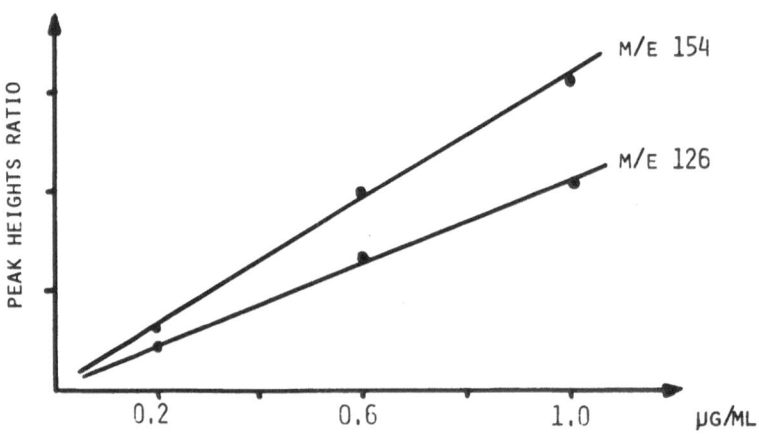

FIG. 6. Spermidine calibration curves.

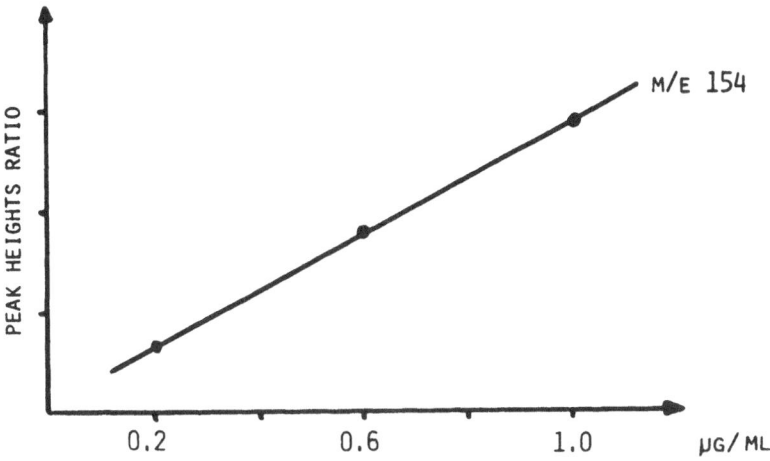

FIG. 7. Spermine calibration curve.

TABLE 5
24 hr urinary excretion of polyamines.

Group	Milk	P	SD	SP	C
			(µg/24hr ± S.E.M.)		
1	Mat.	126 ± 33	45 ± 10	21 ± 6	147 ± 55
2	Nan	68 ± 14	87 ± 37	32 ± 11	35 ± 10
3	Nido	82 ± 25	44 ± 10	12 ± 2	44 ± 12
4	Eledon	62 ± 18	54 ± 24	19	86 ± 19

P = putrescine; SD = spermidine; SP = spermine; C = cadaverine

correlated to the other polyamines, is also given as it is excreted
in appreciable quantities and is determined simultaneously.

DISCUSSION

Age and weight of the subjects. This study was conducted on a
group of neonates born at a gestational age of about 37 weeks. They

were chosen for the homogeneity of their birth weight, and for their
perfect health conditions. Being slightly premature, they were
retained in hospital and thus represent a population sample under
complete clinical control for a certain period of time. At the
time of urine collection, they were about 3 weeks old, still low
in weight and very homogeneous.

Hydrolysis of the samples and N-monoacetylspermidine. As
polyamines may be excreted in a conjugated form (2), hydrolysis
is usually performed before analysis. As our samples were available
in an amount sufficient for only one determination, they were not
hydrolysed, and, therefore only free polyamines were determined.

In some previous studies (10-13), another product related to
spermidine has been detected and identified as a metabolite formed
by N-acetylation of spermidine. N-monoacetylspermidine has been
identified in the urine of normal children (14), but, so far, has
not yet been reported in normal adult unhydrolysed urine.

By hydrolysis of its metabolite, spermidine is formed and
quantified together with the free spermidine present in the sample
before hydrolysis.

Denton et al. (12) compared the chromatograms obtained from the
urine of an adult with acute myelocytic leukemia before and after
hydrolysis. In this particular case a 10-fold increase in spermidine
was detected after hydrolysis. However, these results represent only
one experiment, not performed on a healthy subject, and it seems
that further evidence indicates that even with hydrolysis not all
spermidine is accounted for.

In most of our chromatograms, there was no peak present between
spermidine and spermine which could have corresponded to N-monoacetylspermidine. But the fragment ions we focussed on to detect
the polyamines had a very small abundance in the fragmentation
pattern obtained by Walle (11) and, as the amounts we were
measuring were much lower than those found in adults, we probably
did not have the level of sensitivity necessary to detect this
metabolite in the absence of a synthetic reference molecule.

However, in another series of experiments, polyamines were
determined in the urine of another group of comparable subjects by
gas chromatography and flame ionisation detection (15). In this
case, the sensitivity of the detector was too low to obtain
quantitative results for the polyamines, but large quantities of a
substance chromatographically identified as N-monoacetylspermidine
could be determined. Two samples from the group concerned in our
study could also be analysed, and about 0.1 mg N-monoacetylspermidine
were found to be excreted per 24 hours.

The state of conjugation of the excreted polyamines should be

more thoroughly investigated and the eventual presence of other
conjugates checked before more intensive studies are undertaken.
        Comparison with values from the literature (Table 6). All
the quantitative results found in the literature have been obtained
in hydrolysed samples.

TABLE 6
Values obtained by other authors on hydrolysed samples.

Ref.	Year	Method	N	Age	Putrescine		Spermidine (mg/24hr)		Spermine	
(2)	1971	HV-EP	50		2.7	± 0.5	3.1	± 0.6	3.4	± 0.7
			35		2.9	± 0.5	4.7	± 0.6	3.5	± 0.7
(17)	1973	A.A.A.	10		2.5	± 0.6	2.4	± 0.4	0.4	± 0.2
(16)	1973	GC-FID	12	18-55	1.4	± 0.1	1.3	± 0.2	0.4	± 0.1
					(0.9	− 2.4)	(0.8	− 2.7)	(0.1	− 1.5)
			8	5-48	0.9	± 0.2	0.9	± 0.1	trace	
					(0.3	− 2.4)	(0.4	− 1.4)	(tr.	− 0.2)
(10)	1973	GC-FID	25		1.26	± 0.07	1.06	± 0.09	0.28	± 0.06
(13)	1976	LC			0.89		0.53		0.14	
					(0.02	− 2.55)	(0.02	− 1.32)	(n.d.	− 0.38)
(18)	1978	TLC-Fluor.			3.0	± 0.8	2.2	± 0.4	0.5	± 0.5
					(1.9	− 4.3)	(1.5	− 2.9)	(0	− 1.7)
This work	1976	GC-MS-MID	18	0.4	0.08	± 0.01	0.08	± 0.02	0.04	± 0.01
					(0.02	− 0.17)	(0.01	− 0.42)	(n.d.	− 0.17)

        The techniques used for the determinations have changed with
time, and become more specific and sensitive, and one can observe
that the evolution of the methods is paralleled by an evolution of
the average value found for normal subjects. Our values are about
10 times lower than those observed using comparable methods (GC-
-FID) in the literature. No values for free polyamines are available
in the literature, except for a semi-quantitative result obtained
by Denton et al. (12). Further, the age and weight of the subjects
certainly has an effect on the daily excretion level. Gehrke et al.
(16) also found lower values for 5 and 10 year old children than
for adults.
        Tentative correlation with other parameters. It can be seen
in Table 3 that the subjects can be separated into two groups
according to the level of proteins contained in the milk they were
given. As polyamines are related to protein synthesis, considering
only two groups (and not four) may compensate for the small number

of subjects in each group. In our case, a rationalized way of
expressing polyamine excretion levels is to relate them to the
amount of excreted creatinine which is considered as a very stable
reference. Table 7 shows the amount of polyamines excreted per mg
creatinine. The levels of spermidine and spermine are significantly
higher in the group given a low protein milk ($p<0.1$ and $p<0.05$,
respectively), but not different as far as putrescine is concerned
(Table 8).

TABLE 7
24 hr urinary excretion of polyamines referred
to creatinine.

Group	Milk	P	SD	SP	C
			(µg/mg creatinine±S.E.M.)		
1	Mat.	11 ± 2	4.3 ± 0.8	1.8 ± 0.3	12 ± 4
2	Nan	4.1 ± 0.3	5.0 ± 1.3	2.5 ± 0.9	2.4 ± 0.8
3	Nido	6.2 ± 2.7	2.6 ± 0.6	0.8 ± 0.1	32 ± 12
4	Eledon	8.4 ± 4.3	3.9 ± 0.8	0.7	9.0 ± 4.4

P = putrescine; SD = Spermidine; SP = spermine; C = cadaverine.

TABLE 8
Comparison of the different treatments T tests.

	Mat./Nan	Mat./Nido	Mat.+ Nan/Nido + Eledon
Putrescine	N S	N S	N S
Spermidine	N S	p < 0.05	p < 0.05
Spermine	N S	p < 0.02	p < 0.05
OH-Proline	N S	p < 0.01	p < 0.05

It has been found that in the fetal liver and brain, and,
perhaps in the neonate (especially if it is a premature), the
absence of cystathionase increases the level of methionine. Therefore the synthesis of polyamines, and thereby RNA and DNA may also
be changed (7). This supports our findings, as S-adenosylmethionine
interferes in the polyamine synthesis pathway between putrescine
and spermidine.

Another way of following the development of neonates is to
determine their hydroxyproline index, defined by:

$$I = \text{weight} \times \frac{\text{mM/1 OH-proline}}{\text{mM/1 creatinine}}$$

Again, in the low protein group, this index is significantly higher than in the other group ($p<0.05$) (Table 9).

TABLE 9
Hydroxyproline indices.

Group	Milk	OH-proline index (kg ± S.E.M.)
1	Maternal	2.3 ± 0.1
2	Nan	2.2 ± 0.2
3	Nido	1.5 ± 0.2
4	Eledon	2.3 ± 0.1

To our knowledge, it is the first time that polyamines have been measured in the urine of premature babies.

It is extremely difficult to obtain subjects of about the same weight and gestational age, in perfect health conditions to minimise secondary effects. Also, the collection of samples and the analyses of such small and yet variable quantities add further problems.

This study was a preliminary investigation as part of a more complex research work on infant nutrition. Therefore, the results obtained can only give indications for future work and cannot support any hypothesis.

It can be retained that neonates, considering their weight, seem to excrete large quantities of polyamines, and that the excretion could be related to the protein content of the milk, as well as to the hydroxyproline index.

ACKNOWLEDGEMENTS

The authors gratefully acknowledge the skillful technical assistance of Ms. S. Tanner and Mr. J. Schifferle.

REFERENCES

1) S.S. Cohen, "Introduction to the Polyamines", Prentice-Hall, New-Jersey, 1971.

2) D.H. Russell, C.L. Levy, S.C. Schimpff and I.A. Hawk, Cancer Research, 1971, 31, 1555.

3) H.G. Williams-Ashman, G.L. Cappoc and G. Weber, Cancer Research, 1972, 32, 1924.

4) D.H. Russell, in "Polyamines in Normal and Neoplastic Growth", D.H. Russell, Ed., Raven Press, New York, 1973, p. 1.

5) W.E. Knox, Am. Sci., 1972, 60, 480.

6) J.A. Sturman and G.E. Gaull, Pediatr. Res., 1974, 8, 231.

7) G.E. Gaull, J.A. Sturman and N.C.R. Räikä, Pediatr. Res., 1972, 6, 538.

8) E.J. Herbst and U. Bachrach, Ann. N.Y. Acad. Sci., 1970, 171, 691.

9) S. Nordio, N. Levi and I. Antener, Annales Nestlé, 1977, no 74, p. 47.

10) M.D. Denton, H.S. Glazer, D.C. Zellner and F.G. Smith, Clin. Chem., 1973, 19, 904.

11) T. Walle, in "Polyamines in Normal and Neoplastic Growth", D.H. Russell, Ed., Raven Press, New York, 1973, p. 355.

12) M.D. Denton, H.S. Glazer, T. Walle, D.C. Zellner and F.G. Smith, in "Polyamines in Normal and Neoplastic Growth", D.H. Russell, Ed., Raven Press, New York, 1973, p. 373.

13) H. Adler, M. Margoshes, L.R. Snyder and C. Spitzer, J. Chromatogr., 1977, 143, 125.

14) T. Nakajima, J.F. Zack Jr. and F. Wolfgram, Biochim. Biophys. Acta, 1969, 184, 651.

15) I. Antener, private communication.

16) C.W. Gehrke, K.C. Kuo, R.W. Zumwalt and T.P. Waalkes, in "Polyamines in Normal and Neoplastic Growth", D.H. Russell, Ed., Raven Press, New York, 1973, p. 343.

AN ASSAY FOR SCHIZOPHRENIA: MASS SPECTROMETRIC DETERMINATION OF TWO

CADAVERINE METABOLITES

H. Dolezalova, M. Stepita-Klauco, C. van der Velde and

V. Cassone, Department of Biobehavioral Sciences,

University of Connecticut, Storrs, Connecticut, U.S.A.

INTRODUCTION

Some findings indicate that cadaverine and its acylated metabolites monoacetylcadaverine and monopropionylcadaverine may be
associated with schizophrenia. Monoacetylcadaverine and monopropionylcadaverine were isolated from urine of schizophrenic patients
(1). It was speculated that their presence in the urine may reflect
an abnormal metabolism of cadaverine concomitant with mental
illness (2). We have reported recently that the blood concentrations
of acylated cadaverines are significantly increased in schizophrenic
patients in comparison to those in nonschizophrenic subjects (3).

Cadaverine (1,5-diaminopentane) was the least studied polyamine
in mammals. It has been assumed that this compound was not present
in the brain (4) and that the presence of cadaverine in other
tissues was the result of either microbial metabolism in the intestine
(5) or of tissue putrefaction (6).

The role of cadaverine may be different from that of an exogenous contaminant of the body. Using a mass spectrometric determination having both higher sensitivity and specificity than
previously employed techniques, cadaverine was shown to be present
in the mammalian brain and blood (7, 8). Its biosynthesis from
lysine was demonstrated in the kidney (9). In the brain, an uptake
system has been shown for cadaverine which was inhibited by cyanide
and some polyamines (10). Neither blood nor brain concentrations
of cadaverine are lowered by the absence of bacterial flora (2)
indicating that the brain and blood concentrations of cadaverine
are maintained through mechanisms which are independent of
bacterial action.

Monoacylated cadaverines seem to differ pharmacologically
from their precursor - cadaverine. In contrast to cadaverine, they
can depolarize the cell membrane of neurons in the central ganglia
of molluscs (11). Based on this finding, it was suggested that the
acylation of one amino group of cadaverine could render a compound
which is capable of exerting neuromodulatory effect.

The aim of this study was to decide whether the elevated
concentrations of acylated cadaverines in the human blood could
serve as a biochemical marker of schizophrenia. Our results
suggest that the combined assay of monoacetylcadaverine and monopropionylcadaverine can provide such information.

MATERIALS AND METHODS

Subjects and samples. Blood concentrations of cadaverine,
monoacetylcadaverine and monopropionylcadaverine were measured in
32 subjects: 21 diagnosed schizophrenic patients and 11 nonschizophrenic controls. The patients were selected from in-patient
facility of the Norwich Hospital (Connecticut). They manifested
delusions, hallucinations and/or thought disorder, as well as
inappropriate affect. The patients were free of known organic
disease. Their classification according to DSM-II (12) and medication
taken at the time of sampling are listed in Table 1. The patients
without medication received no drugs during a three month period
prior to their sampling.

The nonschizophrenic controls consisted of hospitalized
nonschizophrenic patients from the same hospital (N=2), volunteer
hospital staff members from the same hospital (N=3), organic neurological hospitalized patients from another hospital (N=4), and
healthy subjects working outside of the hospital environment (N=2).
The controls had no history of mental illness.

Blood samples were collected between 7:00 and 8:00 a.m..
Fasting blood was drawn by venepuncture into siliconized tubes,
the samples were placed on ice, coded, and processed within three
hours by staff members who were unaware of their origin. Samples
from both schizophrenic and nonschizophrenic subjects were processed
simultaneously. All tested individuals are reported in Table 2.

A thin-layer chromatography - mass spectrometry method was
used for determinations. Dansyl derivatives formed in the whole
blood homogenate were separated by thin-layer chromatography,
eluted, and measured by high resolution mass spectrometry.
Dansylation and thin layer chromatography. The blood was
weighed and homogenized in five volumes of 0.2M perchloric acid.
The whole homogenate was then submitted to dansylation, a reaction
with 1-dimethylaminonaphtalene-5-sulfonyl chloride. Dansyl

TABLE 1
Solvent systems for thin-layer chromatographic separation of dansylated cadaverine, monoacetylcadaverine and monopropionylcadaverine.

Compound	First chromatography			Second chromatography		
	Solvents	Runs	Eluant	Solvents	Runs	Eluant
bis-Dns-cadaverine	heptane-acetone (1:1)	one	ethylacetate	chloroform-triethylamine (10:1)	one	ethylacetate
Dns-acetylcadaverine	heptane-acetone (1:1)	one	methanol	benzene-methanol (14:1)	three	ethylacetate-acetone (1:1)
Dns-propionylcadaverine						

TABLE 2

	No:	Diagnosis	Sex	Age	Treatment	Daily Dose in mg	Concentrations in pmoles/g of blood ACCA	PRCA	CA	ACCA + PRCA
CONTROLS	1.	Healthy	F	31	---	-	3.4	4.1	136.7	7.5
	2.	Healthy	F	46	---	-	3.2	3.9	127.4	7.1
	3.	Drug overdose	F	26	---	-	7.2	9.9	37.8	17.1
	4.	Healthy	M	48	---	-	5.7	10.8	117.6	16.5
	5.	Healthy	M	24	---	-	3.3	2.6	109.5	5.9
	6.	Drug overdose	M	31	---	-	4.7	22.0	6.0	26.7
	7.	Stroke	M	72	---	-	3.2	1.1	24.3	4.3
	8.	Healthy	M	35	---	-	2.2	0.7	9.0	2.9
	9.	Stroke	M	69	---	-	3.9	5.6	14.3	9.5
	10.	Stroke	M	56	---	-	3.7	1.1	11.1	4.8
	11.	Stroke	M	69	---	-	3.5	0.0	14.0	3.5
SCHIZOPHRENIC PATIENTS	12.	Schi. Chr.	F	42	Mellaril	800	29.0	54.5	105.2	83.5
	13.	Schi. Chr.	F	22	Haldol	40	27.5	44.8	113.8	72.3
	14.	Schi. Chr.	F	38	Mellaril	800	75.6	63.0	168.7	138.6
	15.	Schi. Chr.	F	48	Haldol	80	66.8	86.4	99,6	153.2
	16.	Schi.	M	43	Thorazine	400	27.3	34.2	33.9	61.5
	17.	Schi. Ac. Cat.	F	56	Nazane	60	69.8	15.4	637.2	85.2
	18.	Schi.	F	24	no medic.	--	18.5	35.2	90.0	53.7
	19.	Schi.	M	32	no medic.	--	22.3	26.0	82.5	48.3
	20.	Schi.	M	31	no medic.	--	13.0	17.5	34.9	30.5
	21.	Schi. Chr.	F	47	no medic.	--	31.3	48.4	124.0	79.7
	22.	Schi. Chr.	F	31	no medic.	--	110.5	57.6	113.9	168.1
	23.	Schi. Chr.	F	41	no medic.	--	3.0	24.2	449.7	27.2
	24.	Schi. Chr. Hall.	M	27	no medic.	--	42.3	75.1	22.7	117.4
	25.	Schi. Acute	M	22	no medic.	--	59.1	100.3	36.7	159.4
	26.	Schi. Acute	F	23	no medic.	--	36.0	89.3	249.9	125.3
	27.	Schi. Acute	M	24	no medic.	--	22.8	37.6	513.1	60.4
	28.	Schi. Ac. Par.	M	19	no medic.	--	37.0	68.8	174.7	105.8
	29.	Schi. Ac. Nondif.	M	29	no medic.	--	29.9	60.7	336.9	90.6
	30.	Schi. Acute	M	32	no medic.	--	8.0	32.7	10.9	40.7
	31.	Schi. Paran.	F	52	no medic.	--	11.7	72.0	315.6	83.7
	32.	Schi. Chr. Ac. Ep.	F	24	no medic.	--	32.0	96.5	202.6	128.5

derivatives of amines were extracted into toluene and separated by thin-layer chromatography on silica gel coated plates (Merck). The fraction which co-chromatographed with standards of the dansylated compound in question was scraped off, eluted, and its content was measured by high resolution mass spectrometry. The solvent systems for chromatography are shown in Table 1.

Reversed peak matching. A modified version of the integrated ion current technique with the peak matching circuit and the internal standard of a dansylated compound having the evaporation profile comparable to that of the compound in question was used. Data conditioning before the quantitation of the recorded molecular ions was performed by correcting the signal for the spectrometer background, assuming a linear change of the background during the sample evaporation, and by machine controlled determination of the beginning and the end of the part of the evaporation curve used for quantitation. The main advantages of this procedure are a higher reproducibility of obtained results and a larger number of biological samples (about sixty) which can be measured in one session without having to wait until the background is reduced.

A known quantity of an internal standard was added to each

eluted chromatographic fraction and the dried mixture was introduced via the probe into the mass spectrometer (AEI MS-902). The list of internal standards and their molecular ions are shown in Table 3. The sample was evaporated over 30-40 seconds by heating it to 350, into the source maintained at 220, and ionized with the electron beam energy 70 eV. The molecular ions corresponding to the dansylated compound in question and to its internal standard were recorded with use of the peak matching circuit of the spectrometer at a resolving power between 2000 and 8000. Their molecular ratio was preset with the accuracy 2ppm. The peaks corresponding to the two molecular ions were consecutively displayed on the screen of a storage oscilloscope and, simultaneously, each individual peak was digitized in a 14-bit analog-to-digital converter (Xerox SDS) and recorded digitally on a magnetic tape. From the ratio between the intensity of the ion of interest and that generated by an internal standard substance, the quantity of the substance of interest in a sample was calculated.

The peak matching technique is the most accurate method available to date for measuring the mass of a molecular ion or fragment. When performed on a high resolution, double focusing mass spectrometer, the mass of an ion of interest can be routinely measured with the accuracy 10 ppm. For quantitation of ions or fragments in biological samples peak matching can be used as the most reliable identification of selected parts of the mass spectrum and thereby of identity of studied components.

The matching of the two peaks (intensities of ions or fragments) has been only subjectively evaluated by the operator in all previous systems for this type of application. In our method, both the quantity and the mass identity (with the accuracy 10 ppm) are objectively recorded. An interference in the spectrum due to impurities or contamination of the sample is therefore objectively demonstrated and such samples are eliminated. Moreover, the method provides an objective proof of the identity of the studied compound, continuously, during the whole evaporation of the substance in the mass spectrometer.

It should be noted that the term peak matching is used in the above description for a procedure which is slightly different from the classical peak matching and could be called, more accurately, a reversed peak matching. While the former implies a stepwise adjustment of a ratio between two masses, in an attempt to find as close as possible numeric value for the actual mass ratio, in the latter procedure, a fixed, calculated six digit ratio is preset; and the difference between the preset (theoretical) ratio and that observed during evaporation of the sample in the mass spectrometer is objectively evaluated. Thus, the disagreement between the expected and observed mass values is evaluated for each pair of two matched mass regions, dynamically, during complete evaporation of the whole

TABLE 3
Mass spectrometric internal standards for quantitative determination
of dansylated cadaverine, monoacetylcadaverine and monopropionylcadaverine.

Compound	Composition	m/e	Standard	Composition	m/e	Ratio
bis-Dns-cadaverine	$C_{29}H_{36}N_4O_4S_2$	568.2178	bis-Dns-hexamethyl-enediamine	$C_{30}H_{38}N_4O_4S_2$	582.2344	1.024666
Dns-acetylcadaverine	$C_{19}H_{27}N_3O_3S$	377.1773	Dns-acetylhexa-methylenediamine	$C_{20}H_{29}N_3O_3S$	391.1930	1.037159
Dns-propionylcadaverine	$C_{20}H_{29}N_3O_3S$	391.1930	Dns-propionylhexa-methylenediamine	$C_{21}H_{31}N_3O_3S$	405.2086	1.035828

sample, in contrast to a subjective evaluation by a mass spectrometer
operator which can be done only for a few sweeps during evaporation
of one sample (due to limitations of the dynamic input range of the
oscilloscope) and rests on a subjective operator's impression rather
than on a recorded measurement.

The evaluation of the accuracy of the peak matching was done
by computer (Xerox Sigma-2). If the mass ratio between the two
matched peak maxima differed at any time during the evaporation
of the sample by more than 40 ppm (due to drift of the instrument,
sample contamination, or electrical interference), the sample was
disregarded.

Calculation of the amount of a substance in each sample was
performed after subtracting the background, by comparing the sums
of the molecular ion maxima collected during the complete evaporation generated by the known amount of the internal standard and
by the unknown amount of the compound in question. The amplitude
of each peak was established and an increase or decrease of the
signal per step for each two consecutive displays was calculated.
The first and the last peak of the evaporation curve used for the
quantitative evaluation were defined as those in which the increase
of the signal per step was for the first time larger than 5% of
the maximal change of the signal per step found during the whole
course of the evaporation of that sample, when searching forward
for the first peak and backward for the last one. The amplitude
of the signal prior to the first peak and after the last was
considered as background. In case these two background values
differed, a linear change of the background during the evaporation
was assumed, and the corresponding background was subtracted from
each peak. From the corrected ratio between the sums of the
molecular ion maxima the actual amount of the compound in question
was calculated.

It is assumed in the course of the calculation procedure that
for the same molar quantities of the two compounds the intensities
of both molecular ions are identical. This is obviously not true
for most compounds. In order to establish the quantitative
relationship between the intensity of the molecular ion of an internal standard and that of the compound in question, a calibration
procedure based on measuring standard samples is performed. These
standard samples contain both the internal standard and the compound
in question in known quantities (two standard samples for each
of about 5 different amounts within the expected operating range).
Assuming the above quantitative relationship, the content of the
compound in question is calculated for each standard sample. Each
standard sample thus gives a calculated and an observed (actual)
value. A regression line is calculated for the whole calibration
set of pairs of expected and observed quantities. From the equation

for the obtained regression line a corrected amount of the internal
standard is calculated. The corrected internal standard is
subsequently used in all experimental samples for calculating the
amount of the compound in question.

The calibration functions of the procedure calculated by linear
regression for 12 samples containing known picomole quantities of
each studied compound had the correlation coefficients 0.9920,
0.9914 and 0.9865, for cadaverine, monoacetylcadaverine and monopropionylcadaverine, respectively. The coefficients of variation
for repeated analyses of identical samples were 12.23%, N=18;
13.85%, N=21; and 12.39%, N=18, for the same sequence of the
studied compounds. The reported concentrations were measured as
quantities more than three times higher than their blanks. Each
sample was analyzed in triplicate (as three separate mass spectrometric samples). The values reported in Tables 2 and 4 (including
blanks) were based on more than 500 mass spectrometric measurements.

TABLE 4

No:	Diagnosis	Comment	Concentrations in pmoles/g of blood				Treatment	Daily Dose	Elapsed Time
			ACCA	PRCA	CA	ACCA + PRCA		in mg	Days
22-1	Schi. Chr.		110.5	57.6	113.9	168.1	No medic.		0
22-2		Unchanged+Med.	10.4	71.1	27.3	81.5	Thorazine	600	2
22-3		Worse+Hallucinations	76.5	49.5	19.8	126.0	Thorazine	600	12
25-1	Schi. Acute	Acute	59.1	100.3	36.7	159.4	No medic.		0
25-2		Improved+Med.	0.0	7.1	212.6	7.1	Thorazine	400	8
25-3		Worse+Hallucinations	11.6	14.2	235.0	25.8	Thorazine	400	20
26-1	Schi. Acute	Acute	36.0	89.3	249.9	125.3	No medic.		0
26-2		Quiet+Med.	26.3	144.9	335.6	171.2	Haldol	30	2
26.3		Calmer+Med.	4.3	16.2	9.7	20.5	Haldol	30	12
27-1	Schi. Acute	Acute	22.8	37.6	513.1	60.4	No medic.		0
27-2		Unchanged+Agitated	11.6	54.7	11.8	66.3	Thorazine	600	2
27-3		Worse+Hall.+Dell.	76.5	49.5	19.8	126.0	Thorazine	600	12
31-1	Schi. Paran.	Acute Episode	11.7	72.0	315.6	83.7	No medic.		0
31-2		Unchanged :Med.	40.9	58.9	534.7	99.8	Prolixin	10	2
31-3		Improved+Med.	6.2	19.4	7.1	25.6	Prolixin	10	12
32-1	Schi. Chron.	Acute Episode	32.0	96.5	202.6	128.5	No medic.		0
32-2	Acute Episode	Unchanged +Med.	32.8	40.4	562.8	73.2	Mellaril	400	2
32-3		Improved+Med.	23.7	31.7	6.1	55.4	Mellaril	400	12
28-1	Schi. Ac. Par.		37.0	68.8	174.7	105.8	No medic.		0
28-2		Calmer+Med.	10.1	75.5	9.2	85.6	Thorazine	1200	2
28-3		Quiet+No hall.	26.0	3.7	8.3	29.7	Thorazine	1200	12
29-1	Sehi. Acute		29.9	60.7	336.9	90.6	No medic.		0
29-2	Nondiff.	Improved	7.2	22.5	24.5	29.7	Stellazine	20	12

RESULTS

The concentrations of cadaverine and acylated cadaverines found in the blood of schizophrenic patients were higher than those in nonschizophrenic subjects (Table 2). The differences between the mean concentrations in nonschizophrenic and schizophrenic subjects were highly significant (t values of the t-test 2.437, 4.124 and 6.080, for cadaverine, monoacetylcadaverine and monopropionyl-cadaverine, respectively; $P<0.05$ for cadaverine, $P<0.001$ for mono-acetylcadaverine and monopropionylcadaverine using the two-tailed t-test). In some patients, only one of the two acylated forms was increased (^23, ^30, ^31-3). The spans of values for each tested compound overlapped between the nonschizophrenic and schizophrenic group (Table 2).

An attempt was made to use elevated blood concentrations of acylcadaverines as a biochemical marker of schizophrenia and to test the relevance of such a marker in individual subjects parti-cipating in this study. For this purpose, the sum of concentrations of both monoacylcadaverines in the blood of each subject was used as his single attribute rather than the concentration of mono-acetylcadaverine or monopropionylcadaverine alone. It should be noted that the sum of the monoacetylcadaverine and monopropionyl-cadaverine concentrations in the blood of each subject represents the concentration of both monoacylcadaverines originating from the same precursor - cadaverine (Fig. 1, Tables 3 and 4).

The ranges of blood concentrations of pooled monoacylcadaverines in the control and in the schizophrenic group were separated from each other. The concentrations of both monoacylcadaverines in the blood of subjects from the control group were between 2.9 and 26.7 pmoles/g, whereas in schizophrenic patients the concentrations were in the region from 27.2 to 168.1 pmoles/g. The difference between the lowest value in schizophrenics (^23 - 27.2 pmoles/g) and the highest value in controls (^6 - 26.7 pmoles/g) was 0.5 pmole/g. The significance of the difference between the schizophrenic and the control group was higher (t=6.260) than that found for monoacetyl-cadaverine (t=4.124) or monopropionylcadaverine alone (t=6.080). All schizophrenic patients has their concentrations of acyl-cadaverines in the blood higher than was the upper limit for the control group (27 pmoles/g).

Some patients were sampled two or three times, after their psychopharmacologic medication was initiated, or if they showed in the opinion of the clinician (C.v.d.V.) an apparent change in the manifestation of their illness. The blood concentrations of those patients are listed in Table 4.

The effect of medication on blood levels of acylcadaverines

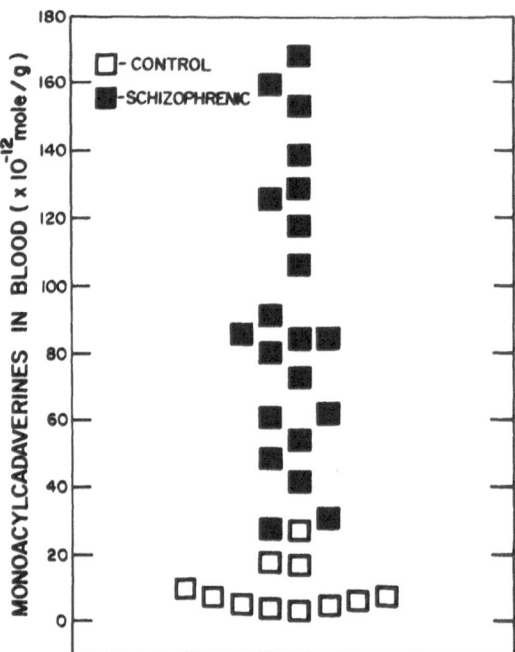

FIG. 1. Concentrations of monoacylcadaverines in
        the blood of studied subjects. The center
        of each square indicates the sum of mono-
        acetylcadaverine and monopropionylcadaverine
        concentrations in the blood of one subject.
        Empty squares - controls, filled squares -
        schizophrenics.

was inconsistent. In some patients (^22, ^25, ^32, ^28, ^29), there
was a decrease in overall concentration of acylcadaverines during
the medication, regardless of the clinical evaluation as improved
(^25-2, ^28-2, ^29-2) or without change (^22-2, ^32-2). In others
(^26, ^27, ^31), the concentrations of acylcadaverines increased
(^26-2, ^31-2) or remained unchanged (^27-2) during the treatment.

     In all patients having medication, the combined concentration
of both acylcadaverines changed with the clinical evaluation in
the same direction: relatively lower values of acylcadaverines were
observed during clinical improvement, and relatively higher values
of acylcadaverines occurred during worsening of clinical symptoms.
In 3 of 8 patients, the concentrations of monoacylcadaverines
were found lower than the upper limit for the control group (27

pmole/g), indicating 'normal' values of monoacylcadaverines, if
their blood was collected at a time of clinical improvement (^25-2,
^25-3, ^26-3, ^31-3).

DISCUSSION

Our evaluation of the nonschizophrenic and schizophrenic group
was done at the time when the clinical diagnoses in both groups
were known. A test of the reliability of the elevated blood
concentrations of monoacylcadaverines as a biochemical marker of
schizophrenia would require determination of reference values on a
group of nonschizophrenic subjects prior to the evaluation of the
tested group; the latter containing nonschizophrenic and schizophrenic subjects with unknown clinical diagnoses at the time of
the testing.

The observed differences in concentrations of blood monoacylcadaverines do not seem to be pharmacologically induced. Only 6 of
21 schizophrenic patients were receiving medication on their first
day of sampling. Moreover, the results obtained on repeatedly
sampled schizophrenics, for which the patients were used as their
own controls, indicate that blood concentrations of monoacylcadaverines follow the clinical change of the illness. Therefore
it is unlikely that their elevation in schizophrenics would be
caused by the diet, dietary habits, hospitalization or other
secondary factors.

In contrast to our previous findings obtained on two groups
of medicated schizophrenics (3), also cadaverine concentrations
were found elevated in some patients (^14, ^17, ^23, ^26, ^27,
^29, ^31, ^32) participating in this study. Only two of these
patients were receiving medication (^14, ^17).

Dramatic changes in the blood concentrations of cadaverine
were observed in all repeatedly sampled patients. However,
cadaverine concentrations did not change consistently with a two-
-day medication, clinical evaluation, or the change in blood
acylcadaverines. With the exception of one patient (^25), all
others had substantially reduced blood levels of cadaverine after
twelve days of pharmacotherapy. It seems therefore that the
cadaverine level in a patient may reflect the history of his
treatment rather than the current status of his illness.

It can be anticipated from our previous results that geographic,
cultural and genetic variables contribute to fluctuations in the
absolute levels of blood acylcadaverines (3). To minimize their
effects, the reference groups of nonschizophrenic subjects should
be selected from the same environment. By assaying a large number
of subjects the range of concentrations of monoacylcadaverines in

both nonschizophrenic and schizophrenic populations will be increased
and could result in an overlap between highest concentrations in
controls and lowest in schizophrenics.

The sites of origin of blood cadaverine and acylcadaverines
are unknown. One can only speculate on their mutual dependence or
significance to mental illness. In the mouse brain, monoacylcadaverines can be elevated by administering monoamine oxidase
inhibitors (3). Thus, in a simplified model, high levels of monoacylcadaverines can be caused by an inefficient monoamine oxidase
system (in the brain or elsewhere), increased acylation rate of
cadaverine, or increased availability of cadaverine. Phenothiazines
and butyrophenones could alleviate this situation by lowering the
cadaverine concentrations (reducing decarboxylation of lysine or
increasing elimination of cadaverine). Since monoacylcadaverines
can activate the nerve cell membrane while cadaverine cannot (11),
it may be the blood level of monoacylcadaverines rather than that
of cadaverine which correlates with the presence of the illness
and its clinical manifestations.

It is concluded from the presented results that the elevated
blood concentrations of monoacylcadaverines seem to be promising
candidates for a combined biochemical marker of certain forms of
mental illness. The possibility that the formation and/or catabolism
of acylcadaverines are connected with the etiopathogenesis of
schizophrenia should be investigated.

ACKNOWLEDGEMENT

We thank Dr. R. Fairweather and M. Thompson for their help in
operating mass spectrometer. This work was supported by PHS Grants
NS 11716 and NS 12482, and by the National Science Foundation
Grant BNS 77-15323.

REFERENCES

1) T.H. Perry, S. Hansen, L. MacDougall, Nature, 1967, 214, 484.
2) M. Stepita-Klauco, H. Dolezalova, Nature, 1974, 252, 158.
3) H. Dolezalova, M. Stepita-Klauco, J. Kucera, in "Mass Spectrometry in Drug Metabolism", A. Frigerio and L. Ghisalberti,
   Eds., Plenum, New York, 1977, p. 201.
4) S.I. Harik, G.W. Pasternak, S.H. Snyder, in "Polyamines in
   Normal and Neoplastic Growth", D. Russel, Ed., Raven Press,
   New York, 1973, p. 307.
5) H. Tabor, C.W. Tabor, Adv. Enzymol. Mol. Biol., 1972, 36, 203.
6) F. Dreyfuss, R. Chayen, G. Dreyfuss, Israel J. Med. Sci.,
   1975, 11, 785.

7) H. Dolezalova, M. Stepita-Klauco and R. Fairweather, Brain Res., 1974, 77, 166.
8) H. Dolezalova and M. Stepita-Klauco, Adavances in Mass Spectrometry in Biochemistry and Medicine, 1976, 1, 206.
9) S. Henningsson, L. Persson and E. Rosengren, Acta Physiol. Scand., 1976, 96, 445.
10) F. Picolli and A. Lajtha, Biochim. Biophys. Acta, 1971, 225, 356.
11) M.W. Miller and M. Stepita-Klauco, Society for Neuroscience Abstracts, 1977, 3, 446.
12) American Psychiatric Association, "DSM-II Diagnostic and Statistical Manual of Mental Disorders", 2nd. ed., Washington, D.C., 1968.

ANALYSIS OF THE URINARY STEROIDS FROM A PATHOLOGICAL PREGNANCY BY LIQUID-GEL CHROMATOGRAPHY AND GAS-CHROMATOGRAPHY-MASS SPECTROMETRY

H. Losty, R.J. Begue, M. Moriniere and P. Padieu

Laboratoire de Biochimie Médicale, Faculté de Médicine

Dijon, France

INTRODUCTION

We have previously described a method (1) employing liquid-gel chromatography on Sephadex LH-20, gas chromatography and gas chromatography-mass spectrometry for the analysis of the principal C_{18}, C_{19}, C_{21} steroids excreted during a normal pregnancy. This paper describes the application of such a method, to the study of the alteration in steroid production and metabolism which occur during a pathological pregnancy. The mother, who had a history of adrenal hyperplasia due to 21-hydroxylase deficiency, was delivered of a normal female infant at term. The steroids excreted during the first, second and third trimester of pregnancy were analysed, identified and quantitated.

MATERIALS AND METHODS

The reference steroids were purchased from Makor Chemicals Ltd (Jerulasem, Israel) and from Steraloids Inc. (Wilton, USA). All solvents were of analytical grade: a)the N-O-bis-(trimethylsilyl)--trifluoroacetamide (BSTFA) was obtained from the Pierce Chemical Co, (Rockford, Ill, USA). The methoxyamine hydrochloride was obtained from Eastman Organic Chemicals (Rochester, N.Y., USA). b) Helix pomatia digestive juice was purchased from "l'Industrie Biologique Française" (Gennevilliers, France). c) Sephadex LH-20 was obtained from Pharmacia Fine Chemicals (Uppsala, Sweden).

Extraction of urinary steroids. 24 h urine samples were collected during the first, second and third trimester and 1/10 of the total volume taken for analysis. The urinary conjugates of

steroids were extracted according to Okerholm et al. (2). The ethyl
acetate extracts were taken to dryness. Acidified ethyl acetate (20
ml) was added and solvolysis (3) allowed to proceed at 39°C for 24
h. Following solvolysis, the ethyl acetate was taken to dryness and
the residue dissolved by sonication in distilled water (20 ml). The
pH was adjusted to 5.20 by addition of acetic acid. Sodium acetate
buffer (2 ml, pH 5.20, 0.1M), Helix pomatia digestive juice (1,000
IU of β-glucuronidase per ml of urine) were added and the sample
incubated at 37°C for 24 h. A second addition of enzyme was made
and the sample incubated for a further 24 h. Following hydrolysis,
the free steroids were extracted twice with ethyl acetate (1:1, by
vol.) and once with diethyl ether (1:1, by vol.). The combined organic
phases were washed with sodium bicarbonate 1M (5 ml) and then with
distilled water until neutral, dried by filtration over anhydrous
sodium sulphate, and evaporated to dryness.

Liquid-gel chromatography. Glass columns, 50 m x 0.7 id, with
a 50 ml solvent reservoir and teflon stopcock were used. Sephadex
LH-20 (5 g) was initially suspended in solvent system S, chloroform-
-heptane-methanol (5:5:1 v/v). The dried urine extract was applied
to the column after sonication in 2.0 ml S and the non-polar steroids
eluted with 50 ml S, under nitrogen, at the rate of 10-15 ml/h.
Fourteen 2.5 ml fraction were collected. The Sephadex LH-20 was resuspended in chloroform-methanol (1:2 v/v) and the polar steroids eluted
in six 5 ml fractions. The internal standard 5β-cholestane-3α-ol
(100 μg) was added to each fraction.

Gas chromatography and gas-chromatography-mass spectrometry.
The eluted fractions were dried under nitrogen. The trimethylsilyl
ether were prepared by heating overnight at 60°C in 150 μl BSTFA.
The methoxime trimethylsilyl derivatives were prepared with 200 μl
of methoxyamine hydrochloride in pyridine (16 mg/ml) by heating
overnight and then silylated as before. The samples were analysed
simultaneously by gas chromatography (Becker 409 Gas Chromatograph)
using twin 3.60 m x 3 mm id glass columns packed with 0.5% OV-1
and 0.5% OV-17. The column temperature was programmed between 180°C
and 250°C at the rate of 1.2°C/min. The temperature of injection
port and detector were 245°C and 275°C respectively.

GC-MS analysis was carried out using an LKB 9 000 instrument
equipped with a 1% OV-1 column. The flow rate of carrier gas (helium)
was 30 ml/min. The temperature of injection port, separator and ion
source were 250°C, 280°C and 290°C respectively. Accelerating
voltage, energy of bombarding electrons and ionizing current were
3.5 kV, 22.5 eV and 60 μA respectively. The column was programmed
from 180°C to 260°C at 1°C/min.

Identification and quantitation of steroids. The steroids were
identified by a comparison of their methylene unit (MU) values,
obtained on OV-1 and OV-17 columns, with those of reference steroids
and by analysis of their mass spectra. The steroids, eluted in each
fraction were quantitated by reference to the internal standard. The
efficiency of extraction of the conjugated steroids, the percentage

recovery of the free steroids after hydrolysis and after chromatography on Sephadex LH-20 and the coefficient of response of the
reference steroids, all previously determined (4) were used to
determine the total amount of steroid present in the original sample.

RESULTS

Gas chromatography and gas chromatography-mass spectrometry
of reference steroids. The MU values on 0.5% OV-1 and 0.5% OV-17
and the coefficient of response as determined on OV-1 are given
including the mass spectra data of the MO-TMS and TMS derivatives
of the principal C_{19} steroids in Table 1. Only the m/e values for
the molecular ion and the most characteristic fragment ions are
represented. Table 2 and Table 3 give the MU values and mass
spectral data for the TMS derivatives of the clinically important
C_{18} and C_{21} steroids.

Analysis of biological samples. - First trimester. The steroids
present in the urine sample taken at 5 weeks were analysed. The gas
chromatogram of those steroids, eluted in Sephadex LH-20 fraction 6,
as their MO-TMS derivatives is illustrated in Fig. 1. The following
steroids were identified by gas chromatography and gas chromatography-
-mass spectrometry: 3α-hydroxy-5α-androstan-17-one (An), 3α-hydroxy-
5β-androstan-17-one (Et); 3α-hydroxy-5α-pregnan-20-one (P) and 3α-
-hydroxy-5β-pregnan-20-one (P3). The other pregnanolone isomer, 3α-
-hydroxy-5β,17α-pregnan-20-one was eluted in the trace amounts. The
mass spectrum of the compound, with a retention time of 40 min and
a methylene unit value of 29.26, is shown in Fig. 2. The molecular
ion M at m/e 521 would indicate a dihydroxypregnandione, bis TMS,
mono MO. The fragment ions at m/e 490: M-31, 431: M-90, 400: M-
-(90+31) and the base peak at m/e 310: M-(2x90+31) are formed by
the loss of the methoxime and trimethylsilyl groups. This compound
is tentatively identified as 3α,17-dihydroxy-5β-pregnan-11,20-dione,
until confirmed by a comparison with the mass spectrum of the
authentic reference steroid. This steroid has previously isolated
in cases of congenital adrenal hyperplasia by David et al. (5).

The gas chromatogram of the steroids, eluted in fraction 10,
as their TMS derivatives is illustrated in Fig. 3. The principal
steroid metabolites were identified as: 3α,11β-dihydroxy-5α-androstan-
17-one (11β-OH-An), 5β-pregnane-3α,20α-diol, 5β-pregnane-3α,17,20α-
triol (pregnanetriol), 3α,17,20α-trihydroxy-5β-pregnan-11-one
(pregnanetriolone) and 5-pregnene-3β,17,20α-triol (pregnentriol).
Pregnanetriol and pregnanetriolone (and 17α-hydroxypregnanolone
eluted in fraction 7) are the characteristic urinary steroids
associated with congenital adrenal hyperplasia (6) and are excreted
in elevated amounts. The mass spectrum of the compound with a
retention time of 25 min and a MU value of 25.59 is shown in Fig.
4 and is identical to that of 3α,6α-dihydroxy-5β-androstan-17-one.

TABLE 1
Principal C_{19} steroids as (a) MO-TMS derivatives (b) TMS derivatives.

Reference compounds	MU/OV-1	MU/OV-17	RC/OV-1	M^+	Predominant peaks
(a) MO-TMS derivatives:					
3α-hydroxy-5α-androstan-17-one	24.94	26.83	1.13	391	213-215-255-270-301-360-376
3α-hydroxy-5β-androstan-17-one	25.18	27.01	1.27	391	213-215-255-270-360-376
3β-hydroxy-5-androsten-17-one	25.56	27.86	1.08	389	129-260-268-284-299-358-374-389
3α-hydroxy-5α-androstan-11,17-dione	25.91	28.59	1.22	405	129-147-284-374-390-405
3α-hydroxy-5β-androstan-11,17-dione	26.07	28.59	0.96	405	261-269-284-300-315-374-390-405
3α,11β-dihydroxy-5α-androstan-17-one	26.84	28.34	0.84	479	125-182-198-213-368-358-448-464-479
3α,11β-dihydroxy-5β-androstan-17-one	27.09	28.54	1.05	479	125-143-182-198-213-215-268-358-448-464-479
(b) TMS derivatives:					
5β-androstane-3α,17β-diol	25.05	24.47	1.80	436	129-215-217-241-256-346
5α-androstane-3β,17β-diol	25.64	26.25	1.32	436	129-148-149-177-218-217-256-331-421
5-androstene-3β,17β-diol	25.78	26.56	1.26	434	129-213-215-289-254-305-329-344-419

TABLE 2
Principal C_{18} steroids as TMS derivatives.

Reference compounds TMS derivatives	MU/OV-1	M^+	Predominant peaks
3-hydroxy-1,3,5(10)-estratien-17-ore	25.34	342	218-244-257-(327)-342
1,3,5(10)-estratriene-3,17β-diol	25.25	416	115-129-(177)-218-232-244-285-326-(401)-416
3,16α-dihydroxy-1,3,5(10)-estratrien-17-one	27.05	430	218-231-244-286-430
3,17β-dihydroxy-1,3,5(10)-estratrien-16-one	27.37	430	129-245-258-271-285-312-415-430
1,3,5(10)-estratrien-3,16α,17α-triol	27.96	504	103-129-(143)-147-205-231-(244)-270-(283)-(285)-297-311-324-345-360-386-414-(489)
1,3,5(10)-estratrien-3,16α,17β-triol	28.33	504	103-129-147-205-231-270-(283)-(285)-297-311-324-345-360-386-414-(489)-504
1,3,5(10)-estratriene-3,15α,16α,17β-tetrol	30.92	592	191-412-502

TABLE 3
Principal C$_{21}$ steroids as TMS derivatives

Reference compounds (TMS derivative)	MU/OV-1	MU/OV-17	RC/OV-1	M$^+$	Predominant peaks
3α-hydroxy-5β-pregnan-20-one	26.11	28.35	0.92	390	215-230-285-300-375
3α,17-dihydroxy-5β-pregnan-20-one	26.96	28.49	0.88	478	215-255-345
5α-pregnane-3α,20α-diol	27.26	27.86	1.10	464	117-269-284-449
5β-pregnane-3α,20α-diol	27.47	28.05	1.19	464	117-269-284-449
3α,6α-dihydroxy-5β-pregnan-20-one	27.55	–	1.04	478	251-298-299-388-478
3α,16α-dihydroxy-5β-pregnan-20-one	27.66	–	1.00	478	96-109-157-159-172-186-255-368-463
5 -pregnene-3β,20α-diol	27.97	28.93	1.09	462	117-129-282-372-462
5β-pregnane-3α,17,20α-triol	28.94	29.87	0.98	480	117-215-255-273
5-pregnene-3β,17,20α-triol	29.49	30.82	0.67	478	117-129-213-253-271-445-463
5-pregnene-3β,16α,20α-triol	29.66	29.94	1.18	550	117-129-141-156-157-460
5α-pregnane-3β,16α,20α-triol	29.70	–	0.95	552	117-141-156-157-462
3α,17,20α-trihydroxy-5β-pregnan-11-one	29.95	31.55	0.71	494	117-287-404

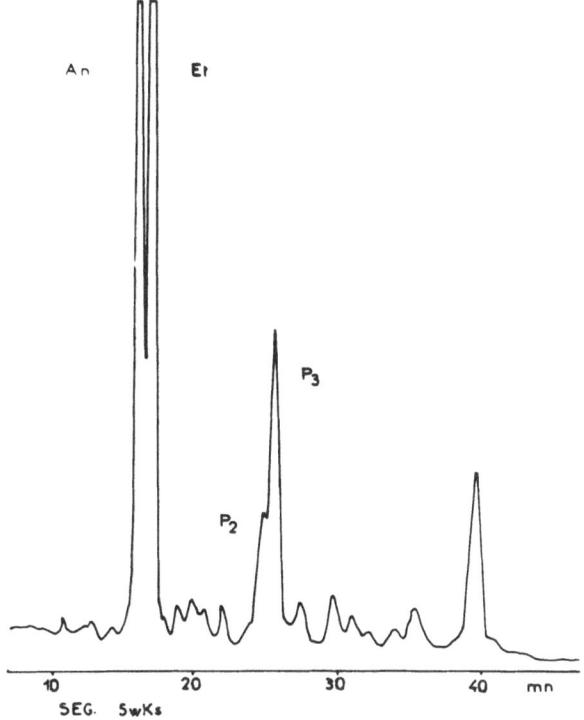

FIG. 1. Gas chromatogram of
first trimester pregnancy
urine (5 weeks), fraction 6.
The steroids were chromatographed as the MO-TMS
derivatives on a 0.5% OV-1
column.

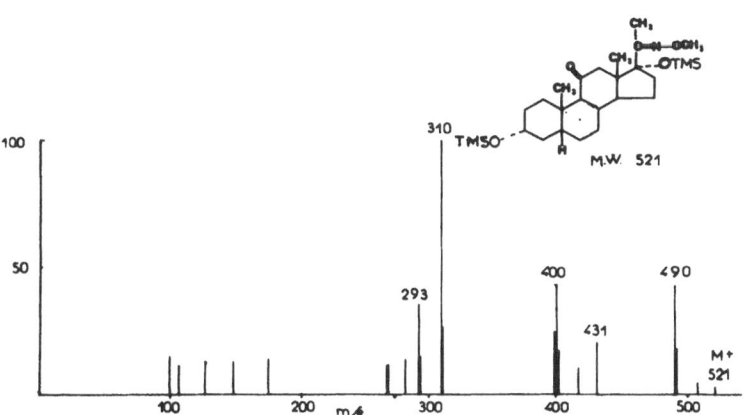

FIG. 2. Mass spectrum of compound identified as:
    3α,17α,dihydroxy-5β-pregnan-11,20-dione MO-TMS.

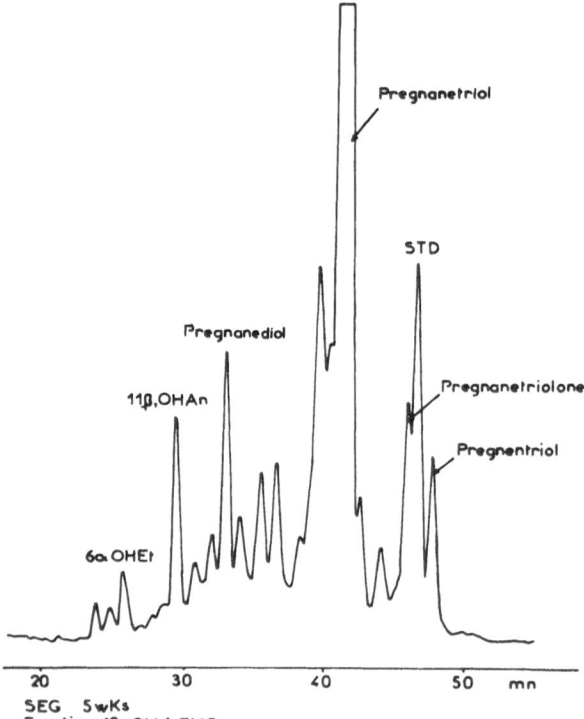

FIG. 3. Gas chromatogram of first trimester pregnancy urine (5 weeks), fraction 10. The steroids were chromatographed as the TMS derivatives on a 0.5% OV-1 column.

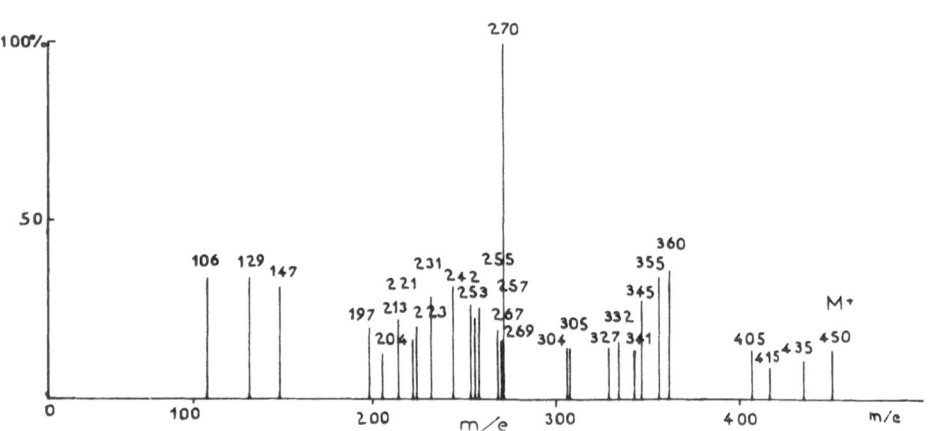

FIG. 4. Mass spectrum of 3α,6α-dihydroxy-5β-androstan-
         -17-one-TMS.

The molecular ion is at m/e 450 and the base peak at m/e 270 with
characteristic fragment ions at m/e 204, 223, 242, 253, 255 and 305.
To our knowledge this steroid 6α-hydroxyetiocholanolone has not been
previously identified in pregnancy urine. In addition, the following
compounds were identified by gas chromatography and gas chromatography-
-mass spectrometry: 5β-androstane-3α,17β-diol, 5-androstene-3β,17β-
-diol, 3α,11β-dihydroxy-5β-androstan-17-one, 5α-pregnane-3α,20α-diol,
3α,16α-dihydroxy-5β-pregnan-20-one and 5α-pregnane-3β,16α,20α-triol.

    In fraction 7, the 3α isomer of 17α-hydroxy-pregnanolone was
isolated. In addition, the 3β isomer was identified by mass spectrometry. The mass spectrum of 3β,17α-dihydroxy-5β-pregnan-20-one as the
MO-TMS derivative is presented in Fig. 5.

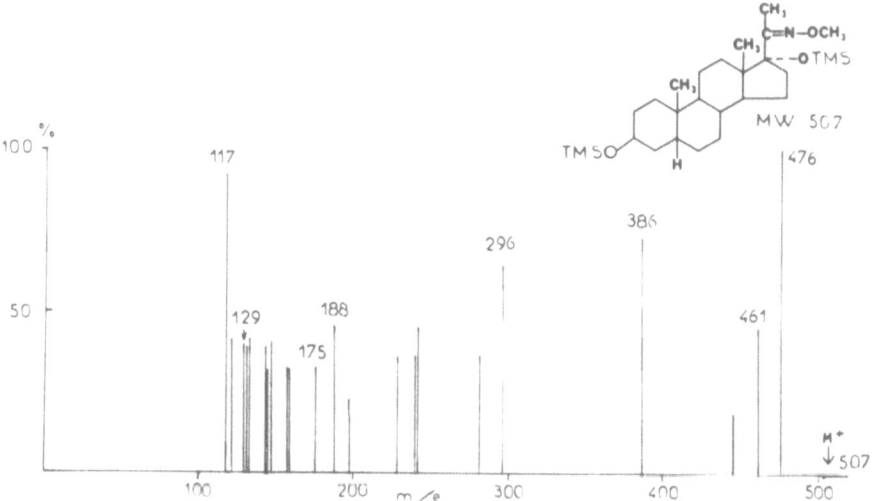

FIG. 5. Mass spectrum of 3β,17α-dihydroxy-5β-
-pregnan-20-one MO-TMS.

The molecular ion M is at m/e 507 and the base peak at m/e 476: (M-31).
The characteristic fragment ions are at m/e 386: (M-90-31), 296:
(M-2x90-31) and 188. The other fragment ions between m/e 117 and 180
are due to an unidentified steroid. The excretion of the 3α and 3β
isomers of 17α-hydroxypregnanolone in the urine of patients with
congenital hyperplasia was previously reported by Viinikka (7).
    Second and third trimester. The steroids excreted during the
second and third trimester were identified and quantitated. The
amounts of estrogens excreted were significantly low.

    The gas chromatogram of fraction 8 of a urine sample from the

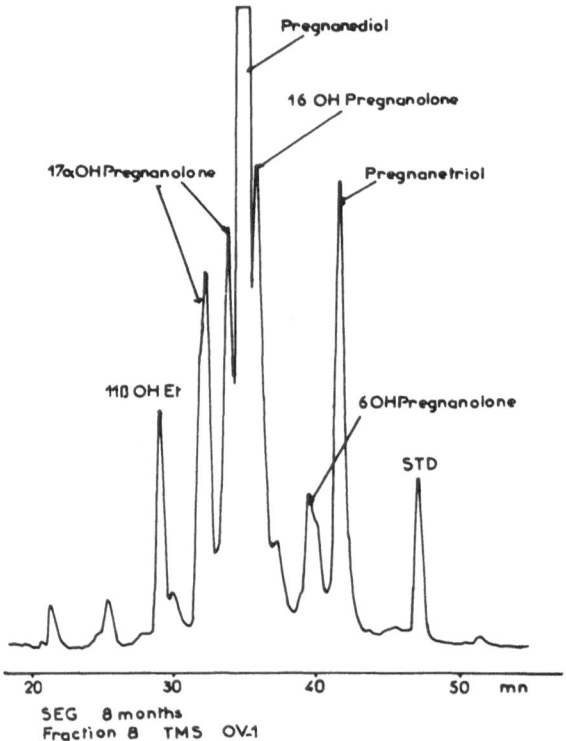

FIG. 6. Gas chromatogram of third trimester pregnancy
        urine (8 months), fraction 8. The steroids were
        chromatographed as the TMS derivatives on a
        0.5% OV-1 column.

eighth month of pregnancy is presented in Fig. 6. The steroids were
chromatographed as their TMS derivatives and identified by gas
chromatography-mass spectrometry as: 3α,11β-dihydroxy-5β-androstan-
-17-one, 5β-pregnane-3α,20α-diol, 3α,16α-dihydroxy-5β-pregnan-20-
-one, 3β,6α-dihydroxy-5α-pregnan-20-one (eluted with a trace of
5-pregnene-3α,16α,20α-triol), 5β-pregnane-3α,17,20α-triol and 5ε-
-pregnan-2ε,3ε,20ε-triol (molecular ion at m/e 552, base peak at
117 and fragment ions characteristic of a 2,3-bis-trimethylsilyl-
oxysteroid at 129, 142 and 143). In addition to pregnanetriol, the
other two steroid metabolites characteristic of congenital adrenal
hyperplasia pregnanetriolone and 17α-hydroxypregnanolone were
isolated in the urine extract. This latter steroid presented two
chromatographic peaks as illustrated in Fig. 6 due to different
products of the silylation reaction (8). The two peaks were
identified as the 3,17 bis-TMS derivative and the 3,20 enol bis-
-TMS derivative.

DISCUSSION

A method developed for urinary steroid analysis was employed
in the investigation of the complex steroids excreted during a
pathological pregnancy where the mother had a history of congenital
adrenal hyperplasia (CAH). In addition to the fetal-placental unit,
the abnormally functioning maternal adrenal was responsible for the
excretion of a range of C_{19} and C_{21} steroids. The versatility and
efficacity of the combination of liquid-gel chromatography, gas
chromatography and mass-spectrometry in isolating, identifying and
quantifying the atypical steroids was demonstrated. In this instance
preliminary separation of the C_{19} and C_{21} steroids by Sephadex LH-20
was indispensable since the total free steroid extract proved to be
too complex to analyse by gas chromatography and for this reason
small (2.5 ml) fractions were collected. An additional means of
identification of the oxygenated steroids was afforded by preparation
of both trimethylsilyl ether derivatives and the methoxime trimethylsilyl derivatives.

In this instance, CAH was attributed to adrenal 21-hydroxylase
deficiency. Cortisol replacement therapy was withdrawn for the first
trimester and the hyperactivity of the maternal adrenal was reflected
in the excretion of pregnanetriol, pregnanetriolone and the 3β isomer
of 17α-hydroxypregnanolone, the characteristic steroid metabolites
associated with 21-hydroxylase deficiency. In addition to the 3β
isomer, 3α,17α-hydroxy-pregnan-20-one was isolated and identified by
mass spectrometry. A steroid, previously isolated in certain cases
of CAH, was tentatively identified by mass spectrometry as 11-oxo-
-17-hydroxypregnanolone; a precursor of pregnanetriolone. The
isolation of these steroid metabolites in significant quantities is
due to the accumulation, in the maternal adrenal, of 17α-hydroxy-
progesterone the immediate steroidal substrate of the deficient
21-hydroxylase enzyme. Alternative means of disposing of the 17α-
-hydroxyprogesterone, by conversion to androgens results in the
excretion of elevated amounts of the following C_{19} steroids:
androsterone, etiocholanolone, their 11-oxygenated and hydroxylated
analogues, androstanediol and androstenediol. The identification
of 6α-hydroxyetiocholanolone represents a novel hydroxylated C_{19}
metabolite. Dehydroepiandrosterone, synthesized by the adrenal in
appreciable amounts, in instances of 21-hydroxylase deficiency, was
identified in the urine in trace amounts only. It is therefore
assumed that this steroid is subject to a rapid turnover in the
maternal circulation.

Urinary steroid analysis of the second and third trimester
reflected the suppression of both maternal and fetal adrenal
activity by 9α-fluorocortisol, the drug used in therapy. The
excretion of C_{19} steroids was restored to normal values, but the
dosage of the drug did not suppress adrenal steroid synthesis
entirely as evidenced by the isolation of 17α-hydroxyprogesterone

metabolites notably pregnanetriol, pregnanetriolone and 17α-hydroxy-
pregnanolone in 8 month pregnancy urine.

The drug, which permeates the placental barrier impairs steroid
synthesis in the fetal adrenal, by mimicking the action of cortisol
in reducing pituitary secretion of ACTH with a consequent reduction
in the conversion of cholesterol to pregnenolone in the adrenal.
The most important effect of this block is on the synthesis of
dehydroepiandrosterone sulphate (DHAS) since placental oestrogen
production is dependent, for the most part, during the second and
third trimester on DHAS and 16α-hydroxyated DHA provided by the
foetus, the abnormally low levels of urinary oestrogens isolated at
8 months gestation are a direct result of reduced DHAS synthesis.
Urinary oestriol levels were only 25% of the normal values and the
other oestrogens, oestrone, oestradiol, 16α-hydroxyoestrone, 16-
-oxo-oestradiol, 15α-hydroxyoestriol, were isolated in trace amounts
only. That such low levels of urinary oestriol were not due to a
deficiency in foetal liver 16 hydroxylase was demonstrated by the
normal excretion of 16α-hydroxypregnanolone.An inverse relationship
between the levels of DHAS and 16α-hydroxylated DHA, in foetal blood,
and the dosage of corticoid drug given to the mother has been
reported (9).

The excretion, at 8 month gestation, of a range of hydroxylated
metabolites of progesterone, is attributed to the vast production
of placental progesterone which is subject to different hydroxylation
reactions. Structural analysis by mass spectrometry of the urinary
pregnanolones, pregnanediols, pregnentriols and pregnanediols,
identifies carbons 2, 3, 16, 17, 20 and 21 as the sites of
hydroxylation.

CONCLUSION

Steroid analysis of maternal plasma by capillary column gas
chromatography would provide additional information on the alterations
to steroid synthsis and metabolism in the maternal + foetal adrenal.
The fact that the infant was normal would suggest that the foetus
was protected against the abnormally high amounts of steroids
secreted by the maternal adrenal and that the different hormonal
modifications, induced by pregnancy, were responsible for increased
peripheral metabolism of these steroids.

ACKNOWLEDGEMENTS

We thank Mrs Nicole Pitoizet for her invaluable technical
assistance and Miss Maroussia Bernot for manuscript realization.

This research was supported by grants from Institut National

de la Santé et de la Recherche Médicale (FRA n° 9), from National de
la Recherche Scientifique (ERA n° 267), Délégation Générale à la
Recherche Scientifique et Technique (ACC n° 76.7.1674.01), Université
de Dijon and Ministère des Affaires Etrangères, Direction des Affaires
Culturelles, Scientifiques et Techniques for a Junior Research grant
for one of us (H.L.).

REFERENCES

1) R.J. Bègue, J. Desgrès, J.Ä. Gustafsson and P. Padieu, J. Steroid.
   Biochem., 1976, 7, 211.
2) R.A. Okerholm, C. Chattoraj, J.L. Pinkus, D. Charles and H.H.
   Wotiz, Steroids, 1970, 16, 66.
3) R. Vihko, Acta Endocrinol., 1966, suppl. 109, 29.
4) R.J. Bègue, M. Dumas, H. Losty, M. Moriniere, C. Perrie and P.
   Padieu, in "Proceedings 9th International Symposium on Gas
   Chromatography and Electrophoresis", A. Frigerio and L. Renoz,
   Eds., Riva del Garda, Italy, 1978, in press.
5) R.R. David, C. Bergada and C. Migeon, J. Clin. Endocrinol. Metab.,
   1965, 25, 322.
6) A.M.Bongiovanni and A.W. Root, New Engl. J. Med., 1963, 268, 1283,
   1342 and 1391.
7) L. Viinikka, O. Jänne, J. Perheentupa and R. Vihko, Clin. Chim.
   Acta, 1973, 48, 359.
8) L. Aringer, P. Eneroth and J.Ä. Gustafsson, Steroids, 1971, 17,
   377.
9) F.L. Mitchell and C.H.L. Shackleton, Adv. Clin. Chem., 1969,
   12, 42.

IDENTIFICATION OF ATYPICAL BILE ACIDS BY GAS CHROMATOGRAPHY-MASS

SPECTROMETRY-COMPUTER TECHNIQUES

F. Stellaard and P.A. Szczepanik-Van Leeuwen

Argonne Bioanalytical Center, Argonne National Laboratory

Argonne, Illinois, U.S.A.

INTRODUCTION

One of the major projects in our laboratory is the study of bile acids. In the field of gasteroenterology and pediatrics, there is an increasing interest in the identification of this category of compounds. The importance of abnormal bile acid composition in biological fluids and tissue as a portent of liver disease is not yet well understood. The appearance of certain atypical bile acids may indicate specific metabolic defects in the pathways leading to bile acid formation, whereas abnormal compositions of normal bile acids may be related to circulation or absorption problems.

BIOLOGICAL AND CLINICAL ASPECTS

Bile acids are synthesized from cholesterol in the liver and are secreted into the bile duct through a network of canaliculi and ductules. After the bile acids enter the bile duct, they are transported to the gallbladder where the bile is stored and concentrated. Following the intake of a meal, the gallbladder is stimulated and its contents emptied in the duodenum. Here the bile acids exert their detergent effect through micelle formation, thus creating a situation to solubilize dietary lipids and permitting action of pancreatic lipase. Absorption of the conjugated and unconjugated bile acids takes place in the jejunum and ileum, leaving only a fraction to be excreted in the feces. In Fig. 1 the structures of the four major bile acids in human bile and serum are presented as the free acids. However, in most biological fluids they are conjugated with glycine or taurine through a peptide bond

chenodeoxycholic acid   cholic acid

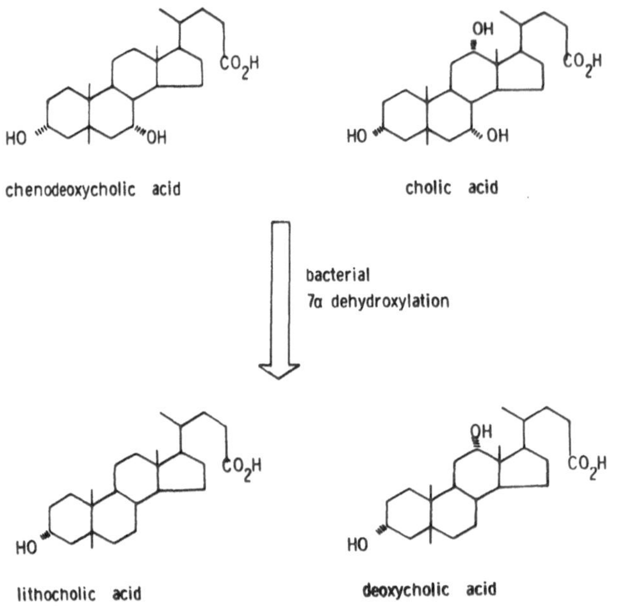

bacterial
7α dehydroxylation

lithocholic acid   deoxycholic acid

FIG. 1. Structures of the major primary and
        secondary bile acids in human bile
        and serum.

at the carboxyl group, and occasionally one or more hydroxyl groups
may be sulfated, predominantly at the 3 position. Recent
investigations (1) indicate that also glucuronidation may occur in
significant amounts in disease states. Of the compounds shown above,
chenodeoxycholic acid and cholic acid are the primary bile acids,
directly synthesized from cholesterol, while lithocholic acid and
deoxycholic acid are their respective metabolites formed through
bacterial 7α dehydroxylation in the colon.

The pathways for the formation of chenodeoxycholic acid and
cholic acid are shown in Fig. 2 as they are currently accepted (2).
They involve two sets of reactions, i.e., the hydroxylation reactions
on the steroid ring (believed to be the result of microsomal
enzyme activity) and the side chain oxidation and cleavage reactions
(believed to take place in the mitochondria).

So far only the metabolism of bile acids in healthy humans has
been discussed. In disease states a number of defects can occur,
related to the synthesis process, to the bile secretion, and to the
bile transport to the gallbladder.

An impaired synthesis in the liver cells may result in a low

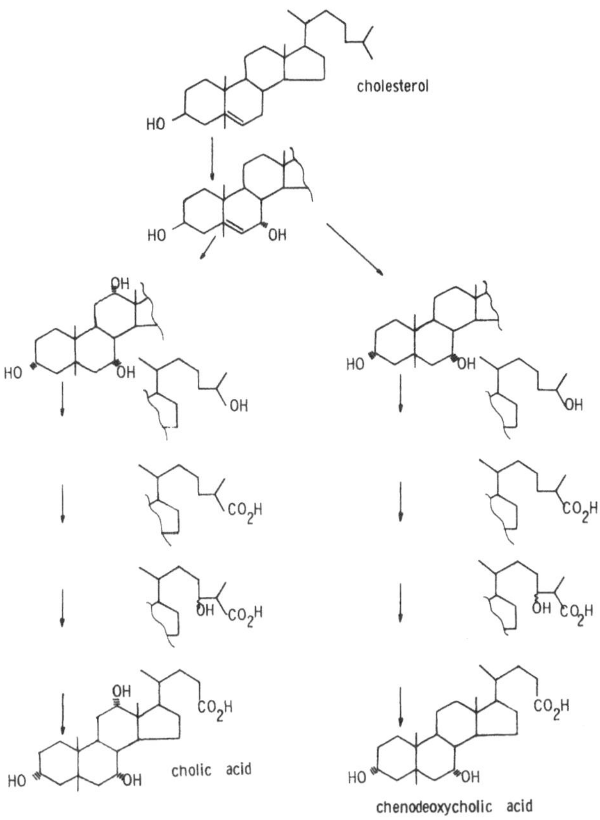

FIG. 2. Pathways for the biosynthesis of chenodeoxycholic
acid and cholic acid in man.

synthesis rate, in an altered pathway, or in a block in the pathway.
An impaired secretion of bile acids into the bile duct will lead
to an accumulation of bile salts in the liver cells. Detoxification
by the liver occurs through enhanced conjugation and thus enhanced
direct excretion into the bloodstream, causing high serum levels of
bile acids. Intrahepatic or extrahepatic obstructions also lead to
elevated bile acid levels in the serum. Gallstone formation may
occur through cholesterol precipitation when cholesterol exceeds
its saturation level in the gallbladder bile. In all these cases
the levels and compositions of bile acids may have a diagnostic
value.

METHODOLOGY

In principle, it is possible to obtain samples from different
sources such as bile, serum, urine, stool and liver tissue. The
preparation of the samples is dependent on the type of sample, but

in all cases the procedure is very extensive.

    In general each procedure consists of the steps discussed
below. An isolation to remove the bile salts from the biological
matrix. This may be done either at the start of the procedure
through a Folch extraction (3) or a petroleum ether extraction, or
it may be performed by a thin layer chromatography step later on
in the procedure (4). A solvolysis step to remove possible sulfate
groups from the steroid ring. This can be carried out with
dimethoxypropane (4), or an acetone/ethanol 9:1 v/v mixture at pH 1
and room temperature (5). A hydrolysis step to cleave the peptide
bond with glycine or taurine. This cleavage occurs under alkaline
conditions and pressure at 120°C (6). In our laboratory 1 N NaOH
is used for this purpose. A newer development in this respect is an
enzymatic hydrolysis process using cholylglycine hydrolase (7).
A derivatization step to volatilize the compound for the GC/MS
analysis. For this purpose the carboxyl groups are methylated using
methanol and dimethoxypropane (1:1 v/v) at pH 1 and room temperature
(4). The hydroxyl groups on the ring system are acetylated using
an acetic acid/acetic anhydride mixture (7:5 v/v) at pH 1 and
0°C (8).

    The final analysis is performed on a BIOSPECT quadrupole mass
spectrometer, interfaced with a VARIAN 1400 gas chromatograph and
a PDP-12 computer or a SIRMID instrument (9), which is a multiple
ion detection device that allows monitoring of up to 6 mass ions in
a single run. The mass fragments are produced by chemical ionization
(CI) using isobutane as the reagent gas. The bile acid methyl ester
acetates are separated on a 1% Poly S-179 on Gaschrom Q column (10).
When atypical compounds are detected, their structures are confirmed
by electron impact mass spectrometry (EI). For this purpose a
PERKIN ELMER 270 double focussing magnetic sector mass spectrometer
is available.

MASS SPECTRA OF BILE ACIDS BY CHEMICAL IONIZATION MASS SPECTROMETRY
(11).

    The CI fragmentation of the methyl ester acetate derivatives
usually results in not more than 2 fragment ions per compound,
which makes the spectra very simple and allows for sensitive
detection (Fig. 3). With the exception of some keto bile acids, a
molecular ion does not exist. The fragments are formed through
cleavage of the acetoxy group only, as very little cleavage of C-C
bonds occurs under CI conditions. Thus the spectra are generally
characterized by the fragments $MH^+ - (n \cdot 60)$, in which 60 stands
for the acetic acid molecule that has been removed in the ionization
process; n is usually 1, 2 or 3.

    The simplicity of the spectra allows for optimal mass separation

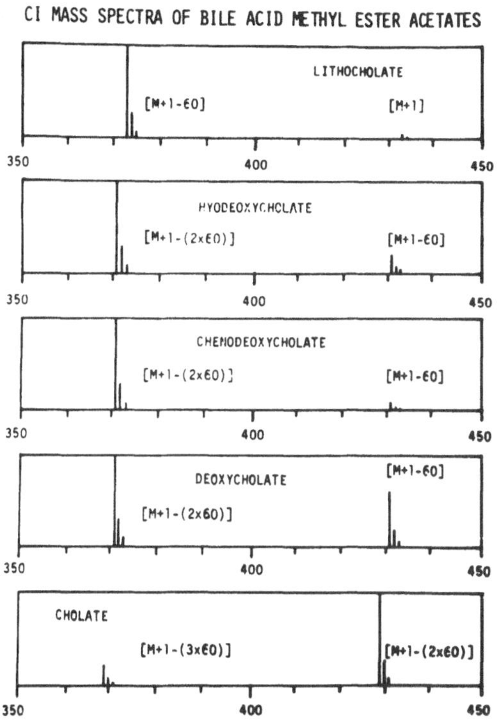

CI MASS SPECTRA OF BILE ACID METHYL ESTER ACETATES

FIG. 3. CI fragmentation patterns for the ethyl
        ester acetate derivatives of some mono,
        di, and trihydroxy bile acids, obtained
        with isobutane as reagent gas.

of the different categories of bile acids according to the number
of hydroxyl groups or keto groups, and the length of the side
chain (Fig. 4). This information, in combination with the GC
retention time data, gives us a powerful tool for screening bile
acid samples by mass chromatographic techniques.

APPLICATION OF THE SINGLE ION MONITORING (SIM) SCREENING PROCEDURE

Clinical problem. In collaboration with Doctors Russel Hanson
of the University of Minnesota and John Watkins of Harvard Medical
School in Boston, we have analyzed bile and urine samples of 5
patients with Zellweger's syndrome (12). One of the characteristics
of this syndrome is an extreme liver dysfunction including bile
duct abnormalities causing cholestasis, structural and functional

CLASS OF COMPOUNDS	M/E
Me C_{24} monoacetoxy	373
Me C_{24} diacetoxy	371, 431
Me C_{24} triacetoxy	369, 429
Me C_{24} monoketo	389
Me C_{24} monoketo monoacetoxy	387, 447
Me C_{24} monoketo diacetoxy	385, 445
Me C_{24} diketo	403
Me C_{24} diketo monoacetoxy	401, 461
Me C_{27} diacetoxy	413, 473
Me C_{27} triacetoxy	411, 471
Me C_{27} tetraacetoxy	409, 469

FIG. 4. Differentiation by m/e value of the
different classes of bile acid methyl
ester acetates.

mitochondrial defects, hepatocellular necrosis and in severe cases,
cirrhosis.

Results. By GC, a number of compounds were detected that have
not been observed in the bile and urine samples of patients with
cholestasis.

After the GC/MS analysis, the mass chromatograms for 20 mass
ions were generated by computer to search for atypical bile acids.
In addition to the normally occurring primary bile acids,
chenodeoxycholic acid and cholic acid, a number of atypical compounds
were detected in the mass ions 409, 471, and 413, representing
tetrahydroxy, trihydroxy, and dihydroxy C_{27} bile acids, respectively
(Fig. 5). Together with the GC retention time data, the compounds
were identified as trihydroxycoprostanic acid (m/e 413), dihydroxycoprostanic acid (m/e 471), and varanic acid (m/e 409) (Fig. 6).

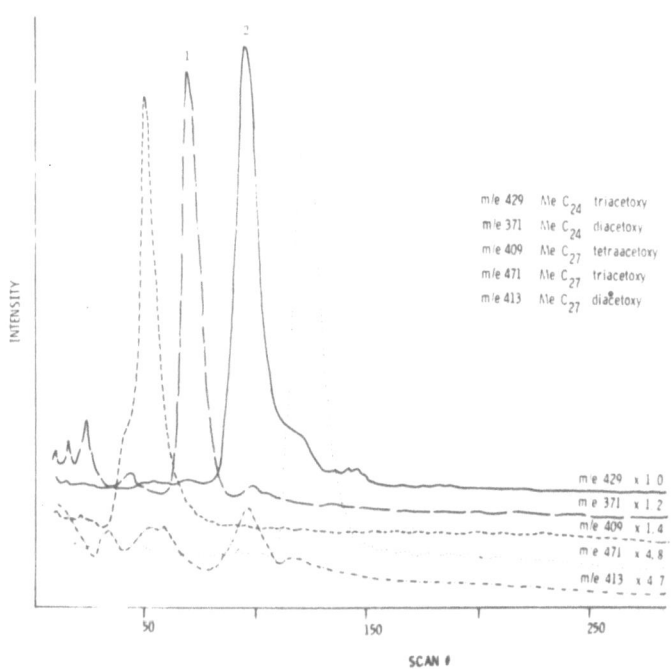

FIG. 5. Mass ion profile of a urine sample of a patient
with Zellweger's syndrome. 1- chenodeoxycholic
acid, 2- cholic acid. Scan speed 10.7 sec/scan.

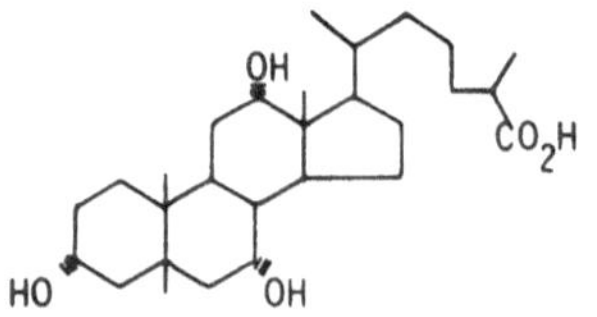

TRIHYDROXYCOPROSTANIC ACID

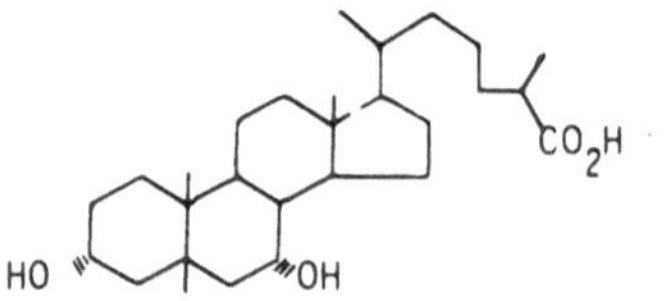

DIHYDROXYCOPROSTANIC ACID

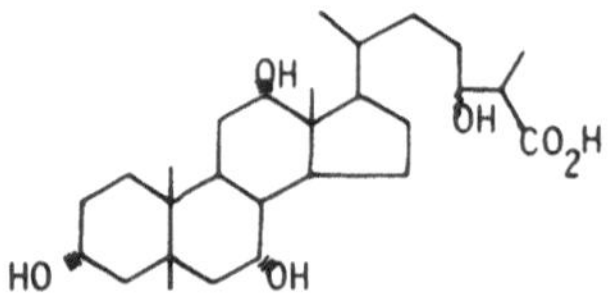

VARANIC ACID

FIG. 6. Structures of trihydroxycoprostanic acid,
dihydroxycoprostanic acid, and varanic acid
identified in the urine of patients with
Zellweger's syndrome.

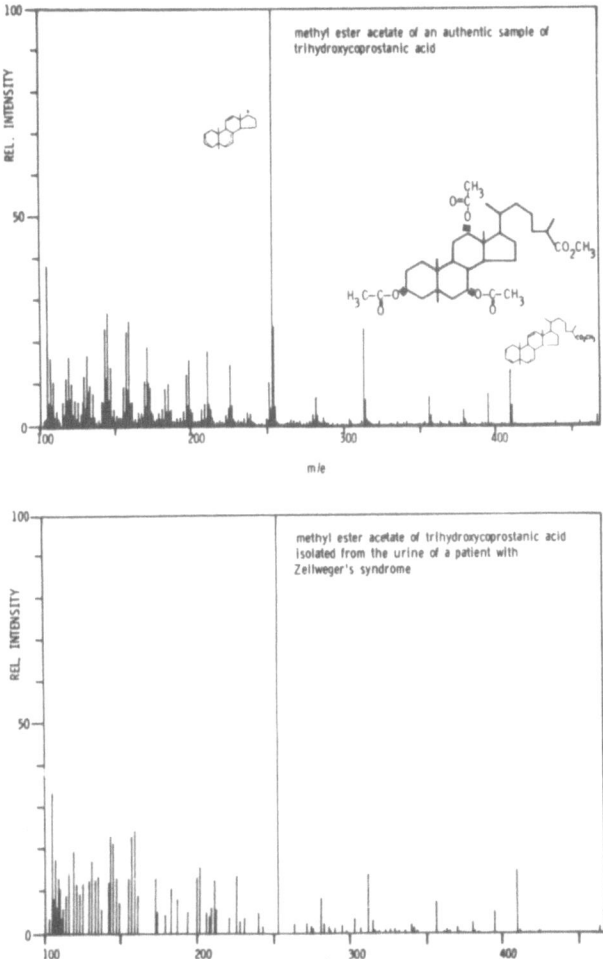

FIG. 7. Structure confirmation of trihydroxycoprostanic acid detected in urine through
comparison of the EI spectra with that of
an authentic sample.

The spectra of varanic acid and trihydroxycoprostanic acid
have been confirmed with EI mass spectrometry. The spectra for
trihydroxycoprostanic acid isolated from urine and an authentic
sample are shown in Fig. 7. Mass ions at m/e 253 (bare steroid
skeleton), 313, 356, 379, 395, and 410 (skeleton + side chain) are
identical and give evidence for the structure.

CONCLUSION

The bile acids identified are rarely detected in human specimens although they supposedly play a role in the pathways as shown in Fig. 2. Trihydroxycoprostanic acid and varanic acid are intermediates in the cholic acid synthesis, while dihydroxycoprostanic acid appears in the chenodeoxycholic acid pathway. Since the oxidation and cleavage of the side chain are known to take place in the mitochondrial part of the liver cell, the results indicate that there is a correlation between the appearance of C_{27} bile acids and the mitochondrial defects observed.

REFERENCES

1) W. Fröhling, A. Stiehl, et al., in "Bile Acid Metabolism in Health and Disease", G. Paumgartner and A. Stiehl, Eds., MTP Press Limited, Lancaster, 1977, p. 101.
2) R.F. Hanson and M. Pries, Gastroent., 1977, 73, 611.
3) J. Folch, M. Lees and G.H. Sloane-Stanley, J. Biol. Chem., 1957, 226, 497.
4) S.S. Ali and N.B. Javitt, Can. J. Biochem., 1970, 48, 1054.
5) R.H. Palmer and M.G. Bolt, J. Lipid Res., 1971, 12, 671.
6) D.H. Sandberg, J. Sjövall, K. Sjövall and D.A. Turner, J. Lipid Res., 1965, 6, 182.
7) O.J. Roseleur and C.M. van Gent, Clin. Chim. Acta, 1976, 66, 269.
8) J. Roovers, E. Evrard and H. Vanderhaeghe, Clin. Chim. Acta, 1968, 19, 449.
9) P.D. Klein, J.R. Haumann and D.L. Hachey, Clin. Chem., 1975, 21, 1253.
10) P.A. Szczepanik, D.L. Hachey and P.D. Klein, J. Lipid Res., 1978, 19, 280.
11) P.A. Szczepanik, D.L. Hachey and P.D. Klein, J. Lipid Res., 1976, 17, 314.
12) R.F. Hanson, P.A. Szczepanik, et al., Clin. Res., 1978, 26, 320.

THE EFFECTS OF TRIISOPROPYLACETIC ACID ON THE METABOLIC PROFILE OF

URINARY ORGANIC ACIDS

T.Kuhara,Y.Hirokata,S.Yamada,I.Matsumoto,J.A.MacPhee° and

J.-E.Dubois°;Kurume University School of Medicine,Kurume,

Fukuoka,Japan;°Université Paris VII, Paris, France

INTRODUCTION

Metabolic profiling, introduced by Horning and Horning (1) for
the analysis of complex mixtures of physiological fluids, is now
mostly applied to screen metabolic disoders. Gas chromatography-mass
spectrometric techniques have proved to be one of the most suitable
methods for these purposes. Recent advances in this field made it
possible not only to screen metabolic disorders easily and precisely
but also to resolve many unknown relationships between an original
defect by gene mutation and secondarily caused biochemical or
clinical abnormalities. As the biological reactions are so closely
interrelated and one compound may act on multiple reactions, it is
advisable to analyze as many endogenous compounds as possible in
biological samples. Metabolic profiling, in this sense, is useful
not only in the study of inborn errors of metabolism but also in
the study of drug action in humans. For drugs acting on the central
nervous system, such as anesthetics, hypnotics or psychotropic drugs,
metabolic profiling in the brain is desirable. It is our final goal
to understand the action mechanism of these drugs by the analysis
of metabolic profiles in brain tissue or cerebrospinal fluid. In
this respect we have only begun to see normal profiles in brain
tissue from rats or cerebrospinal fluid from humans. At present,
however, profiling of organic acids or amino acids from serum and
urine have been established.

For the past several years, we have been studying the metabolism
of sodium dipropylacetate (2-4). Concerning the action mechanism,
Godin et al. (5) showed the increase in γ-aminobutyric acid (GABA)
level by DPA in rats. In their experiments with brain homogenate
of rats, DPA inhibited the GABA metabolizing enzyme, 4-aminobutyrate:

307

2-oxoglutarate aminotransferase. However, the inhibitory concentration
was too high to be considered physiologically significant.

In order to investigate the effect of DPA in the human body, we
studied the metabolism in normal adults. In urine, within 2 hrs
after administration of a single therapeutic dose, we detected
considerable amounts of β-oxidative-intermediates in addition to
the glucuronide conjugate and ω-oxidation products. We also analyzed
urinary acids from a DPA-treated patient. The gas chromatograms are
shown in Fig. 1.

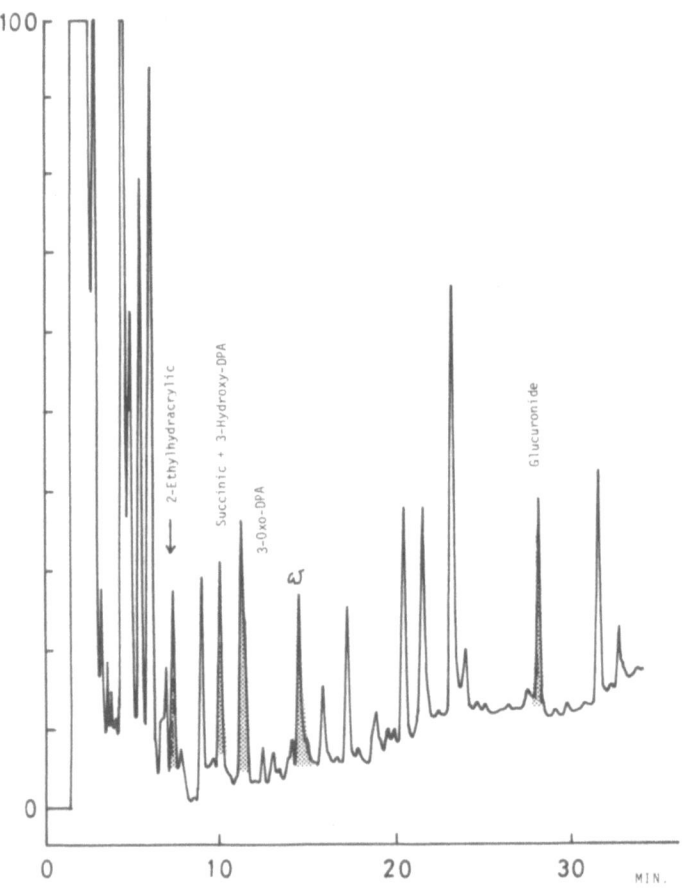

FIG. 1. Gas chromatogram of urinary acids from a DPA-
-treated patient. ω; 2-n-propylglutaric acid
(ω-oxidation product). A marked excretion of
2-ethylhydracrylic acid is clearly shown. The
peak is hardly detected in control urine.

Recently Gompertz et al. (6) reported his experience that urinary
metabolites of DPA in man are potential sources of interference
with organic acid screening procedures. In this sense, metabolic
study of drugs in detail is useful also for screening inborn errors
of metabolism.

As shown in Fig. 1, we confirmed the marked increase in 2-ethyl-
-hydracrylate, in urine of DPA-treated child. 2-Ethylhydracrylate
was identified as a minor but normal constituent of human urine in
1974 by Mamer et al. (7). It was also demonstrated as a catabolic
intermediate of isoleucine (8). Accordingly DPA competes with 2-
-methylbutyryl CoA for the enzyme system responsible for the β-
-oxidation. It is believed that branched chain amino acids are
catabolized in mitochondria of extrahepatic tissues and serve as
energy sources (9). From these reasons, it is clear that DPA enters
mitochondria. The reason why various enzymic reactions in mitochondria
are not confused seems to be rapid metabolism of DPA by β-oxidation.
Also rapid conversion into water-soluble compounds by either ω-
-oxidation or glucuronide conjugation may decrease its distribution
in mitochondria.

In an attempt to see the effect of branched chain fatty acids,
we synthesized a compound which seems difficult to undergo biotransformation. The compound, triisopropylacetic acid, cannot be metabolized via β-oxidation from its high branching. This compound, however,
is expected to enter mitochondria because it is lipid soluble, small
in molecular weight and neutral as well as acidic like DPA. It may
affect the mitochondrial function directly as TIPA is not metabolized
in mitochondria, and this may be reflected on the profiles of urinary
organic acids. In the present study we describe the effect of TIPA
on the profiles of urinary organic acids.

$$CH_3-CH_2-CH_2 \diagdown$$
$$CH-COOH$$
$$CH_3-CH_2-CH_2 \diagup$$

$$(CH_3)_2-CH \diagdown$$
$$(CH_3)_2-CH-C-COOH$$
$$(CH_3)_2-CH \diagup$$

Dipropylacetic acid (DPA)   Triisopropylacetic acid (TIPA)

MATERIALS AND METHODS

Animals. Male Wister albino rats weighing 200 ± 20 g were used.
Aqueous TIPA solution was prepared with 1 N Na_2CO_3 and adjusted to
pH 8, 9 and 10. In some experiments TIPA was suspended in 1% Tween
80. Fifteen rats were loaded with 6 mg per kg body weight of TIPA
by stomach intubation after 14 or 38 hr fasting. 24-Hour urine
specimens were collected individually from rats in metabolic cages.

Bacterial growth was prevented by adding toluene (0.05 ml) to the
collecting flasks. The 24 h urine specimens from rats receiving 0.9%
NaCl, 1% Tween-80 or Na_2CO_3 solution (pH 8, pH 9 or pH 10) were
used as controls.

Extraction. Two ml of 24 hr urine made up to 15 ml with
deionized water was taken into a centrifuged tube and acidified to
pH 1 with 6 N HCl. After adding 1 g of sodium chloride, and 80 µg
of glutaric acid and nonadecanoic acid as internal standards, the
solution was finally made up to 5 ml with deionized water. The
extraction was done twice with 10 ml of ethyl ether and ethyl acetate,
successively. The combined organic phase was then dehydrated with
anhydrous sodium sulphate and evaporated to dryness. The methoxime
TMS derivatives were prepared by addition of 10 mg methoxylamine
hydrochloride in 100 µl of pyridine and then 200 µl of N,O-bis-tri-
methylsilyltrifluoroacetamide. One or two µl of the reaction mixture
was injected onto GLC or GC-MS.

Gas chromatography and gas chromatography-mass spectrometry.
Gas chromatography was performed by JEOL JGC 1100 equipped with
FID. Glass column packed with 3% OV-17 on 80 - 100 mesh Gas Chrom Q
was used. The column temperature was programmed at 6°C from 80°C
per min. GLC peak area relative to an internal standard; nonadecanoic
acid was determined with an on-line Shimazu Chromatopac 4B. GC-MS
analysis was carried out by JEOL JMS D-100 on-lined to JEC 2000
data system. The separator was maintained at 280°C and the ion source
had an electron energy of 75 eV.

RESULTS AND DISCUSSION

The administered triisopropylacetic acid was detected in the
urine. TMS ester of TIPA appears after the succinate peak and has
methylene unit value of 14.25. In the mass spectrum of TIPA, the
molecular ion peak is not observed but the peak at $/M-15/^+$ at m/e
243 is clearly detected. The McLafferty rearrangement peak is also
found at m/e 216. As compared with the control, remarkable increase
in several peaks were observed in the gas chromatogram of TIPA-
-treated rats. GC-MS analysis of these peaks was done in order to
investigate whether the increase is due to a normal component or
an abnormal compound. The mass spectrum of the abnormally elevated
peak, which elutes immediately after methyladipate is shown in Fig.
2. The molecular ion is not observed but the peak at m/e 349 is
found clearly which is due to the favored loss of methyl group
linked to silicon in the molecular ion. The ion at m/e 247 corresponds
to $/M-COOTMS/^+$ and the ion at m/e 321, $/M-43/^+$ is produced by the
loss of carbon oxide through rearrangement decomposition of cyclic
oxonium $/M-15/$ ion (10). From these observations, we estimated
the compound as 2-hydroxyglutaric acid tri-TMS and confirmed this by
comparison with authentic mass spectra (11). The mass spectrum of the
other large peak is shown in Fig. 3. This is assigned to 2-oxo-

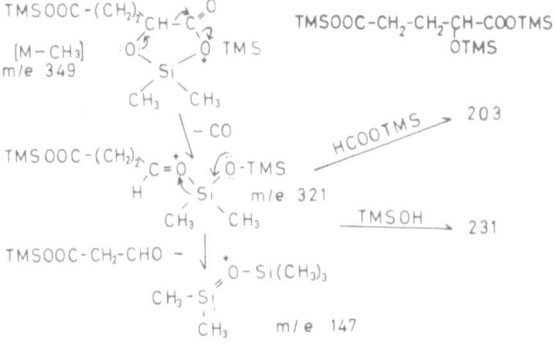

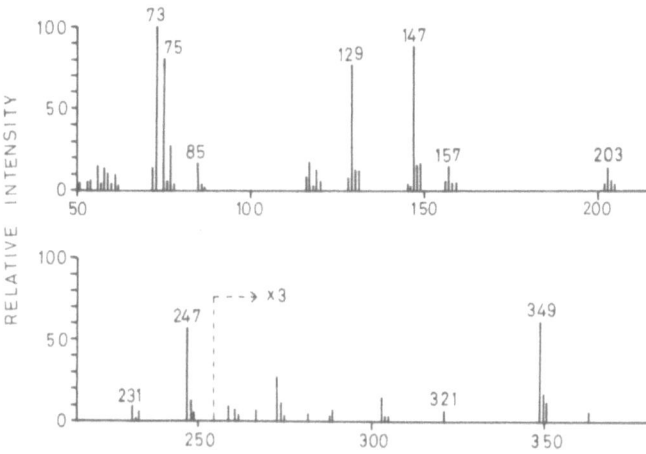

FIG. 2. Mass spectrum of TMS derivative of 2-
-hydroxyglutaric acid obtained from
TIPA-treated rats.

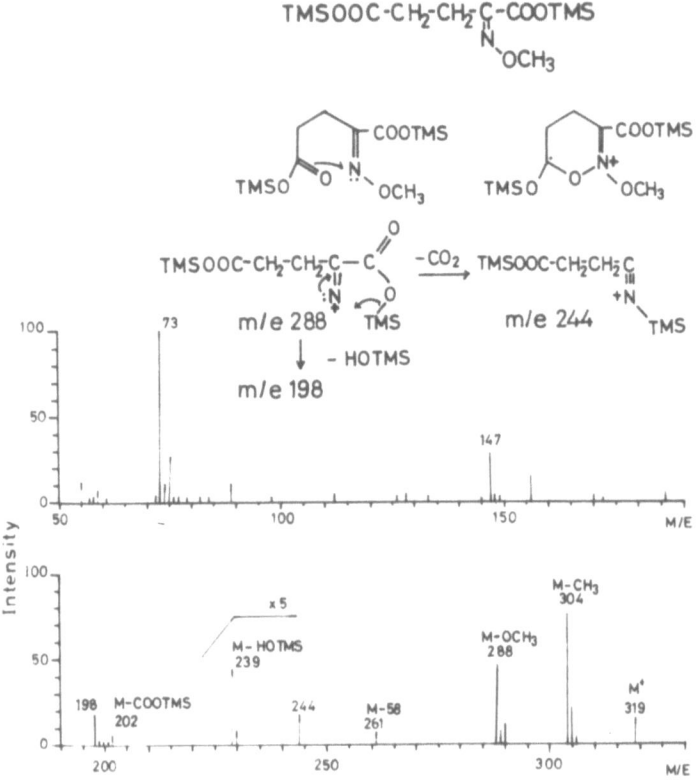

FIG. 3. Mass spectrum of methoxime TMS ester of
2-oxoglutaric acid obtained from TIPA-
-treated rats.

glutarate methoxime TMS ester. The highest mass number at m/e 319
due to molecular ion is found significantly. It is stabilized by
the formation of a cyclic 6-membered transition state (12). Loss of
methoxy radical from methoxime gives ion at m/e / M-31_7. From the
ion at m/e 288, carbon dioxide is eliminated to give the ion at
m/e 244. The ions due to loss of trimethylsilanol or carbotrimethyl-
silyloxy group are also found at m/e 229 or m/e 202.

By GC-MS analysis the increase in succinate and citrate was
demonstrated in addition to huge peaks due to 2-hydroxyglutarate and
2-oxogluturate. All of them are the intermediate of TCA cycle. The
increase in 2-oxogluturate was always shown in repeated experiments
and this effect induced by TIPA was most remarkable. The data are
summarized (in Table 1). As compared with the average value of

TABLE 1.
Increased urinary acids in TIPA-treated rats.

	Control	TIPA-Treated	
Citrate	100	180	(6 fold)¨
2-Oxogluturate	100	281	(7 fold)¨
2-Hydroxygluturate	100	247	(5 fold)¨
Succinate	100	202	(4 fold)¨

n=10   n=15

¨Maximum Increase

control (10 rats) urinary excretion of 2-oxoglutarate was elevated three-fold, and maximum level was seven-fold. Increase of succinate was two-fold on the average and maximum value was four-fold. As for citrate, the increase was two-fold but the highest value was six-fold with respect to the control. Massive excretion of 2-oxoglutarate in a patient with a deficiency of 3-methylcrotonyl CoA carboxylase was reported (13) but the mechanism of the increase of 2-oxoglutarate was not clear.

The increase in 2-oxoglutarate does not seem to be due to inhibition of 2-oxoglutarate dehydrogenase because succinate and other intermediates of TCA cycle were also elevated. It seems to be due to stimulation of the TCA cycle as a whole. In the cycle the enzyme responsible for the formation of 2-oxoglutarate has a characteristic nature. The oxidation of isocitrate to 2-oxoglutarate is catalyzed by NAD-linked isocitrate dehydrogenase in mitochondria. The NAD-linked enzyme is an allosteric enzyme that requires ADP as a specific effector (14). Isocitrate dehydrogenase does not appear to be the primary rate-controlling step in the liver: however, it may be involved in a secondary regulatory role. In other tissues such as muscle, it is believed to act as the primary regulatory site.

From this characteristic nature of the enzyme, the decrease in ATP concentration seems to be the reason for the increase in 2--oxoglutarate followed by concomitant increase in its reduced form, 2-hydroxyglutarate, observed in the present study.

As is well known, electron transport from NADH formed in the TCA cycle to oxygen is the direct source of the energy used for the coupled phosphorylation of ADP. Uncoupling agents allow electron transport to continue but prevent the phosphorylation of ADP to ATP. Lehninger et al. showed that fatty acids were able to uncouple oxidative phosphorylation. Wojtczak and Lehninger (15) showed the presence of an endogenous factor capable of uncoupling oxidative phosphorylation and causing the swelling of mitochondria. They concluded that this uncoupling factor is a fatty acid or a mixture of fatty acid since these effects can be replaced by fatty acids and can be prevented by serum albumin which is capable of binding fatty acid. Accordingly, it is reasonable to consider that TIPA is capable of uncoupling oxidative phosphorylation, and that the decrease of ATP formation accelerates TCA cycle, especially iso-citrate dehydrogenase activity.

There is still much doubt as to the exact mechanism by which barbiturates produce their pharmacological effects. Various experimental data showing the site of action for the barbiturates have been accumulated. However, no enzyme which can be specifically inhibited by a reasonable concentration of the drug has been reported. Brody and Bain (16) demonstrated that barbiturates are effective uncoupling agents. At present a number of chemical agents are known as uncoupling factors. Most of them are lipid soluble substances containing an acidic group.

In summary TIPA causes a marked excretion of 2-oxoglutarate and its reduced form, 2-hydroxyglutarate in urine of TIPA-treated rats. These findings together with concomitant increase in other intermediates of the TCA cycle suggest that TIPA acts as an un-coupling agent in mitochondrial oxidative phosphorylation. In vitro experiments are now under study.

REFERENCES

1) E.C. Horning and M.G. Horning, Clin. Chem., 1971, 17, 802.
2) T. Kuhara and I. Matsumoto, Biomed. Mass Spectrom., 1974, 1, 291.
3) T. Kuhara, Y. Iwai, S. Haraguchi, T. Shinka and I. Matsumoto, in "Recent Developments of Mass Spectrometry in Biochemistry and Medicine", A. Frigerio, Ed., Plenum Press, New York, 1978, p.191.
4) I. Matsumoto, T. Kuhara and M. Yoshino, Biomed. Mass Spectrom., 1976, 3, 235.
5) Y. Godin, L. Heiner, J. Mark and P. Mandel, J. Neurochem., 1969, 16, 869.
6) D. Gompertz, P. Tippett, K. Bartlett and T. Baillie, Clin. Chim. Acta, 1977, 74, 153.
7) O.A. Mamer and S.S. Tjoa, Clin. Chim. Acta, 1974, 55, 199.
8) O.A. Mamer and S.S. Tjoa, Biomed. J., 1976, 160, 417.
9) L.L. Miller, in "Amino Acid Pools, Distribution, Formation and Function", J.T. Holden, Ed., New York, 1962, p. 708.
10) G. Petersson, Tetrahedron, 1970, 26, 3413.
11) S.P. Markey, W.G. Urban and S.P. Levine, Eds., "Mass Spectra of Compounds of Biological Interest", National Technical Information Service, Virginia, 1974.

12) A.M. Lowson, R.A. Chalmers and R.W.E. Watts, Biomed. Mass
    Spectrom., 1974, 1, 199.
13) M.D.A. Finnie, K. Cottrall, J.W.T. Seakins and W. Snedden,
    Clin. Chim. Acta, 1976, 73, 513.
14) A.L. Lehninger, in "Biochemistry", 2nd ed., Worth Publishers
    Inc., New York, 1975, p. 456.
15) L. Wojtczak and A.L. Lehninger, Biochem. Biophys. Acta, 1961,
    51, 442.
16) T.M. Brody and J.A. Bain, Proc. Soc. Exp. Biol. Med., 1951, 77,
    50.

APPLICATION OF CHEMICAL IONIZATION MS TO THE STUDY OF REGIONAL BRAIN

CATECHOLAMINE METABOLISM IN (S)-α-METHYLDOPA TREATED RATS

D. Karashima, R.L. Cockerline, K.L. Melmon and

N. Castagnoli, Jr., University of California,

San Francisco, California, U.S.A.

INTRODUCTION

Following the early studies of Oates and co-workers (1) establishing the antihypertensive properties of (S)-α-methyldopa $/$ (S)-α-MD $/$, many workers have investigated the mechanistic features underlying the pharmacological properties of this clinically useful drug (2-4). The currently most widely accepted hypothesis, the central false neurotransmitter theory, proposes that (S)-α-MD is transported into the brain where it undergoes enzymatic decarboxylation to α--methyldopamine (α-MDA) which is oxidized to α-methylnorepinephrine (α-MNE). These metabolites displace the endogenous central catecholamines, dopamine (DA) and norepinephrine (NE), and act as "central false neurotransmitters" (5).

Assessment of the interactions of (S)-α-MD and its amine metabolites with endogenous central biogenic amines has often relied on analysis of these compounds using somewhat insensitive and non-specific fluorometric assays (6-8). While these approaches have proved useful in the past, limitations in specificity have encouraged workers in this field to develop more sensitive and specific assays which provide more reliable quantitative estimations of catecholamines (9, 10).

We have employed chemical ionization mass spectrometric techniques to study (S)-α-MD metabolism in an effort to obtain structural information (11) and, with the aid of stable isotopically labelled internal standards, quantitative information relating to the mode of action of this drug in the brain (9, 12). This report summarizes our studies on the quantitative relationships between

brain catecholamine and (S)-α-MD levels and blood pressure responses
in normotensive rats treated with (S)-α-MD.

METHODS

Male Sprague-Dawley normotensive rats were infused with (S)-α-
-MD in normal saline solution at a constant rate for 24 to 30 hours
(5-30 mg/kg per hour) via a cannula in the right jugular vein. The
blood pressure (B.P.) in the left carotid arterial catheter was
monitored with a Statham pressure transducer interfaced with a Grass
polygraph. After infusion, rats were sacrificed by decapitation and
the corpus striatum (C.S.), hypothalamus, and brain stem (B.S.)
were isolated according to the method of Glowinski and Iversen (13)
and weighed. Internal standards (Fig. 1), synthesized in our
laboratory, and sodium metabisulfite solution were added to the
tissues.

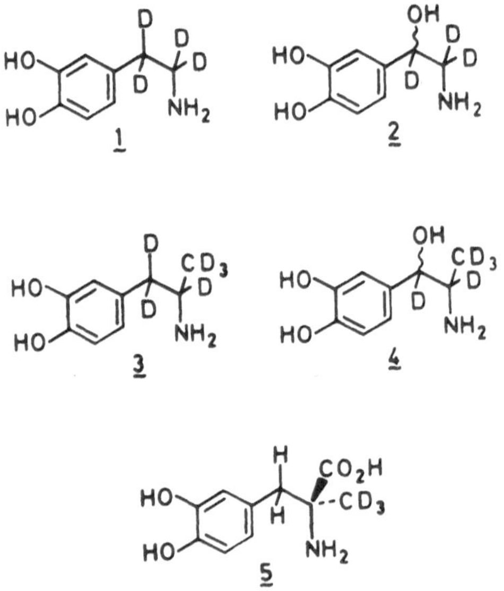

Fig. 1. Deuterated internal standards used in the CIMS
analysis of (S)-α-MD and catecholamines in the
rat brain. 1= dopamine-d_4; 2= Norepinephrine-d_3;
3= α-methyldopamine-d_6; 4= α-methylnorepinephrine-
-d_5; 5= (S)-α-methyldopa-d_3.

After homogenization of the tissues in 5% trichloroacetic acid and
centrifugation, the decanted supernatants were purified by ion
exchange chromatography. The desired amines and amino acid were
eluted with ethanolic HCl from the sodium and hydrogen forms,
respectively, of Dowex-50 columns and were derivatized according to
the scheme outlined in Fig. 2.

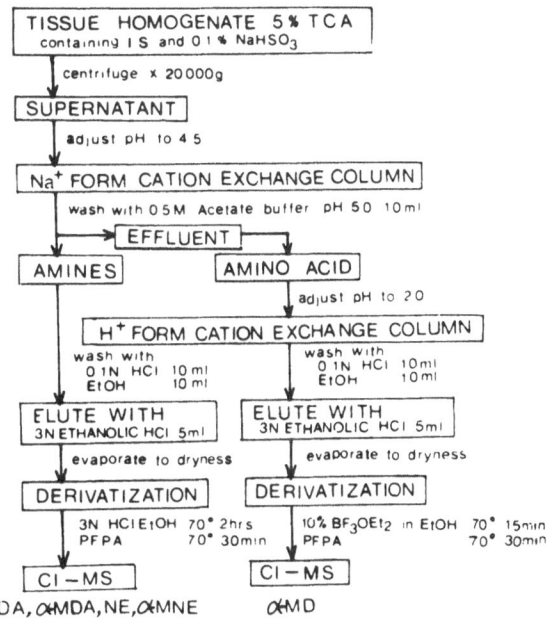

FIG. 2. Scheme of analysis of (S)-α-MD and the
       catecholamines. ("I.S. = Internal Standard).

The derivatives (Fig. 3), dissolved in pentafluoropropionic anhydride
(PFPA), were analyzed (direct insertion probe technique) for deuterated and non-deuterated amines and (S)-α-MD by multiple scans on an
AEI MS-902 modified for chemical ionization (CI) mass spectrometry
(14). The reagent gas was isobutane at 0.4 torr, and the probe
temperatures were 150°C for the amines and 180°C for (S)-α-MD.

RESULTS AND DISCUSSION

The tris-pentafluoropropionyl derivatives of α-MDA and DA gave
abundant protonated molecular ions (MH+) at m/e 606 and 592, respect-

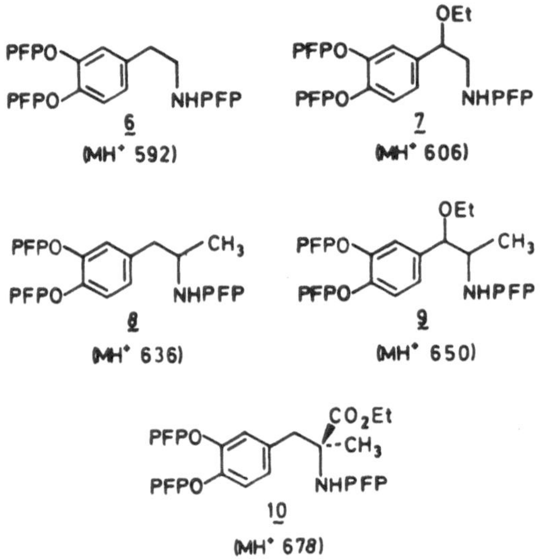

FIG. 3. Structures and m/e values for derivatives of
         (S)-α-MD and the catecholamines: 6= tris-(PFP)-
         -DA; 7= tris-(PFP)-NE β-ethyl ether; 8= tris-
         -(PFP)-α-MDA; 9= tris-(PFP)-α-MNE β-ethyl ether;
         10= tris-(PFP)-(S)-α-MD ethyl ester.

ively, and those of the β-ethyl ethers of α-MNE and NE at m/e 650
and 636, respectively. The corresponding pentafluoropropionyl
derivative of (S)-α-MD ethyl ester displayed a good MH$^+$ ion at m/e
678 (Fig. 3). These high molecular weight derivatives proved
satisfactory in terms of volatility and absence of significant
background ions in the mass region of interest.

   Fig. 4 shows the isobutane CI mass spectrum of the derivatized
internal standards and the amines obtained from the hypothalamus
of an (S)-α-MD treated rat. No interfering ions within the mass
region of interest were observed except at m/e 593. This nominal
mass corresponds to the fragment ion arising from the loss of
ethanol from the protonated pentafluoropropionyl derivative of
norepinephrine-d$_3$ β-ethyl ether. Since the internal standard for
NE contains about 10% NE-d$_2$, we calculated the NE-d$_3$/NE-d$_2$ ratio
and used this information to correct the m/e 592 peak height for the
determination of DA. The m/e 588 peak shown in Fig. 4 has not been
identified but is not derived from the tissues. Standard curves
constructed for the analysis of the five compounds were linear in
the ranges used. The recoveries of the amines were between 70% and
80%. The recovery of (S)-α-MD was about 50%. Our control values for

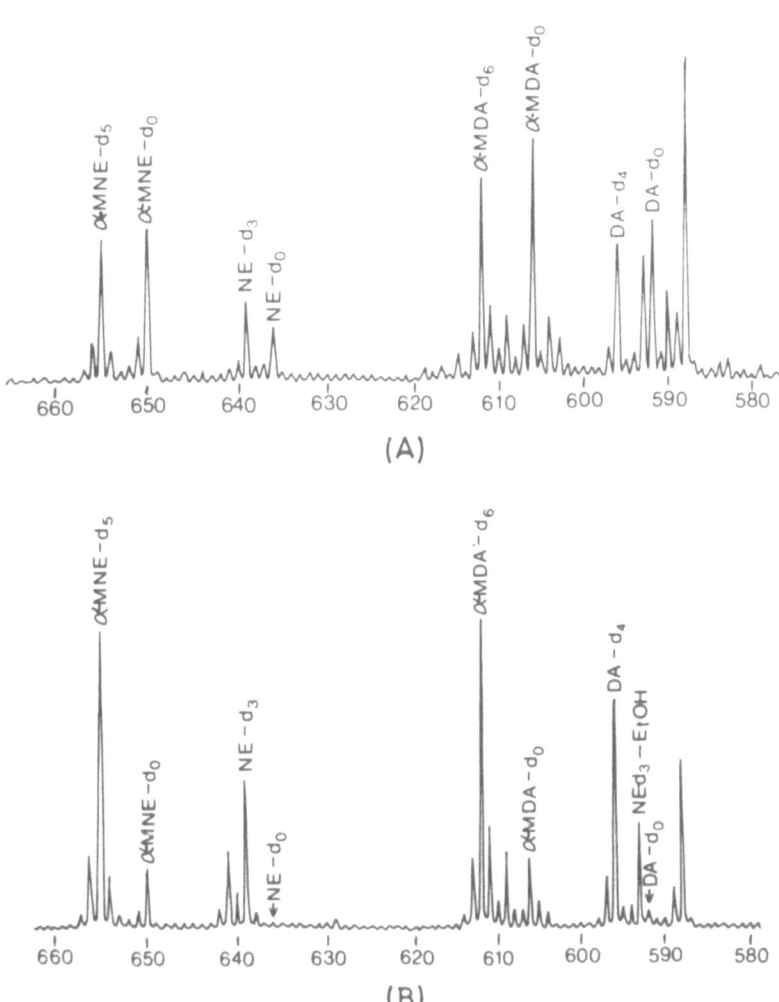

FIG. 4. Partial isobutane CI-mass spectrum of (A) a
        mixture of catecholamines and their deuterated
        internal standard and (B) amines isolated from
        the hypothalamus of a rat treated with (S)-α-MD
        (25 mg/kg per hour) for 24 hours. All samples
        were passed through ion exchange columns.

DA and NE are in agreement with literature values (9, 15).

    After 2 hours of constant infusion of (S)-α-MD the blood pressure,
following its initial increase, began to devrease rapidly until it
stabilized after 6 hours of treatment. Although the blood pressure
appeared to remain constant between 6 and 8 hours, a small decrease
was usually observed over the next 16 hours (Fig. 5).

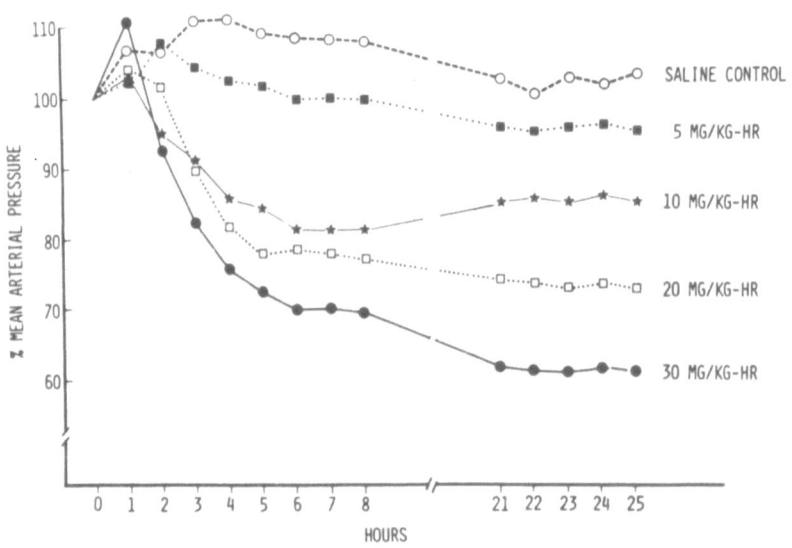

FIG. 5. Blood pressure response during the constant
        intravenous infusion of (S)-α-MD. Drugs were
        dissolved in saline and infused via right
        jugular venous catheter. The infusion rate
        was 0.85 ml/hr. Blood pressure is expressed
        as the % mean arterial pressure compared to
        the pretreatment value. Data shown are mean
        values of 8 rats.

Except in rats given 30 mg/kg of (S)-α-MD per hour, the final blood
pressure was not different statistically from the 8-hour blood
pressure levels (t-test for difference between means, P > 0.10).

    Fig. 6 shows the levels of (S)-α-MD, α-MDA, and α-MNE in the
hypothalamus of rats treated with 25 mg/kg of (S)-α-MD per hour for
8, 21, 24, and 30 hours. The levels of amines at these times were

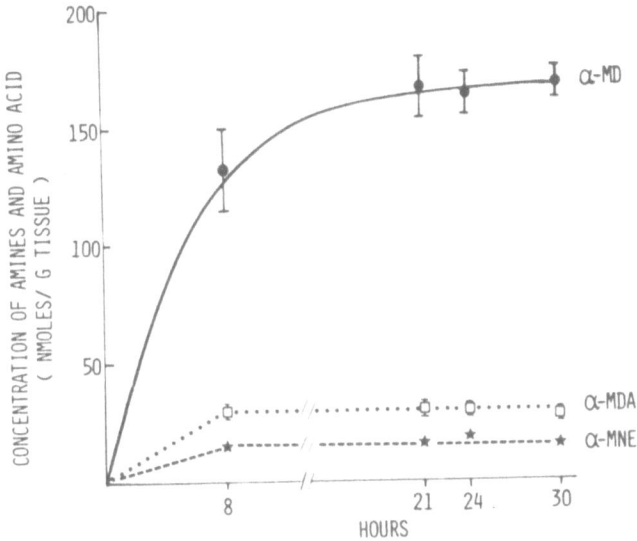

FIG. 6. Concentration of (S)-α-MD and its metabolites
        measured in rat hypothalamus after different
        durations of infusion of (S)-α-MD. Data are
        shown as mean ± S.E.M. of 6 rats.

not significantly different from each other (t-test, p > 0.10).
However, (S)-α-MD levels continued to increase until 21 hours, after
which they became constant. Similar results were found for tissue
levels of (S)-α-MD and its amine metabolites in the brain stem and
corpus striatum. Therefore, steady-state levels of α-MNE appeared
to be reached by 8 hours and of (S)-α-MD by 21 hours. We have
assumed that steady-state levels will be reached by these times for
all doses of (S)-α-MD. Based on these considerations, it was decided
to study the hypotensive responses of (S)-α-MD infused for 24 hours
(steady-state conditions). Good correlation between the blood pressure
response and the 24-hour infusion dose of (S)-α-MD was observed
(Fig. 7).

      Fig. 8 shows the concentrations of amines in various brain
regions following 24-hour infusion of different doses of (S)-α-MD.
The DA concentration in the corpus striatum decreased in a dose-
-dependent manner. On the other hand, the α-MDA concentration
increased in the corpus striatum and hypothalamus. At the higher
infusion doses, the levels of the α-methylated compounds were
significantly higher than the control levels of the endogenous

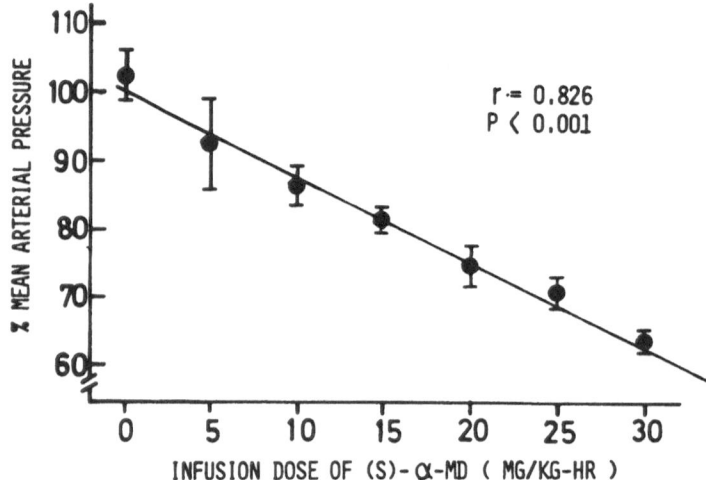

FIG. 7. Per cent mean arterial blood pressure change
versus infusion dose of (S)-α-MD. Mean arterial
pressure shown is percentage of the pretreatment
control value. Data shown are mean ± S.E.M. of
8 rats.

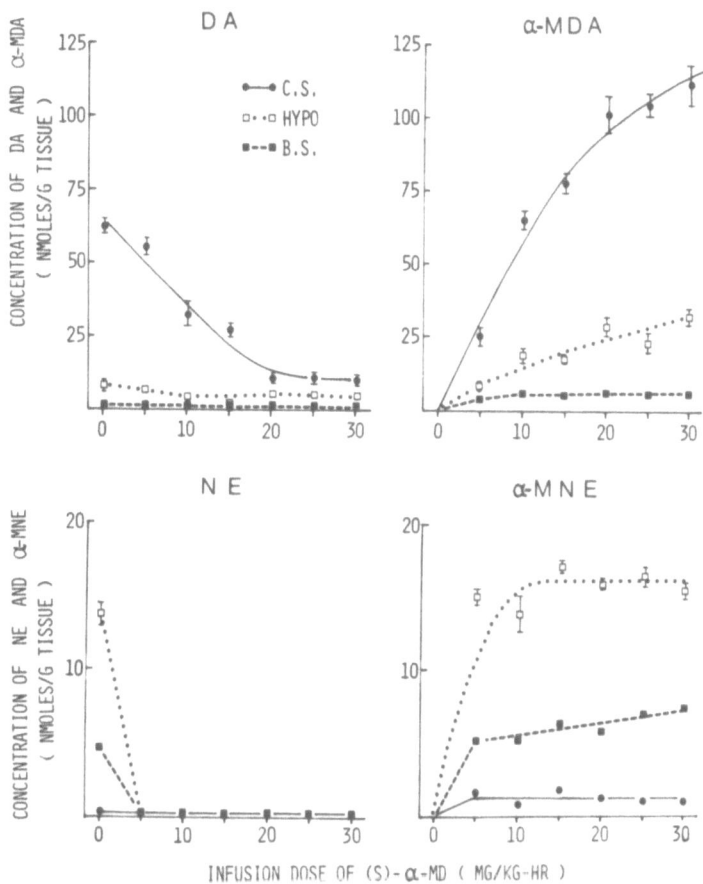

FIG. 8. Concentrations of amines in various brain regions
        following constant intravenous infusion of different
        doses of (S)-α-MD (24 hours). Data are shown as
        mean ± S.E.M. of 6 rats.

catecholamines. The brain stem DA levels did not change significantly
when (S)-α-MD was administered; the levels of α-MDA attained were
constant regardless of the doses of (S)-α-MD given. NE concentrations
in the hypothalamus, brain stem, and corpus striatum fell to
essentially zero even at the lowest infusion dose of (S)-α-MD (5
mg/kg per hour). Concentrations of α-MNE reached a maximum level at
relatively low infusion doses.

   This replacement of endogenous central catecholamines with
α-methylated amines following administration of (S)-α-MD has been
reported previously (4, 7, 8). The mechanistic significance of this
replacement is still not completely understood. With the hope of
differentiating between the blood pressure regulating effects of
α-MDA and α-MNE more clearly, we compared the levels of amines in the
corpus striatum, hypothalamus, and brain stem with the blood pressure
responses. Figs. 9-11 show that reasonably good linear correlations
were found between the blood pressure responses and the 24-hour
levels of DA, α-MDA, and (S)-α-MD in the three brain regions studied.

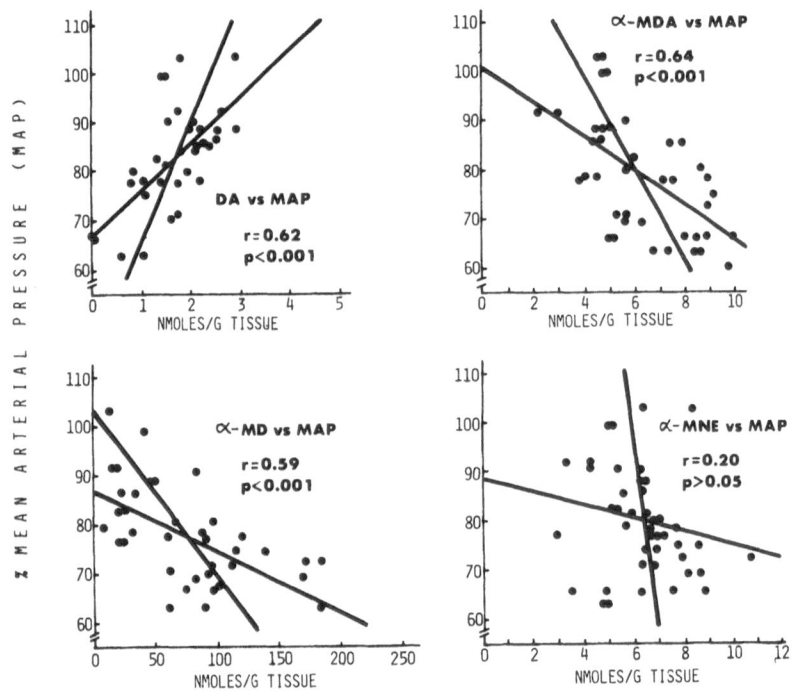

FIG. 9. Scatter plot of concentration of DA, α-MDA,
        α-MNE and (S)-α-MD in brain stem and blood
        pressure response following constant
        intravenous infusion of (S)-α-MD (24 hours).
        Blood pressure is expressed as percentage
        mean arterial pressure (MAP) of pretreatment
        control value. "Least squares regression" lines
        are drawn for estimating the mean % MAP from
        the concentration and estimating the mean
        concentration from the % MAP.

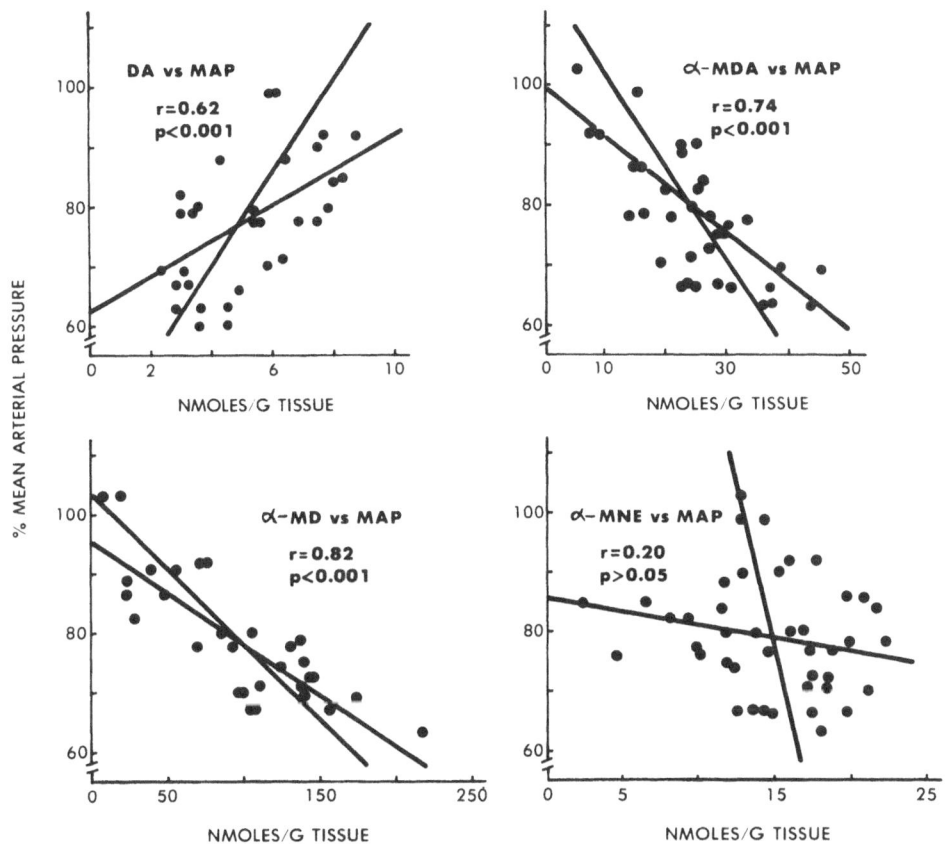

FIG. 10. Scatter plot of concentration of DA, α-MDA, α-MNE and (S)-α-MD in hypothalamus and blood pressure response following constant intravenous infusion of (S)-α-MD (24 hours). (See explanation in legend of Fig. 9).

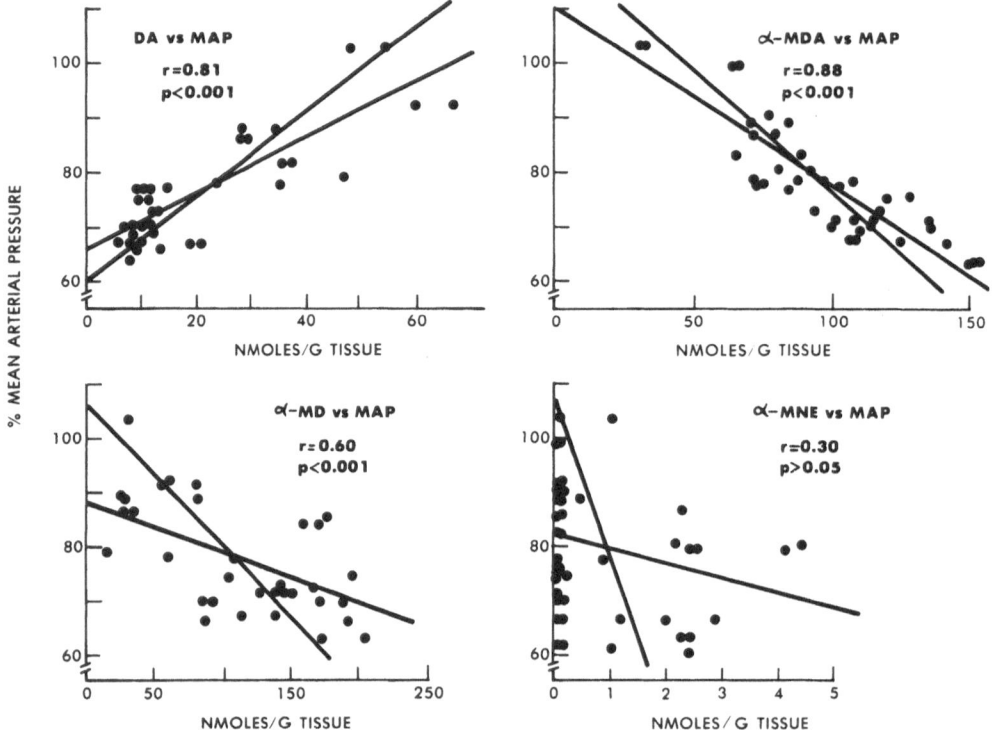

FIG. 11. Scatter plot of concentration of DA, α-MDA,
         α-MNE, and (S)-α-MD in corpus striatum and blood
         pressure responses following constant intravenous
         infusion of (S)-α-MD (24 hours). (See
         explanation in legend of Fig. 9).

The NE levels were too low to be quantitated. On the other hand,
no correlation was observed between α-MNE levels and blood pressure
responses (Figs. 9-11). In the hypothalamus and brain stem essentially
the same results were observed as in the corpus striatum. Waldmeier
et al. (8) reported similar correlations between α-MDA levels in
whole brains and the corresponding blood pressure response in renal
hypertensive rats treated with a single subcutaneous dose of (S)-
-α-MD.

Previous reports have suggested that α-MNE may be the critical
central metabolite responsible for the blood pressure regulating
properties of (S)-α-MD. For example, injection of α-MNE into the
medulla oblongata or hypothalamus of rats produced a dose-dependent
fall in blood pressure (6, 17). Intracerebroventricular injections
of α-MDA caused a dose-dependent hypotensive response in rats and
cats. This effect, however, was blocked by pretreating the animals

with selective central dopamine α-hydroxylase inhibitors (18-20).

The data from the present study may be explained in two different ways. First, only a small fraction of the total α-MNE may be responsible for regulating blood pressure. NE may exist in two compartments -- a "functional compartment" and a "main storage compartment" (21). Only freshly synthesized NE is released from the "functional compartment" alone upon neural stimulation (22, 23). If this applies to α-MNE, the amount of α-MNE released upon neural stimulation might be proportional to the amount of α-MDA in the tissue. The studies on α-MNE reported here do not differentiate between the two pools. Even though no correlation between α-MNE levels in the brain and hypotensive response was observed, the correlation between the hypotensive response and the brain tissue levels of α-MDA is consistent with a mechanism of action dependent upon the rate of de novo synthesis of α-MNE.

An alternate interpretation is that α-MDA itself may have a direct hypotensive effect. Heise (24) has reported that intracerebroventricular perfusion of DA in cats produced dose-dependent hypotensive responses which could be inhibited by the dopaminergic blocking agents, haloperidol and pimozide. Others also have suggested that central dopaminergic sites may be important in regulating blood pressure (25-27). Although it may be dangerous to relate α-MDA levels in these brain regions to blood pressure control mechanisms, the correlation of α-MDA levels and blood pressure response is impressive particularly when contrasted with the lack of such a correlation with α-MNE levels. The present data are consistent with the proposal that central dopaminergic receptors may contribute to the mediation of the antihypertensive activity of (S)-α-MD.

ACKNOWLEDGEMENTS

(S)-α-methyldopa was generously supplied by Dr. J. Baer of Merck Sharp and Dohme, Westpoint, Pa. This work was supported in part by NIH Grant GM 16496 and the Earl C. Anthony Fund (U.C.S.F.).

REFERENCES

1) J.A. Oates, L. Gillespie, S. Udenfriend and A. Sjoerdsma, Science, 1960, 131, 1890.
2) E. Muscholl, Ann. Rev. Pharmacol., 1976, 6, 107.
3) M. Henning, Acta Physiol. Scand. Suppl., 1969, 322, 1.
4) F.P. Nijkamp and W. De Jong, Prog. Brain Res., 1977, 47, 349.
5) C.C. Porter, M.L. Torchiana and C.A. Stone, in "Antihypertensive Agents. Handbook of Experimental Pharmacology, Ser. 2, Vol. 39", F. Gross, Ed., Springer-Verlag, New York, 1977, p. 263.
6) M. Henning and P.A. van Zweiten, J. Pharm. Pharmacol.,1968,20,409.

7) N.J. Uretsky, J. Pharmacol. Exp. Ther., 1974, 189, 359.
8) P. Waldmeier, P.R. Hedwall and L. Maitre, Naunyn-Schmiedeberg's Arch. Pharmacol., 1975, 289, 303.
9) C.R. Freed, R.J. Weinkam, K.L. Melmon and N. Castagnoli, Anal. Biochem., 1977, 78, 319.
10) S.H. Koslow, F. Cattabeni and E. Costa, Science, 1972, 176, 177.
11) K.S. Marshall and N. Castagnoli, Jr., J. Med. Chem., 1976, 16, 266.
12) M.M. Ames, K.L. Melmon and N. Castagnoli, Jr., Biochem. Pharmacol., 1977, 26, 1757.
13) J. Glowinski and L.L. Iversen, J. Neurochem., 1966, 13, 655.
14) W.A. Garland, R.J. Weinkam and W.F. Trager, Chem. Instrumentation, 1974, 5, 271.
15) M. Holzbauer and D.F. Sharman, in "Catecholamines, New Series", H. Blaschko and E. Muscholl,Eds., Springer-Verlag, New York, 1972, p. 110.
16) F.P. Nijkamp and W. De Jong, Eur. J. Pharmacol., 1975, 32, 361.
17) H.A.J. Struyker Boudier, G. Smeets, G. Brouwer and J.M. van Rossum, Arch. Int. Pharmacodym., 1975, 213, 285.
18) M.D. Day, A.G. Roach and R.L. Whiting, Europ. J. Pharmacol., 1973, 21, 271.
19) L. Finch and G. Hausler, Brit. J. Pharmacol., 1973, 47, 217.
20) L. Finch, A. Hersom and P. Hicks, Brit. J. Pharmacol., 1975, 54, 445.
21) J. Glowinski, Brain Res., 1973, 62, 489.
22) I.J. Kopin, G.R. Breese, K.R. Krauss and V.K. Weise, J. Pharmacol. Exp. Ther., 1968, 161, 271.
23) J. Glowinski, M.J. Bresson, A. Cheramy and A.B. Thierry, Adv. Biochem. Psychopharmacol., 1972, 6, 93.
24) A. Heise, in "New Antihypertensive Agents", A. Scribine and C.S. Sweet, Eds., Spectrum, New York, 1976, p. 135.
25) P. Bolme and K. Fuxe, Eur. J. Pharmacol., 1971, 13, 168.
26) D.B. Calne and P.F. Teychenne, Prog. Brain Res., 1977, 47, 331.
27) M.W. Osborne, in "New Antihypertensive Agents", A. Scriabine and C.S. Sweet, Eds., Spectrum Publications, New York, 1976, p. 105.

CONCURRENT EXTRACTION AND GC-MS ASSAY OF ENDOGENOUS TP, T, IAA,

5HT AND 5HIAA IN SINGLE RAT BRAIN SAMPLES

F. Artigas and E. Gelpí

Instituto de Biofísica y Neurobiología. C.S.I.C.,

Barcelona, Spain

INTRODUCTION

The indoleamine Serotonin (5-hydroxytryptamine) is presently considered as one of the few established central neurotransmitters, its metabolism having been related to physiological functions of the body as well as several psychiatric disorders, such as schizophrenia and depression. Consequently, various analytical methods have been developed for serotonin (5HT) as well as for its precursor tryptophan (TP) and its acidic metabolite, 5-hydroxyindole--3-acetic acid (5HIAA) in brain tissue (1-3).

Tryptamine, the non hydroxylated analog of 5HT, also seems to be related in some way to neurotransmission processes. The accumulation of T in rat brain, after inhibition of MAO produces mydriasis, decreased locomotor activity, body tremors and convulsions, and when infused into humans it increases blood pressure, dilates pupils and produces changes in sensation and preception similar to those of the LSD-like hallucinogens (4).

However, until recently, there was no adequate analytical methodology for the detection of T. This is mostly due to the extremely low endogenous levels of this amine in mammalian brain. Philips et al. (5), using a high resolution mass spectrometric technique have shown that T is present in rat brain at levels of 0.5 ng/gr. Higher values have been reported by other authors (6, 7), ranging from 25 to 70 ng/g, but Martin et al. (8) failed to detect endogenous T with both GC and fluorimetric methods at a sensitivity limit of 5 ng/g. Also during the course of this work, a mass fragmentographic (MF) method has been published (9), though the

authors did not detect T in rat brain at an assay sensitivity of
0.5 ng.

Furthermore, nothing was known of the postulated acidic
metabolite of T, indole-3-acetic acid (IAA) in mammalian brain
until a very recent publication on this subject in which Warsh et
al. (10) showed that IAA is present in rat brain at levels of about
12 ng/g. However, most of the reported analytical methods for TP
and its metabolites have the following common characteristics:
a) they concentrate on the assay of one, or two metabolites and
b) in general, they are relatively complex, involving several
extractive and/or chromatographic procedures, which in some cases
causes drastic reductions of the recovered amounts, basically due
to the high lability and the low amounts of some of these compounds
(specially T and IAA).

These considerations coupled to our interest in studying the
possible neuroactive role of T, per se, or related to 5HT metabolism
led to the development of a simple and sensitive method based on
the concurrent extraction of the indolic compounds from brain
tissue on a short XAD-2 column and further MF analysis of the resin
eluate.

MATERIAL AND METHODS

Glass material. Glass material was cleaned with a commercial
laboratory detergent, rinsed with dilute HCl, tap water, distilled
water and allowed to dry in an oven.
Chemicals. Solvents and reagents (analytical and chromatographic
grades) were supplied by Carlo Erba, Milano, E. Merck, Darmstadt
and Xpectrix Int., Barcelona, and were used as received from the
supplier, except methano) that was distilled once in an all-glass
apparatus. Tryptamine (T) hydrochloride, serotonin (5HT) hydrogen
oxalate, Tryptophan (TP), 5-hydroxyindole-3-acetic acid (5HIAA)
and 5-methoxytryptamine (5MeOT) were purchased from Regis Chemical
Co. Indole-3-acetic acid (IAA) was obtained from Sigma Chemical Co.
Tryptamine α-d$_2$, β-d$_2$ (T-d$_4$) and Serotonin α-d$_2$, β-d$_2$(5HT-d$_4$) were
a gift of Dr. John Shaw, National Institute of Health (Bethesda,
Md. USA). Indole-3-acetic acid α-d$_2$(IAA-d$_2$) and 5-hydroxyindole-
-3-acetic acid (5HIAA-d$_2$) were synthesized in the laboratory (11).
Tryptophan β-d$_2$(TP-d$_2$) was purchased from MSD (München). ^{3}H-Tryptamine
(1,65 Ci/mmol) was from Amersham (Radiochemical Center). Stock
solutions of all compounds were prepared in distilled methanol and
stored at -10°C. Concentrations were individually adjusted within
the range of 0,5-3 x the expected endogenous levels of each compound (max. 100 ng/µl). Isotopic purity of deuterated standards
was found to be constant for at least three months at these
conditions.

Boron trichloride (BCl_3) in methanol (10% W/V) and pentafluoropropionic anhydride (PFPA) were purchased from Xpectrix Int. and were
used without further purification.

Chromatographic material. The XAD-2 resin (300-1000 μ) was
obtained from Serva, Heidelberg. Prior to its use it was washed with
methanol in a Soxhlet during 12-18 h, allowed to dry at 60°C and
stored in glass beakers. Glass columns (30 x 0.4 cm i.d.) were
filled with the XAD-2 resin suspended in methanol and washed with
30 ml of HCl/H_2O at pH 1.5-1.7, before adding the sample.

The stationary phases used for GC-MS analysis were 5% OV-17
and 3% Dexsil 300 on gas Chrom Q 100/120 mesh. The average efficiency
of GC columns was about 1200-1500 plates/m.

Instrumentation. The combined gas chromatography-mass spectrometry (GC-MS) system consisted of a Perkin Elmer model 3920 gas
chromatograph coupled to a Hitachi RMU-6H mass spectrometer through
a gold lined jet separator. The mass spectrometer is equipped with
a four channel multiple ion detection (MID) system of our design
(12) which allows the monitoring of up to four m/e values for mass
fragmentographic determinations. Recoveries of the ^{3}H-T added to
the brain tissue were checked on an Intertechnique model SL30 liquid
scintillation spectrometer.

METHODS

A full description of the analytical method is to be published
(13). Briefly, it is carried out as follows: adult Sprague-Dawley
rats of either sex were decapitated, the brains rapidly taken up
in ice, blotted with filter paper, weighed and homogenized in 10
volumes of ice-cold freshly prepared 0.1N HCl containing 1N KCl
and 0.2% $NaHSO_3$. The deuterated standards used for quantification
as well as 5MeOT used as a carrier were added to the HCl medium
before homogenization. The homogenate was centrifuged at 100.000
x g for 20 min, and the supernatant was saved. The pellet was
rehomogenized with 10 volumes of the homogenization medium and
centrifuged at 100.000 x g for 20 min. The pooled supernatants
were then adsorbed on a short column of a XAD-2 (usually 0.5 grs),
the column washed with 20 bed volumes of dilute HCl at the pH of
the sample (1.4-1.6), the washings discarded, and the column eluted with 6 ml of distilled methanol.

One drop of formic acid and 10 μg of carrier (5MeOT) were
added to the eluate and it was evaporated to dryness in a Rotary
film evaporator. The dry residue was then derivatized by combined
methylation-perfluoroacylation according to described procedures
(14, 15), with some slight modifications: the methylation reaction
was carried out at 85°C for 15 min to achieve a better methylation
of TP and the volume of methylating agent (BCl_3/MeOH) was set to

20 µl per vial. The acylation was performed with a solution of 100 µl PFPA/100 µl acetonitrile saturated with $NaHSO_3$, during 2 hr at 60°C. After evapoarting the excess of reagent, the dry residue was dissolved in 20 µl of C_6H_6/PFPA (95/5) and 2-3 µl taken up for the MF analysis.

The analysis of samples of cerebrospinal fluid was performed as described omitting the homogenization and centrifugation steps.

GC-MS assay. The routine assay was performed in a 5% OV-17 column at 210-220°C. The ions selected for verification of the identity of endogenous compounds by MID were those recorded at the following m/e values (14, 15) 276 and 347 for TP, 276 and 289 for T, 276 and 335 for IAA, 438 and 451 for 5HT and 438 and 497 for 5HIAA. Quantitation was achieved by comparison of the height of the peaks corresponding to the endogenous compounds with the height of the corresponding deuterated analogs at the following m/e values: 276 for IAA, T and TP, 278 for $IAA-d_2$, $T-d_4$, $TP-d_2$, 438 for 5HIAA, 5HT and 440 for $5HIAA-d_2$, $5HT-d_4$. The isotopic purity of the deuterated standards which was individually checked by monitoring both m/e values of the protium and deuterium forms after injecting only the deuterated compounds, was found to be better than 98% in all cases. Nevertheless, the contributions of the ions from the deuterated standard to those from the endogenous counterparts were subtracted from the latter.

RESULTS AND DISCUSSION

The starting point of this work was a previous report (15) in which we described the use of the resin XAD-2 as a excellent tool for the one-step extraction of these metabolites of TP from physiological fluids. The procedure allows the concurrent extraction of the parent aminoacid together with the amine and acid metabolites, probably through a non-polar interaction between the polystyrene matrix of the resin and the indole ring of all these metabolites. However, initially the application of the previous method to brain tissue led to serious problems, as recoveries were much lower than those obtained with urine of CSF. Briefly, the reasons for the discrepancies observed were to be found in the higher amounts of interfering substances (obviously, the complexity of a homogenate of one brain is higher than that of 2 ml of urine or CSF) and the lower amounts of TP metabolites, specially IAA and T, which are present in rat brain at the low nanogram range. These facts, coupled to the high lability of these compounds and the need for additional sample preparation steps, such as homogenization, would affect the overall sensitivity of the method, as shown below.

To summarize, the additional analytical precautions that have to be taken are: 1) use of a mild homogenization medium, as the use of $HClO_4$ or trichloroacetic acid results in almost quantitative losses for T and IAA; 2) use of a double homogenization step, to counteract the binding of T to lipidic fractions (the binding of 6 ng ^3H-T to the first pellet was calculated to be around 25%); 3) use of low amounts of XAD-resin and tissue: the amounts of XAD-2 resin and tissue were adjusted, after a detailed study, to 0.5 grs of resin per no more than 1 gr of brain tissue. Apparently, the use of higher amounts of tissue lead to lower recoveries of IAA and T, as shown in Fig. 1.

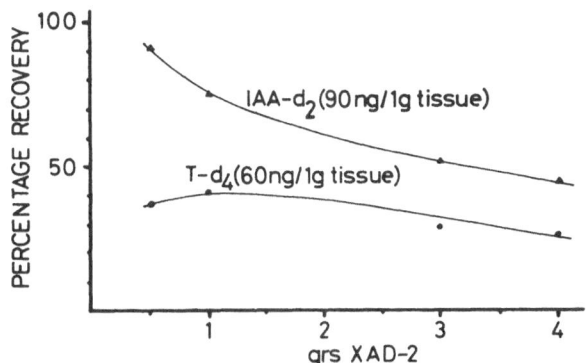

FIG. 1. Procedural recoveries of the stated amounts of T-d_4 and IAA-d_2 added to 1 gr of brain tissue VS XAD-2 resin bed.

This apparently contradictory data can be easily explained by taking into consideration the yield of the derivatization step: that is, for compounds being in the nanogram range, the competitive and/or catalytic effects exerted by non-indolic compounds also adsorbed on the XAD-2 resin during the derivatization step, lead to lower derivatization yields. Actually, the adsorption of indoles on the XAD-2 resin increases with the volume of resin bed. For instance, the recovery of about 6 ng of ^3H-T added to one brain homogenate is 60%, while the recovery of the process when tested with similar levels of T-d_4 (that is, including the derivatization step) is only 38%. Thus, for a fixed amount of T and IAA, the level of derivatization-interfering substances has to be kept to a minimum,

either by lowering the amount of resin or the amount of brain tissue
per column. Nevertheless this effect was not observed at the same
degree for compounds present in larger amounts, such as 5HT, 5HIAA
or TP or even for IAA and T when added at levels ten times higher.
For this reason larger resin beds (up to 3-4 gr) can be used only
if one is interested just in the levels of 5HT, 5HIAA or TP.

TABLE 1
Procedural recoveries of deuterated Tryptophan
and indolic metabolites on 0.5 gr of XAD-2 resin.

Compounds	Amount	Apparent[a] recovery (%)
IAA-d$_2$	25 ng	54.7 ± 3.3 (n = 5)
T-d$_4$	12 ng	38.1 ± 1.4 (n = 5)
TP-d$_2$	3.7 µg	57.5 ± 4.1 (n = 5)
5HIAA-d$_2$	830 ng	57.9 ± 8.6 (n = 5)
5HT-d$_4$	400 ng	34.7 ± 5.4 (n = 5)

Final recoveries of the stated amounts of deuterated
standards added at the beginning of the analytical
procedures to a brain homogenate. Results are expressed as mean ± standard deviation of five experiments.

[a]The term "apparent" refers to the data obtained by
mass fragmentographic evaluation, as described in the
text, and thus it is a combination of the MeOH eluate
recovery plus the yield of derivatization of this
eluate.

Table 1 gives the final recoveries of the stated amounts of IAA-d$_2$,
T-d$_4$, TP-d$_2$, 5HIAA-d$_2$ and 5HT-d$_4$ added as described to each of five
samples of brain tissue (0.9 grs each). However, these recovery
values can be improved by extraction of higher amounts of deuterated
standards, as can be seen by comparing the values of Table 1 with
those of Fig. 1. In the latter case the illustrated values were
obtained with 90 ng of IAA-d$_2$ and 60 ng of T-d$_4$ per gram of tissue
adsorbed on the stated amounts of XAD-2 resin. Recoveries were
calculated by comparing the peak height of the MF traces (278 for
indoles and 440 for 5-hydroxyindoles) corresponding to a) the five
brain samples and b) two control samples containing only the stated
amounts of the deuterated standards plus the carrier compound
(5MeOT). (Data shown in Fig. 1 were obtained without adding the
carrier compound).

In any case, taking into account the < 100% derivatization

yield, the absolute recoveries in the MeOH eluate can be calculated
as about 90% for IAA, TP and 5HIAA, 60% for T, and 50% for 5HT.
4) Use of distilled methanol as eluant of the indoles adsorbed on
the resin: this was deemed an absolute necessity due to the importance
of degradative processes (losses of T and IAA larger than 40% at the
low nanogram range) probably caused by the presence of oxidizing
agents in the commercial product.
5) Use of a carrier compound: the carrier effect on the derivatization
yield is shown in Table 2. As can be seen, the response as well the
precision of the measurements were remarkably improved.

TABLE 2
Carrier effect of 5-methoxytryptamine on the
derivatization of deuterated Tryptamine and
indole-3-acetic acid.

	without carrier (n=4)		with carrier (n=4)	
	$IAA-d_2$	$T-d_4$	$IAA-d_2$	$T-d_4$
$\bar{x}$	88	38	239	88
$s/\bar{x}$	0.11	0.18	0.04	0.04

Data represent peak heights at m/e 278 of $IAA-d_2$
12.7 ng/vial and $T-d_4$ 5.7 ng/vial standards derivatized alone or in the presence of 2.5 µg of 5-Metho-
xytryptamine (5MeOT) per vial. Number of vials
given in brackets.

RAT BRAIN SAMPLES

Identification and quantification of IAA, T, TP, 5HIAA and 5HT.
The identification of all endogenous compounds rests in a) the
coincidence of their respective GC retention times with those of
the standards, b) the MS response at selected masses and c) the
correct matching of the abundance ratios of both of the ion
pairs monitored for the endogenous and the deuterated authentic
compounds given in Table 3. Fig. 2 shows a schematic example of
expected MF(MID) profiles for TP, T and IAA, illustrating how these
identification criteria would be applied in practice.

The gas chromatographic separation was performed on a 5% OV-17
column. On this column the peaks corresponding to IAA, T, TP, 5HIAA
and 5HT were located on the mass fragmentographic profiles with the
aid of their respective standards. However, due to an occasional
contribution of an interfering unidentified component to the
abundance of the molecular ion of IAA-1PFP at m/e 335, often

TABLE 3
Abundance ratios of endogenous and standard
compounds.

Ion ratio	Abundance	Compounds
$\dfrac{m/e\ 276}{m/e\ 335}$	1.69	Endogenous IAA
$\dfrac{m/e\ 278}{m/e\ 337}$	1.60	Standard IAA-d_2
$\dfrac{m/e\ 276}{m/e\ 289}$	non detectable	Endogenous T
$\dfrac{m/e\ 278}{m/e\ 292}$	0.85	Standard T-d_4
$\dfrac{m/e\ 276}{m/e\ 347}$	3.68	Endogenous TP
$\dfrac{m/e\ 276}{m/e\ 347}$	3.72	Standard TP
$\dfrac{m/e\ 278}{m/e\ 348}$	4.45	Standard TP-d_2
$\dfrac{m/e\ 438}{m/e\ 497}$	1.97	Endogenous 5HIAA
$\dfrac{m/e\ 440}{m/e\ 499}$	2.00	Standard 5HIAA-d_2
$\dfrac{m/e\ 438}{m/e\ 451}$	0.48	Endogenous 5HT
$\dfrac{m/e\ 440}{m/e\ 454}$	0.49	Standard 5HT-d_4

Values are mean of three determinations a)Explana-
tion of the anomalous value obtained for standard
TP-d_2 is given in the text.

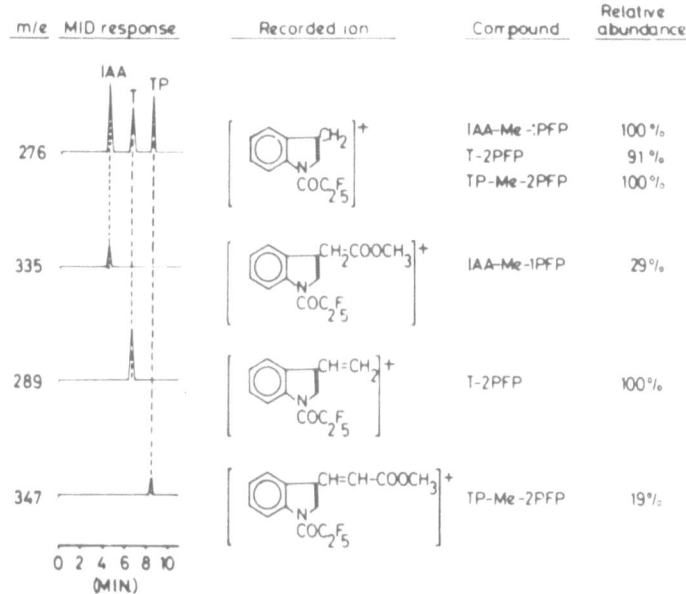

FIG. 2. Theoretical mass fragmentographic responses
for IAA-Me-1PFP, T-2PFP and TP-Me-2PFP as
predicted from the mass spectra of these
compounds.

observed in the mass fragmentogram obtained with the OV-17, the
identity of the presumed IAA peak was also verified on a different
glass GC column packed with 3% Dexsil on Gas Chrom Q, 100/120
(Fig. 3). The ion abundance ratio at m/e 276 and 335, measured with
this second GC column, fits exactly the abundance ratio of the
deuterated analog (Table 3). Furthermore, the level of IAA quantified at m/e 335 and 337 on Dexsil 300 was the same as when
calculated using the ions at m/e 276 and 278, with the OV-17 column.

At this point it is also interesting to note that while the
ion abundance ratios of endogenous and deuterated standard compounds
are, within the experimental error, identical in every case, there
is an appreciable difference in the case of endogenous and deuterated
Tryptophan. However, this effect was also determined for the ratio
between the authentic protium and deuterium forms of TP, as indicated
in Table 3. This could be accounted for by the fact that as the
ion at m/e 347 is known to arise in indoleamines by β-cleavage with
rearrangement of a hydrogen atom (14) to the carbonyl oxygen of the

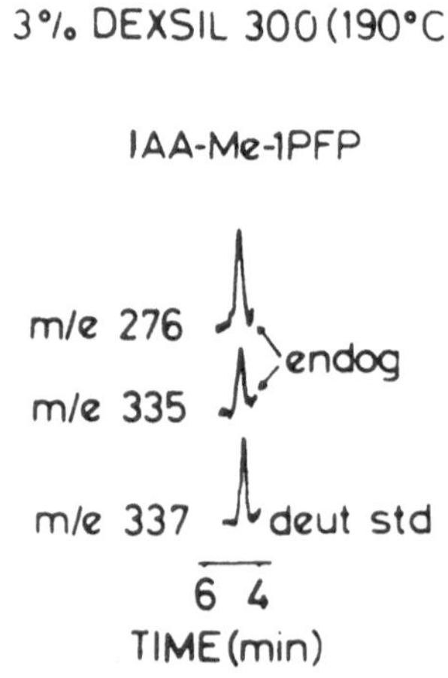

3% DEXSIL 300 (190°C)

IAA-Me-1PFP

m/e 276
                     endog
m/e 335

m/e 337   deut std

6 4
TIME (min)

FIG. 3. Mass fragmentograms (MF) obtained from an
        extract corresponding to 150 mgrs of brain
        tissue. The MF traces at m/e 335 and 337
        correspond to the molecular ions of
        IAA-Me-1PFP and IAA-d_2-Me-1PFP respectively
        (14) while the trace at m/e 276 corresponds
        to the α-cleavage on the side chain of the
        endogenous IAA. Gas chromatographic and
        mass spectrometric conditions were: 3%
        Dexsil 300 column (3m x 2 mm i.d.) at 190°C.
        He flow= 30 ml/min. Injector and gold-lined
        separator temperatures were 250 and 280°C,
        respectively. Chamber voltage: 70eV. Trap
        current: 80 μA. Nominal accelerating voltage:
        1800 V. Chamber temperature: 180°C.

perfluoroacyl moiety, giving a characteristic M-163 fragment
(HO-C(=NH)C_2F_5), the corresponding rearrangement of one of the
deuterium atoms on the β^-carbon atom of TP (TP-d_2) leading to m/e
348 could be destabilized by the interfering driving force of the
neighboring TP carboxyl group on the α^-carbon of the aminoacid.

───────────

  According to nomenclature used for 3-substituted indoles instead
of the current aminoacid labelling of C atoms used herein for
TP-d_2.

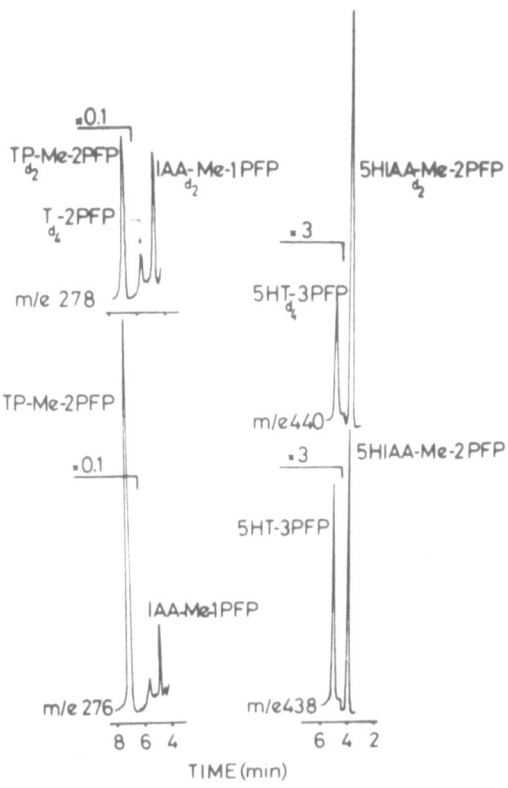

FIG. 4. Mass fragmentographic profiles from an
injection of a brain extract (equivalent
to 140 mgr of tissue) obtained on a 5%
OV-17 column (2.5 m x 2.5 mm i.d.). The
responses at m/e 276 and 438 identify the
endogenous compounds and the peaks
appearing at m/e 278 and 440 are the
corresponding deuterated standards. GC
and MF conditions were the same as in
Fig. 3 except for the column temperature,
which was set at 210°C.

Fig. 4 shows the mass fragmentographic profiles of the ions at
m/e 276, 278, 438 and 440. The responses shown would correspond
to an injection of a sample of 140 mgr of brain tissue. Although
the m/e 276 profile contains a small peak at the same retention
time as that of authentic T-2PFP ($\sim$ 5 min, 30 sec) it did not
produce the expected ion abundance of the corresponding signal at
m/e 289, predicted from the ratio m/e 276/m/e 289 in the mass
spectrum of T-2PFP. In fact, the very small peak measured at m/e

289 would fix an upper limit of T in rat brain at the level of
350-400 pg/gr. Table 4 shows the endogenous levels of IAA, T, TP,
5HIAA and 5HT found in whole rat brain.

TABLE 4
Whole rat brain content of IAA, T, TP, 5HT and 5HIAA.

IAA......................	13.1 ± 2.0 ng/gr	(n=6)
T........................	< 380 pg/gr	(n=6)
TP.......................	4.16 ± 0.23 µg/gr	(n=6)
5HIAA....................	442 ± 24 ng/gr	(n=6)
5HT.....................	526 ± 81 ng/gr	(n=5)

Results are expressed as $\bar{x}$ ± s.d. Number of animals
is given in brackets. Data refers to gr of wet tissue.

These results confirm a recent finding (5, 9) showing that the
rat brain level of T is $\leq$ 500 pg/gr. Also, the values of IAA agree
with the only data available up to date on this subject: the 11.6
ng/gr reported by Warsh et al. (10). On the contrary, the results
obtained for T by Saavedra et al. (6) and Snodgrass et al. (7)
differ markedly being 50-200 times higher than those reported
herein, most likely as a consequence of the lower specificity of
the separation and detection techniques used by these authors.

According to these data the low, practically undetectable,
levels of endogenous T inferred in adult rat brain seem to question
the possible classic neurotransmitter role of this amine in CNS.
Also, the ratio IAA/T is significantly much higher than that of
5HIAA/5HT, suggesting a higher availability of T to MAO, although
it has also been reported that T is found to some extent in
particular fractions (16). To summarize, it could be concluded
that the question whether T exerts some physiological action in
normal metabolic conditions remains still largely unsolved in spite
of the interesting body of the data available. This is mostly
because some of the conclusions established in this regard must
be considered with caution as they have been based on results
obtained with analytical methods not always suitable for the
specific and unequivocal detection of such low levels of T. In other
words, what has been reasonably labelled as tryptamine may not be
all tryptamine in some cases due to lack of the necessary methodological specificity.

The high sensitivity achieved with this method (about 40 pg
of T and IAA give a signal to noise ratio of 2) as well as the
concurrent extraction and detection of various significative

members of the metabolic pathway of TP open the way to a serious
study of TP metabolism in small amounts of tissue (at the low
milligram range). As indicated above in relation to the experimental
recoveries of the sample clean up on XAD-2, the lower the quantity
of tissue per column bed the higher the apparent recovery that
could be attained, so that 1 gr of neural tissue represents the
upper optimal limit established under the conditions herein
described. Thus, for smaller weights of tissue, the apparent
recoveries could be expected to be even higher than reported in
Table 1.

CSF samples. Samples of CSF from patients affected by psychiatric
disorders are being analyzed with this method, and the full results
will be published elsewhere. Nevertheless, to illustrate the
analytical capabilities of the method, we wish to present the results
corresponding to a case of a deep schizophrenic patient (F.R., male,
28 y. old), who had been taking several doses of LSD per week over
1 ½ years.

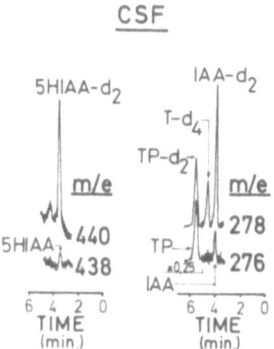

CSF

FIG. 5. MF responses corresponding to a sample of
        CSF from a schizophrenic patient. Upper and
        lower profiles correspond to the deuterated
        standards and endogenous compounds respec-
        tively (for details see text).

Fig. 5 shows the mass fragmentographic profile corresponding to 0.2
ml of CSF (2 ml of CSF were processed as described, finally dissolved
in 20 µl of C_6H_6: PFPA and 2 µl taken up for GC-MF analysis). The
levels found for 5HIAA, and IAA were extremely low (890 and 920
pg/ml, mean of two determinations), and the TP level was found to
be 376 ng/ml. The amines, T and 5HT were not detectable at a
sensitivity of 200 and 500 pg/ml, respectively. These results, when
compared to data from the literature corresponding to healthy
individuals (17, 18) reflect, in this case,a deep alteration of the
brain TP metabolism.

In conclusion these results demonstrate that besides being
highly specific and sensitive, one of the most significant and
remarkable assets of this method lies in its wide scope as it
readily allows a simultaneous determination of various members of
a metabolic pathway, regardless of their different chemical
properties (acids, bases and amphoterous products) and at high
sensitivity ($\approx$ 50 pg of T and IAA at a S/N of 2).

Capitalizing on this useful property, work is presently under
way to extend the method to the investigation of the possible occurrence of other indolic compounds of neurochemical interest in
these extracts, whose analysis could provide a complete screening
of all neuroactive metabolites of TP in normal as well as in
pathological conditions.

ACKNOWLEDGEMENTS

The authors wish to acknowledge receipt of samples of $T-d_4$
and $5HT-d_4$ from Dr. J. Shaw, National Institute of Health,
Bethesda Md. USA, and samples of CSF from the Preventorio
Psiquiátrico Municipal, (Drs. Bulbena and Costa) Barcelona, Spain.
This work was carried out with the support of Research Grant from
the Comisión Asesora de Investigación Científica y Técnica,
project n° 1287.

REFERENCES

1) W.D. Denkla and H.K. Dewey, J. Lab. Clin. Med., 1967, 69, 160.
2) G. Curzon and A.R. Green, Br. J. Pharmacol., 1970, 39, 653.
3) O. Beck, F.A. Wiesel and G. Sedvall, J. Chromatogr., 1977,
   134, 407.
4) W.R. Martin and J.W. Sloan, Psychopharmacologia, 1970, 18, 231.
5) S.R. Philips, D.A. Durden and A.A. Boulton, Can. J. Biochem.,
   1974, 52, 447.
6) J.M. Saavedra and J. Axelrod, J. Pharmacol. Exp. Ther., 1972,
   182, 363.

7) S.R. Snodgrass and A.S. Horn, J. Neurochem., 1973, 21, 687.
8) W.R. Martin, J.W. Sloan, S.T. Christian and T.H. Clements, Psychopharmacologia, 1972, 24, 331.
9) J.J. Warsh, D.D. Godse, H.C. Stancer, P.W. Chan and D.V. Coscina, Biochem. Med., 1977, 18, 10.
10) J.J. Warsh, P.W. Chan, D.D. Godse, D.V. Coscina and H.C. Stancer, J. Neurochem., 1977, 29, 955.
11) J. Segura, J. Veiguela and E. Gelpí, in preparation.
12) F. Artigas, E. Gelpí, M. Prudencio, J.A. Alonso and J. Ballart, Anal. Chem., 1977, 49, 543.
13) F. Artigas and E. Gelpí, Anal. Biochem., in press.
14) E. Gelpí, E. Peralta and J. Segura, J. Chromatogr. Sci., 1974, 12, 701.
15) J. Segura, F. Artigas, E. Martínez and E. Gelpí, Biomed. Mass Spectrom., 1976, 3, 91.
16) A.A. Boulton and G.B. Baker, J. Neurochem., 1975, 25, 477.
17) L. Bertilsson and L. Palmer, Science, 1972, 177, 74.
18) S.J. Dencker, V. Malm, B.E. Roos and B. Werdinius, J. Neurochem., 1966, 13, 1545.

A NEW METHOD FOR SIMPLIFIED PROSTAGLANDIN PROFILING BY SELECTED ION MONITORING

J. Roselló and E. Gelpí

Instituto de Biofísica y Neurobiología,

Barcelona, Spain

INTRODUCTION

We have lately been interested in a feasibility study of the experimental requirements that would apply to the simultaneous assay of various classes of prostaglandins, like PGAs, PGBs, PGEs and PGFs, by combined gas chromatography-mass spectrometry (GC-MS). The aim of such a study was the development of a "simple" (in terms of sample handling and clean up procedures) and at the same time practical extraction and derivatization approach suitable for GC-MS applications. In general, many of the reported analytical procedures were deemed rather cumbersome (1), not readily reproducible in our hands or too restrictive in scope, as they are useful only for the determination of a given prostaglandin. In this sense one of the first operational requirements that had to be satisfied due to the analytical constraints imposed by the limited sample preparation procedure was the use of the high selectivity inherent to the specific ion detection techniques. In other words, any possible lack of sufficient sample purification, due to the simplification of the enrichment process as well as to any lack of gas chromatographic resolution prior to the MS analysis, would be compensated by resorting to a very highly selective method of detection. The work undertaken in this direction has resulted to date in 1) a simple and efficient two step extraction procedure for PGs in human seminal or blood plasma, recently reported (1) and 2) an equally simple one step derivatization based on the direct silylation of all PGs with mixtures of a silylating agent/non-aromatic nitrogen heterocycle (2, 3), preferably BSTFA/piperidine, according to the overall behaviour and stability of the derivatives thus formed (4).

In contrast, all of our previous attempts to achieve a concurrent assay of these prostaglandins by using standard methoximation-
-silylation procedures (5) led to highly complex gas chromatographic
profiles, difficult to deal with and interpret (4). The methoximation-
-silylation procedure, as described (5) gives two SIN and ANTI
isomers (6) for each component member of series I and series II in
PGA, PGB and PGE.

On the other hand our recently reported simultaneous derivatization of PGs, while remarkably simplifiying such profiles and
producing derivatives very suitable for Selected Ion Monitoring
(SIM) (1-4), still does not give a sufficiently well resolved GC
profile for the fully silylated PGF (TMS)$_4$ and 9-enol PGE (TMS)$_4$
derivatives. (PGBs give the corresponding di-TMS derivatives and
PGAs give the new 11-piperidyl, 9-enol PGA (TMS)$_3$ derivatives
(4, 7)). Although the observed lack of GC resolution between the
PGE and PGFs derivatives can be compensated by the inherent
specificity of the SIM technique, it nevertheless interferes   with
the direct quantification of both PGs, which appear to be the most
significant PGs from a physiological stand point.

This communication presents a new method for PG profiling by
selected ion monitoring. The method is based on the combined
derivatization of these Prostaglandins, first with N-butyl boronic
acid (5, 6) and then with the BSTFA/piperidine (1:1) mixture (2, 4).

MATERIALS AND METHODS

Reagents. Methanol, dimethoxypropane (DMP), n-butyl boronic
acid (NBBA) and N,O bis (trimethylsilyl) trifluoroacetamide (BSTFA),
either analytical or chromatographic grade were used as received
from local suppliers.

Samples of prostaglandins A, B, E and F (series I and II) were
generously supplied by Dr. Pike of Upjohn Co. Kalamazoo, Mich., USA.
Gas chromatographic conditions. Silanized glass columns were
used throughout this work. The columns were 1.8 m long and 2.5 mm
i.d. and they were filled with 3% OV-17 on Gas Chrom Q, 100-120
mesh.

Before use they were conditioned for 48 hours at 300°C and
silylated by repeated injections of ½ μl of BSTFA. Samples were
directly injected into a Perkin Elmer series 900 Gas Chromatograph,
equipped with dual FID detectors, under the following conditions:
injector and manifold temperatures 300°C; column temperature 260°C;
helium carrier flow rate 30 ml/min.
Gas chromatographic-mass spectrometric conditions. The SIM
profiles were obtained with a Hitachi-RMU-6H mass spectrometer

connected to a 3920 Perkin Elmer-Gas Chromatograph through a
"molecular jet separator" under the following set of operating
conditions: electron energy 70 e.v.; accelerating voltage 1800 V.
Emission current 80 μA; separator and source temperatures 280°C and
180°C respectively; column temperature 260°C and helium carrier flow
rate at 35 ml/min.

The mass spectrometer was equipped with a four channel multiple
ion detection unit designed and built in our laboratory (8).

Derivatization procedures. Quantities of 100-120 μl of ethanol
PG-solutions (1-0.5 μl/μl) of PGA_1, PGA_2, PGB_1, PGB_2, PGE_1, PGE_2,
$PGF_{1\alpha}$ and $PGF_{2\alpha}$ were used. The dry residues were derivatized as
follows. N-Butyl boronation. The boronated $PGF_{1\alpha}$ and $PGF_{2\alpha}$ derivatives
were prepared according to the methods described in literature (9-11).
Briefly, the method involves the formation of the 9,11-cyclic butyl-
boronate with a 100 μl solution of NBB/DMP (2.5 mg NBB/1ml DMP) at
60°C. After 10-20 min the solution is taken to dryness and the
residue silylated according to the procedure described below.
Direct silylation. Silylation of the previously boronated PG mixture
was accomplished by reaction with a mixture 1:1 of BSTFA/piperidine
(1h at 60°C). The samples were directly injected into the GC.

RESULTS AND DISCUSSION

Fig. 1 shows the kind of GLC profile obtained from a mixture
of prostaglandins A, B, E, and F (series I and II), directly
derivatized with a mixture of BSTFA/piperidine (1:1) (Fig. 1A) or
previously n-butylboronated (Fig. 1B). As already reported, the
direct approach leads in a single reaction to the formation of the
derivatives listed in Table 1 (2-4, 7).

The coelution, for instance, of both PGA_1 and PGA_2 derivatives
in a single GC peak (ret. index of 3185, Table 1), while simplifying
the GLC profile also precludes their individual quantitative assay.
However, this would not be a problem by selected ion monitoring
since the two components of the gas chromatographically unresolved
peak can be readily differentiated by monitoring the appearance of
those ions specific to the derivatives of PGA_1 at m/e 464 and PGA_2
at m/e 462, as illustrated in Fig. 2. The same reasoning would apply
to PGB_1 and PGB_2, readily detectable by monitoring the ions at m/e
480 and 478, respectively (2) (Fig. 2).

The mass spectral patterns of all of the derivatives included
in Table 1 have been reported in literature (2-4) and will not be
further discussed here. A summary of the most suitable specific ions
that can be used for SIM in each case is also included in Table 1.

However, in the case of PGEs and PGFs the small ΔI differences
observed in the retention indices of the four derivatives may easily
lead to the coelution of both classes of PGs, as shown in Fig. 1A.

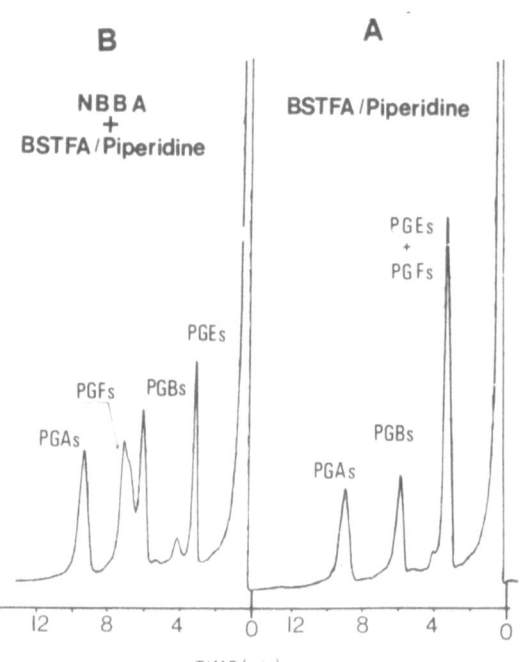

FIG. 1. Gas chromatographic profiles of a synthetic mixture
of eight prostaglandins (PGE$_1$, PGE$_2$, PGF$_{1\alpha}$, PGF$_{2\alpha}$,
PGB$_1$, PGB$_2$, PGA$_1$ and PGA$_2$) derivatized A) with a
1:1 mixture of BSTFA/piperidine (1 hr at 60°C) and
B) same as A but with prior derivatization of the
PG mixture with n-butylboronic acid (10-20 min
at 60°C). Both profiles were obtained on a 2 m
long x 2 mm i.d. glass column packed with 3% OV-17
on Gas Chrom Q 100/120 at 260°C. Under these
conditions the peak corresponding to the NBB-TMS
derivative of both PGFs is wider than the rest due
to the partial separation of the mixed PGF$_{1\alpha}$ and
PGF$_{2\alpha}$ derivatives.

TABLE 1
Derivatives of prostaglandins by direct reaction with BSTFA/piperidine

Prostaglandin	Derivative	Retention index"	Specific ion pair^
$PGF_{1\alpha}$	$PGF_{1\alpha}$ (TMS)$_4$	2807	483(55) 464(21)
$PGF_{2\alpha}$	$PGF_{2\alpha}$ (TMS)$_4$	2780	481(48) 462(33)
PGE_1	9-enol PGE_1 (TMS)$_4$	2774	481(38) 462(49)
PGE_2	9-enol PGE_2 (TMS)$_4$	2774	479(63) 460(100)
PGB_1	PGB_1	3006	480(48) 381(100)
PGB_2	PGB_2	3006	478(32) 379(100)
PGA_1	11-piperidyl, 9-enol PGA_1 (TMS)"$_3$	3185	464(100)
PGA_2	11-piperidyl, 9-enol PGA_2 (TMS)"$_3$	3185	462(100)

: Data obtained on 3% OV-17 at 260°C.

^ Figures shown represent the m/e values of the selected ions. Their relative abundance in each mass spectra are given in parenthesis.

" The nomenclature of these derivatives has been recently revised (7) from 11-piperidyl PGA(TMS)$_3$ to 11-piperidyl, 9-enol PGA(TMS)$_3$ as indicated here in accordance to their structural features.

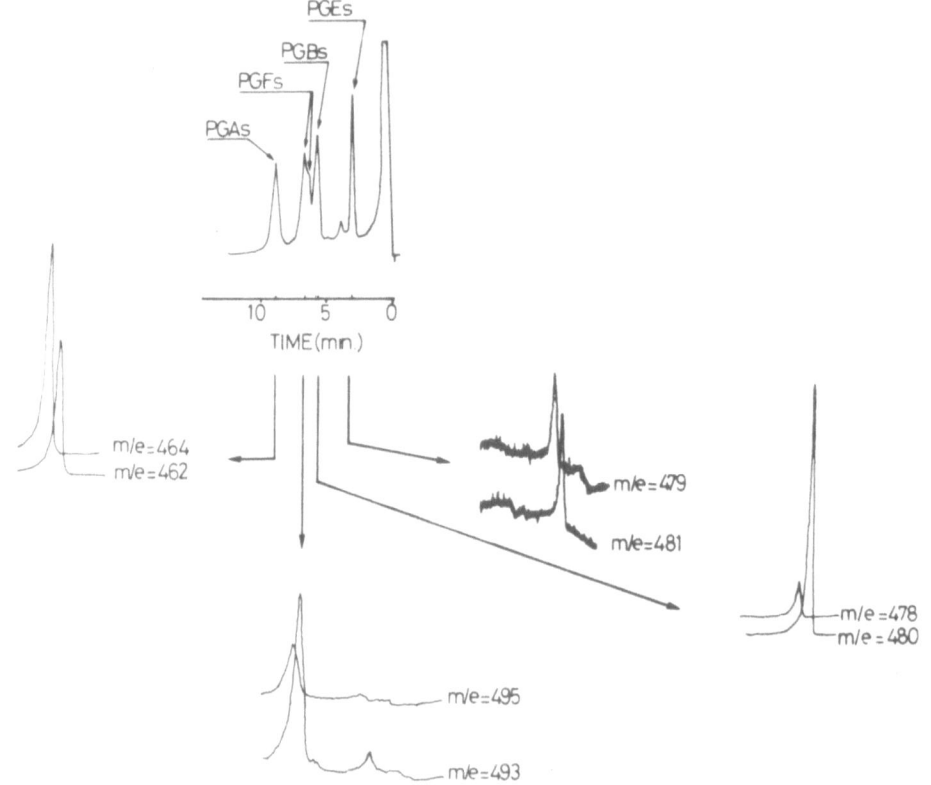

FIG. 2. Detection of the eight PGs by selected ion
monitoring. As illustrated every GC peak can
be resolved into its two components by
selective recording of the specific ions
indicated in Table 1.

The four component GLC peaks cannot be readily resolved by SIM as the
mass spectral features of all four PGs are essentially identical
(see for instance the specific ions of $PGF_{2\alpha}$ (TMS)$_4$ and 9-enol PGE_1
(TMS)$_4$ in Table 1).

Nevertheless, partial resolution of the $PGF_{1\alpha}$ can be obtained
on more efficient GC columns (2). This is illustrated by the double
peak at ≈6 minutes on the m/e 464 profile in Fig. 3.

On the other hand, PGEs and PGFs have now been completely
separated in these profiles by previous boronation of the mixture
of PGs. This only affects the PGFs by forming the corresponding
9,11 cyclic NBB $PGF_{1\alpha}$ and $PGF_{2\alpha}$ derivatives without altering the

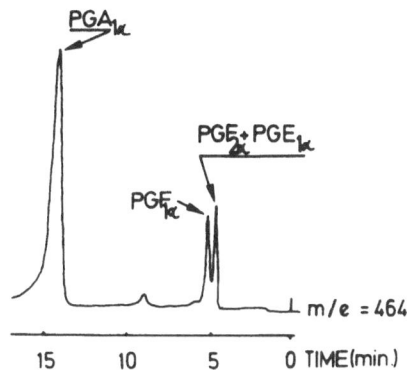

FIG. 3. Prostaglandin profile obtained by selected
        ion monitoring of the ion at m/e 464. The
        sample thus analyzed was an extract (1) of
        human seminal plasma. The gas chromatographic
        phase was also an OV-17 with similar
        characteristics to the one used for the
        analysis in Fig. 1 although its efficiency
        was much higher.

rest of PGs, which can be fully silylated next, as described. The
procedure of combined boronation-silylation results in an improved
GLC profile in the sense that it is now possible to have four
distinct peaks each representing a different PG class as shown in
Fig. 1B.

The mass spectra of the NBB-(TMS)$_2$ derivatives of PGF$_{1\alpha}$ and
PGF$_{2\alpha}$ thus obtained from the PG mixture are identical to those
reported by Wolfe and Pace Asciak (9). The molecular ions appear at
m/e 566 and m/e 564, respectively, and the M-71 ions (m/e 495 and
493) corresponding to the loss of the $C_{16}-C_{20}$ fragment are very
abundant in both cases, which is useful for their detection by SIM
as illustrated in Fig. 2. With a four channel multiple ion detection
unit the GLC PGE peak (Fig. 1B) can be resolved into its two
components, PGE$_1$ and PGE$_2$, by monitoring the specific ions at m/e
481 and 479, respectively, while the PGF peak of the NBB-(TMS)$_2$
derivative, now eluting after PGBs, as seen in Fig. 1B, can also be
resolved into its two PGF$_{1\alpha}$ and PGF$_{2\alpha}$ components by monitoring the

specific ions at m/e 495 and 493, respectively.

To summarize, these four classes of PGs can be concurrently analyzed by combined GC-MS (SIM) techniques after synthesis of their mixed n-butylboronated-TMS (PGFs), piperidyl-TMS (PGAs) and fully silylated (PGEs and PGBs) derivatives.

REFERENCES

1) J.M. Tusell, J. Roselló and E. Gelpí, Prostaglandins, 1978, 15, 219.
2) J. Roselló, J.M. Tusell and E. Gelpí, J. Chromatogr., 1977, 130, 65.
3) J. Roselló, C. Suñol, J.M. Tusell and E. Gelpí, Biomed. Mass Spectrom., 1977, 4, 237.
4) J. Roselló and E. Gelpí, J. Chrom. Sci., 1978, 16, 177.
5) F. Vane and M.G. Horning, Anal. Lett., 1969, 2, 357.
6) B.S. Middleditch and D.M. Desiderio, Prostaglandins, 1972, 2, 115.
7) J. Roselló and E. Gelpí, Biomed. Mass Spectrometry, in press.
8) F. Artigas, E. Gelpí, M. Prudencio, J.A. Alonso and J. Baillart, Anal. Chem., 1977, 49, 543.
9) L.S. Wolfe and C. Pace-Asciak, J. Chromatogr., 1971, 56, 129.
10) A.G. Smith and C.S.W. Brooks, Biomed. Mass Spectrom., 1977, 4, 258.
11) R.W. Kelly, Anal. Chem., 1973, 45, 2079.

ANALYSIS OF URINARY STEROIDS BY LIQUID-GEL CHROMATOGRAPHY AND GAS

CHROMATOGRAPHY - MASS SPECTROMETRY. CLINICAL APPLICATIONS

R.J.Begue, M.Dumas, H.Losty[+], M.Moriniere, C.Perrier

and P.Padieu

Centre Hospitalier Régional Universitaire,Dijon,France;

[+]University of Glasgow, Glasgow, Scotland, U.K.

INTRODUCTION

In clinical chemistry extensive gas phase investigations (GLC and GC-MS) of urinary steroids should be considered as comprehensive analyses carried out on metabolic families of compounds. Such a procedure must cope with their physiopatho-logical consistency. First, such assays apply to urine specimens of extreme diversity due to age, sex, pregnancy and clinical status etc.... Therefore, the selection of metabolic groups of steroids should give a clear cut clinical significance. Second, in order to develop these assessments to a useful purpose they should not only compare with results obtained by more conventional methods but also be explicit and complete them.

The high complexity by number and by species of the steroids of urine samples cannot be resolved by the separative capacity of GLC columns. In order to get a satisfactorily separative and quantitative chromatography for steroid metabolites one must not only meet the above requirements but should also offer a workable procedure which can be handled routinely in the clinical laborato-ry with the aim of achieving better diagnostic and therapeutic uses. In this article, we will describe such a method operating already in the laboratory of clinical chemistry and devised for steroid multigroup assays. It will be illustrated by several clinical case applications.

355

MATERIAL

The following chemicals were used:

(7-^{3}H)-dehydroepiandrosterone (SA 22Ci/mmol) was purchased from the Radiochemical Centre (Amersham,England). Except for 5β-cholestane-3α-ol, obtained from Sigma (St.Louis,USA) and for 3α, 16α -dihydroxy-5β-androstan-17 one, 3α,16α-dihydroxy-5α-androstan-17-one, 3α ,6α-dihydroxy-5β-androstan-17-one, 3α,15α-dihydroxy-5β-androstan-17-one and 3β, 16β -dihydroxy-5-androsten-17-one obtained from Steraloids Inc., (Wilton,USA) all other reference steroids were purchased from Ikapharm (Ramat-Gan,Israel).

All solvents of analytical grade and anhydrous sodium sulfate were purchased from Merck (Darmstadt,Germany).

Solution of dipyridinium sulfate (1N) for extraction of steroid conjugates (ref.1) was prepared by mixing equal volumes of 2N pyridine and 2N sulfuric acid. The acidified ethyl acetate for solvolysis was prepared by saturating redistilled ethyl acetate (10 vol.) with 2M sulfuric acid (1 vol.) in a separatory funnel (ref.2).

All other chemicals of analytical grade were: N-O-bis-(trimethylsilyl)-trifluoro-acetamide (BSTFA) obtained from the Pierce Chemical Co., (Rockford,Ill.USA); methoxyamine hydrochloride obtained from Eastman Organics Chemicals (Rochester,N.Y.USA); Helix pomatia digestive juice purchased from l'Industrie Biologique Française,(Gennevilliers,France) and Sephadex LH-20 obtained from Pharmacia Fine Chemicals,(Uppsala,Sweden).

The solvent mixtures used for Sephadex LH-20 chromatography were S1: chloroform-heptane-methanol (5:5:1 by vol) and S2: chloroform-methanol (3:1 by vol.).

METHODS

Column chromatography on Sephadex LH-20. Glass columns, 50cm x 0.7cm i.d., with a 50 ml solvent reservoir and teflon stopcock were used. Sephadex LH-20 (5g) was suspended in 25 ml of S1. The mixture was stirred for 15 min, added to the column, and let stand overnight. Solutions of reference standards (100μg of each) were used to determine the elution volumes and the percentage recovery of individual steroids after chromatography on Sephadex LH-20 as estimated by GLC. The neutral steroids were eluted with 50 ml of S1, and the polar steroids with 50 ml of S2. During chromatography, the pressure was adjusted to maintain a flow rate of approximately 10-15 ml/h. Fractions of 5 ml were collected and 100μg of 5β-cholestane-3α-ol (Std) was added to each fraction. After evaporation to dryness and TMS or MO-TMS derivatization,the steroids of each fraction were identified by GLC.

Gas liquid chromatography and gas chromatography-mass
spectrometry. A Becker 409 gas chromatograph equipped with a flame
ionization detector was used. The glass columns were 3.60 x 3 mm
with 0.5% OV-1 or 0.5% OV-17. The flow rate of carrier nitrogen
gas was 30 ml/min. On OV-1 and OV-17, the temperature was programmed from 180°C to 250°C at a rate of 1.2°C/min. The temperature
of injection port and detector were 245°C and 275°C respectively.

GC-MS analysis was carried out using an LKB 9000 instrument
equipped with a 1% OV-1 column. The flow rate of carrier helium
gas was 30 ml/min. The temperature of injection port, separator
and ion source were 250°C, 280°C and 290°C respectively. Accelerating voltage, energy of bombarding electrons and ionizing current
were 3.5 kV, 22.5 eV and 60 μA respectively. The column was
programmed from 180°C to 260°C at 1°C/min.

Procedure for the analysis of steroids in urine. The urinary
conjugates of steroids were extracted according to Okerholm et al.
(ref.1). The ethyl acetate extracts were taken to dryness.
Acidified ethyl acetate (20 ml) was added and the solution was
incubated at 39°C for 24 h. Following this solvolysis, the ethyl
acetate was taken to dryness and the residue dissolved by sonication in distilled water (20 ml). The pH was adjusted to 5.20 by
addition of acetic acid. Sodium acetate buffer (2 ml, pH 5.20,
0.1M), Helix pomatia digestive juice (1,000 IU of β-glucuronidase
per ml of urine) were added and the sample incubated at 37°C for
48h. A further 1,000 IU of β-glucuronidase was added after the
first 24 h. Following hydrolysis, the free steroids were extracted
twice with ethyl acetate (1:1 by vol.) and diethyl ether (1:1 by
vol.). The combined organic phases were washed with sodium
bicarbonate 1M (5 ml) and distillated water until neutral, dried
by filtration on anhydrous sodium sulfate, and evaporated to dryness. The residue was dissolved by sonication in 0.5 ml of S1 and
the solution transferred to the Sephadex LH-20 column. The steroids
were eluted with 50 ml of S1. The polar steroids excreted during
pregnancy and various pathological states were eluted with 50 ml
of S$_2$ (ref.3). Chromatography on Sephadex LH-20 and subsequent
gas-chromatographic analyses were carried out as previously
described.

Identification and quantitation of steroids. The steroids
were identified by a direct comparison of MU values (4) obtained on
OV-1 and OV-17 columns and on the basis of identity of a mass
spectrum with that of the reference steroid. The steroids present
in each fraction were quantitated by addition of a known amount of
5β-cholestane-3α-ol (25 to 150 μg) to the samples prior to TMS or
MO-TMS derivatization. The total concentration of each identified
metabolite was calculated after the following parameters had been
determined: i. initial sample volume as a percentage of the total
24 hour urine; ii. efficiency of extraction of the conjugated
steroids (1) and the percentage recovery of the free steroids after

hydrolysis; iii. percentage recovery after chromatography on
Sephadex LH-20; iv. coefficient of response of the reference
steroids (5).

RESULTS

Gas liquid chromatography and gas chromatography-mass
spectrometry of reference steroids. The MU values on 0.5% OV-1 and
0.5% OV-17 and the coefficient of response as determined on OV-1
are given including the mass spectral data of the MO-TMS derivatives of eleven 17-oxosteroids in Table 1. Only the m/e values
for the molecular and the most characteristic fragment ions are
represented. The five 3,16-dihydroxy-17-oxosteroids were analysed
as TMS derivatives, since each isomer produces two distinct peaks
when the MO-TMS derivative is prepared. Their chromatographic and
mass spectral characteristics are presented in Table 2. Table 3
and Table 4 give the MU values and mass spectra data for the TMS
derivatives of the clinically important C_{19} and C_{21} steroids.
Table 5 lists elution volumes of the steroids and their percentage
recoveries after chromatography on Sephadex LH-20 as performed
under the previously described experimental conditions. The
percentage recovery determined as the mean from three experiments
was of the order of 82-97%. Chromatography on Sephadex LH-20 and
gas chromatography appear to be good complementary processes for
separation of steroids, as evidenced by the different order of
elution of reference steroids on each type of column. Where
separation was not achieved by gas-phase analysis, the individual
steroids were separated on the liquid gel column.

Analysis of biological samples. Figs.1,2 and 3 illustrate the
gas chromatographic separation on OV-1 and mass spectra of
steroids excreted by a 33-year old woman with minor virilism (case
1) associated with arterial hypertension.

The excretion of 17-oxosteroids, determined colorimetrically
was 14.80 mg/24h. The testosterone plasma level was normal, 400
ng/l. Fig 2 illustrates the gas chromatogram of LH-20 fraction 3.
Analysis by GC-MS identified the peak at 24.73 as 3α-hydroxy-5-
androsten-17-one and the mass spectra is presented in Fig.2. The
identity of the remaining three peaks was confirmed by mass
spectrometry as 3α-hydroxy-5α-androstan-17-one (MU = 24.92), 3α-
hydroxy-5β-androstan-17-one (MU = 25.13) and 3β-hydroxy-5-androsten
-17-one (MU = 25.53). The amount of each steroid excreted was
calculated and is included in Table 4.

Fig. 3 illustrates the gas chromatogram obtained from
fraction 4. The steroids excreted were identified as 3α-hydroxy-
5β-androstan-11,17 dione (MU = 26.03), 3α,11β-dihydroxy-5α-
androstan-17-one (MU = 26.83), a trace amount of 5α-pregnane-3α,
20α-diol, 5α-pregnane-3α, 20α-diol (MU = 27.47) and 5-pregnene-3α,
16α, 20α-triol (MU = 28.61).

TABLE 1

Methylene unit values (MU) on 0.5% OV-1 and 0.5% OV-17 response coefficient (RC) on OV-1, m/e values of molecular ion (M⁺) and of characteristic fragment ions (predominant peaks) of 17-oxosteroids as MO-TMS derivatives

Reference compounds (MO-TMS)	MU OV-1	MU OV-17	RC OV-1	M⁺	Predominant peaks
3α-hydroxy-5α-androstan-17-one	24.94	26.83	1.13	391	213-215-255-270-301-360-376
3α-hydroxy-5β-androstan-17-one	25.18	27.01	1.27	391	213-215-255-270-360-376
3β-hydroxy-5-androsten-17-one	25.56	27.86	1.08	389	129-260-268-284-299-358-374-389
3β-hydroxy-5α-androstan-17-one	25.60	27.83	0.82	391	213-215-255-270-343-345-360-376
3α-hydroxy-5α-androstan-11,17-dione	25.91	28.59	1.22	405	129-147-284-374-390-405
3α-hydroxy-5β-androstan-11,17-dione	26.07	28.59	0.96	405	261-269-284-300-315-374-390-405
3α,6α-dihydroxy-5β-androstan-17-one	26.48	27.89	0.84	479	96-211-213-252-253-268-358-448
3β-hydroxy-5α-androstan-11,17-dione	26.71	29.58	0.89	405	109-129-147-284-344-374-390
3α,15α-dihydroxy-5β-androstan-17-one	26.73	27.89	0.82	479	133-168-184-201-268-332-358-448-464-479
3α,11β-dihydroxy-5α-androstan-17-one	26.84	28.34	0.84	479	125-182-198-213-268-358-448-464-479
3α,11β-dihydroxy-5β-androstan-17-one	27.09	28.54	1.05	479	125-143-182-198-213-215-268-358-448-464-479

TABLE 2

Data of the same parameters as in Table 1 but for 3,16-dihydroxy-17-oxosteroids as TMS derivatives

Reference compounds (TMS)	MU OV-1	MU OV-17	RC OV-1	M^+	Predominant peaks
3α,16α-dihydroxy-5β-androstan-17-one	26.04	27.70	0.91	450	117-162-201-216-306
3α,16α-dihydroxy-5α-androstan-17-one	26.16	26.79	0.86	450	106-117-190-201-216-306
3β,16α-dihydroxy-5-androsten-17-one	26.99	28.90	0.64	448	117-129-175-196-199-214-304
3β,16α-dihydroxy-5α-androstan-17-one	27.14	29.00	0.84	450	106-117-201-216-306
3β,16β-dihydroxy-5-androsten-17-one	27.32	29.20	0.66	448	117-129-175-196-199-214-304-433

TABLE 3

Data of the same parameters as in Table 1 for $C_{19}O_2$ and $C_{19}O_2$ steroids as TMS derivatives

Reference compounds	MU OV-1	MU OV-17	RC OV-1	M^+	Predominant peaks
5β-androstane-3β,17β-diol	24.84	25.20	1.43	436	129-215-217-241-256-346-421
5α-androstane-3α,17β-diol	25.05	25.47	1.80	-	129-148-215-217-241-256-331-346
5β-androstane-3α,17β-diol	25.05	24.47	1.80	-	129-215-217-241-256-346
5α-androstane-3β,17β-diol	25.64	26.25	1.32	-	129-148-149-177-215-217-256-331-421
5-androstene-3β,17β-diol	25.78	26.56	1.26	434	129-213-215-289-254-305-329-344-419
3β,17β-dihydroxy-5-androsten-16-one	27.32	29.19	0.88	448	129-143-171-214-304-358-433-448
5-androstene-3β,11β,17β-triol	27.65	29.26	0.82	522	129-211-213-237-252-327-342
5-androstene-3β,16α,17β-triol	28.29	29.61	1.18	522	129-147-213-239-329-432
5-androstene-3β,16β,17β-triol	28.44	28.80	1.07	522	129-147-213-239-329-432

TABLE 4

Data of the same parameters as in Table 1 but for the C_{21} steroids as TMS derivatives

Reference compounds (TMS)	MU OV-1	MU OV-17	RC OV-1	M^+	Predominant peaks
3α-hydroxy-5β-pregnan-20-one	26.11	28.35	0.92	390	215-230-285-300-375
3β-hydroxy-5-pregnan-20-one	26.67	–	0.73	388	129-241-259-283-298-333-375
3α,17-dihydroxy-5β-pregnan-20-one	26.96	28.49	0.88	478	215-255-345
5α-pregnane-3α,20α-diol	27.26	27.86	1.10	464	117-269-284-449
5β-pregnane-3α,20α-diol	27.47	28.05	1.19	464	117-269-284-449
3α,6α-dihydroxy-5β-pregnan-20-one	27.55	–	1.04	478	251-298-299-388-478
3α,16α-dihydroxy-5β-pregnan-20-one	27.66	–	1.00	478	96-109-157-159-172-186-255-388-463
5-pregnene-3β,20α-diol	27.97	28.93	1.09	462	117-129-282-372-462
3β,16α-dihydroxy-5-pregnen-20-one	28.50	30.22	0.99	476	129-141-157-159-172-461
5β-pregnane-3α,17,20α-triol	28.94	29.87	0.98	480	117-215-255-273
5-pregnene-3β,17,20α-triol	29.49	30.82	0.67	478	117-129-213-253-271-445-463
5-pregnene-3β,16α,20α-triol	29.66	29.94	1.18	550	117-129-141-156-157-460
5α-pregnane-3β,16α,20α-triol	29.70	–	0.95	552	117-141-156-157-462
3α,17,20α-trihydroxy-5β-pregnan--11-one	29.95	31.55	0.71	494	117-287-404

TABLE 5
Elution volume and percentage recovery of unconjugated steroids
after chromatography on Sephadex LH-20.

Reference compounds	Effluent volume (ml)	Percentage recovery
5-cholestene-3β-ol	5 - 15	96
5α-hydroxy-5β-pregnan-20-one	10 - 15	96
3β-hydroxy-5-pregnen-20-one	10 - 15	89
3α-hydroxy-5β-androstan-17-one	10 - 20	97
3β-hydroxy-5-androsten-17-one	10 - 20	90
3α-hydroxy-5β-androstan-11,17-dione	15 - 20	97
3α,17-dihydroxy-5β-pregnan-20-one	15 - 25	-
3α,6α-dihydroxy-5β-pregnan-20-one	15 - 25	92
5α-pregnane-3α,20α-diol	15 - 25	98
5β-pregnane-3α,20α-diol	15 - 25	97
5-pregnene-3β,20α-diol	15 - 25	94
5-pregnene-3β,20β-diol	15 - 25	92
3α,11β-dihydroxy-5α-androstan-17-one	20 - 30	95
3β,17β-dihydroxy-5-androsten-16-one	20 - 30	98
3β,16α-dihydroxy-5-pregnen-20-one	20 - 30	88
3β,16β-dihydroxy-5-androsten-17-one	20 - 30	93
5β-pregnane-3α,17,20α-triol	20 - 30	92
5α-androstane-3α,17β-diol	25 - 30	82
5-androstene-3β,17β-diol	25 - 30	91
5β-androstane-3α,17β-diol	25 - 30	85
5α-androstane-3β,17β-diol	25 - 30	86
5-pregnene-3β,16α,20α-triol	30 - 35	94
5-pregnene-3β,17, 20α-triol	30 - 35	78
5β-androstane-3α,11β,17β-triol	40 - 45	97
5-androstene-3β,16α,17β-triol	45 - 55	92
5-androstene-3β,16β,17β-triol	45 - 55	94
5-androstene-3β,11β,17β-triol	50 - 60	87

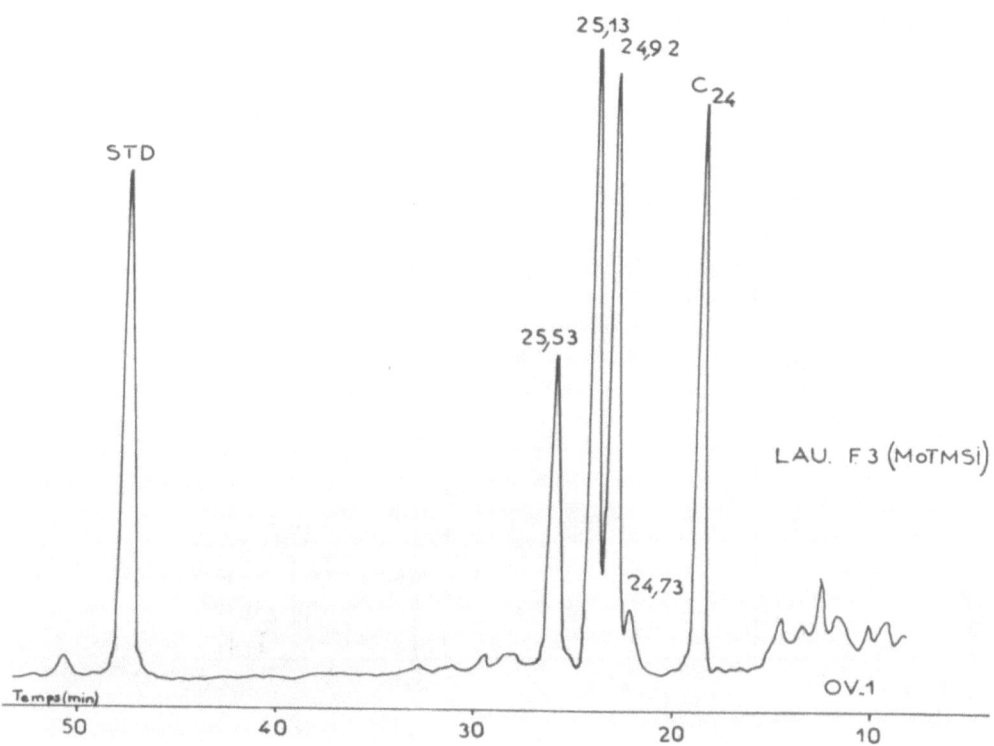

Fig.1. Case report 1: **gas** chromatogram on OV-1 of MO, TMS-steroids eluted in LH-20 fraction 3 and numbered by their MU values. They were identified by GLC and GC-MS (see text for their nomenclature names).

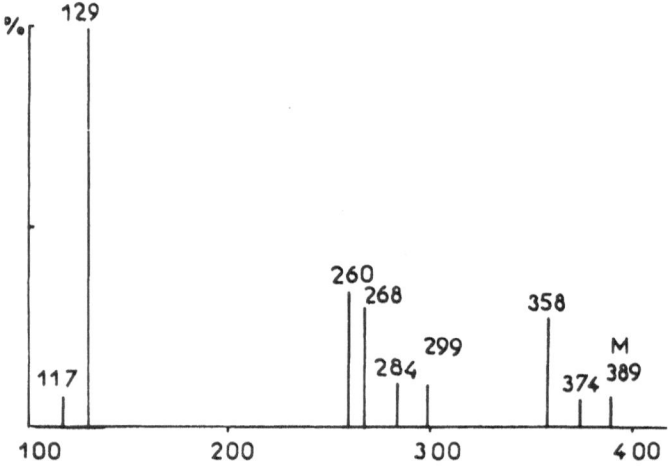

Fig. 2. Mass spectrum of compound with an MU = 24.73 on OV-1
eluted in Sephadex LH-20 fraction 3 and identified by GC-MS as MO,
TMS 3α-hydroxy-5-androsten-17-one

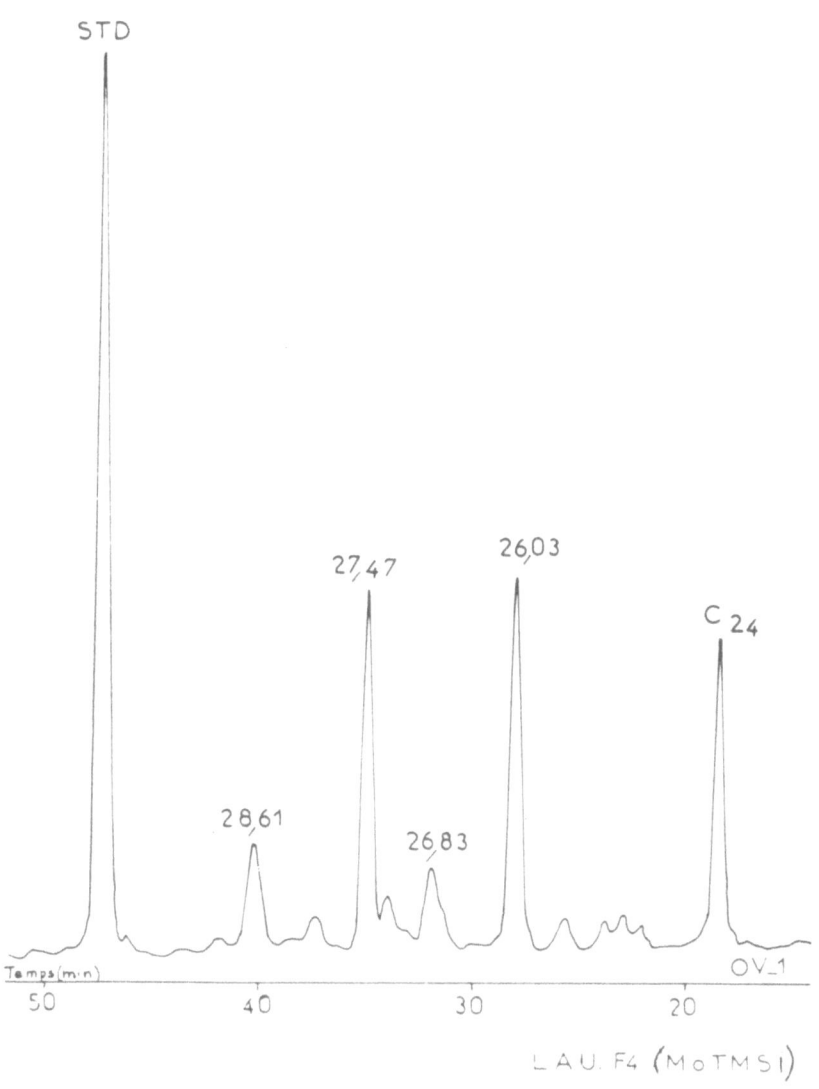

FIG. 3. Case report 1: same as in Fig. 1 but in the case
of steroids eluted in fraction 4 from the
Sephadex LH-20 column.

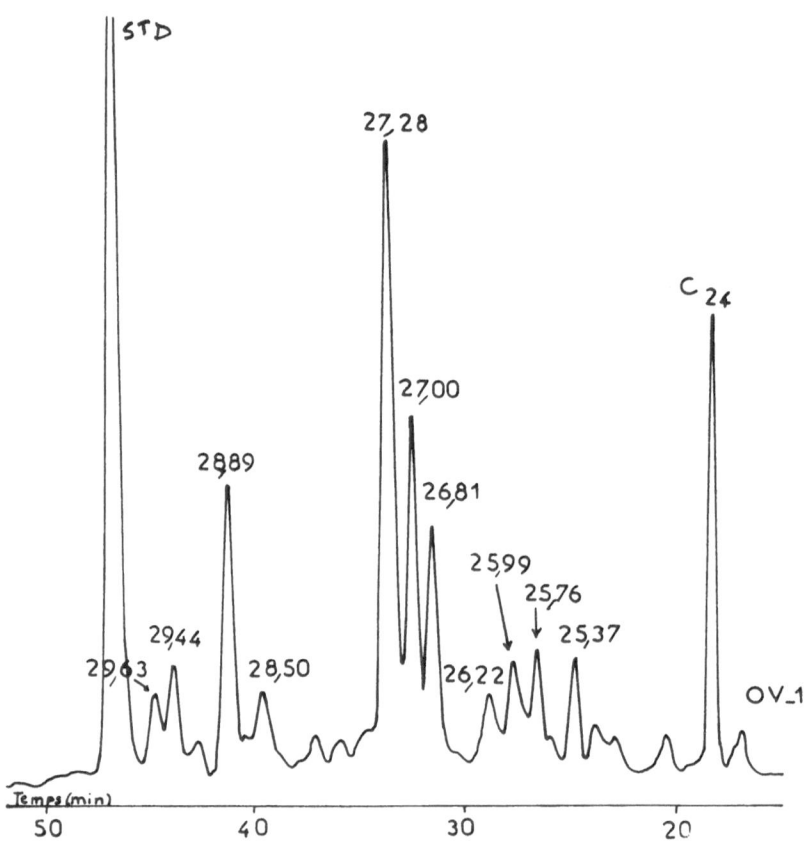

Fig. 4   Case report 1: same as in Fig.1 but in the case of
steroids eluted in fraction 5 from the Sephadex LH-20 column.

Fig.4 illustrates the chromatogram obtained from fraction 5.
The nine compounds were identified as: 5β-adrostane-3α,17β-diol
(MU = 25.37), 5-androstene-3β, 17β-diol (MU = 25.76), 3α,11β-di-
hydroxy-5α-androstan-17-one (MU = 26.81), 3α,11β-dihydroxy-5β-
androstan-17-one (MU = 27.00), 3β, 16α-dihydroxy-5-androsten-17-one
(MU = 27.28) the mass spectrum of which is shown in Fig. 5,
5-pregnene-3α,16α,20α-triol (MU = 28.50), 5β-**pregnane-3α,17,20α-tr**
-triol (MU = 28.89), 5-pregnene-3β,17, 20α-triol (MU = 29.44) and
5α-pregnane-3β,16α,20α-triol (MU = 29.63).   Analysis of the
remaining fractions indicates that the total neutral steroids were
contained in an elution volume of 15-30 ml.

Repeated analysis of urine samples proved that the elution
pattern of each steroid was reproducible.   However in some cases,

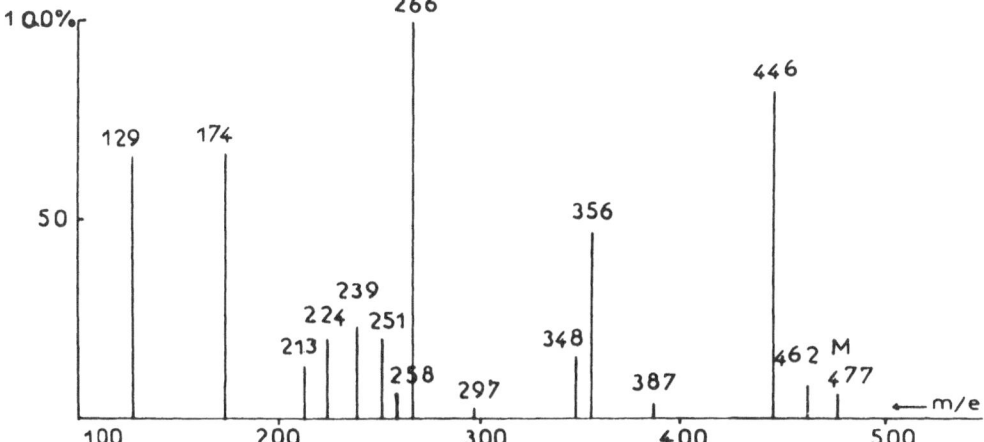

Fig. 5. Case report 1: mass spectrum of compound with an MU = 27.28
on OV-1 eluted in Sephadex LH-20 fraction 5 and identified by GLC
and GC-MS as MO, TMS 3β,16α-dihydroxy-5-androsten-17-one.

compounds (often of therapeutic origin) interfered during elution
and could retard the occurrence of some steroids in the group
separated on the lipophilic gel, as can clearly be seen by
comparison of Fig.3 and Fig.6.

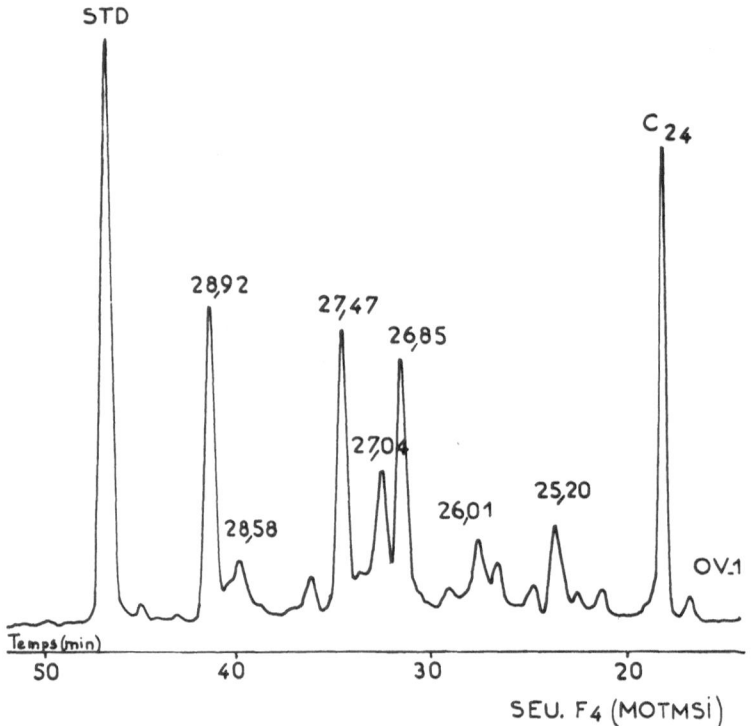

Fig. 6 Case report 2: gas chromatogram on OV-1 of MO,TMS-steroids
eluted in Sephadex LH-20 fraction 4 and numbered by their values.
They were identified by GLC and GC-MS (see text for their
nomenclature names).

    Fig. 6 illustrates the gas chromatogram obtained from the
fraction 4 of a urine extract of a 32-year old women (case 2).
Urinary 17-oxosteroid excretion was 16.50 mg/24 h and plasma
testosterone was elevated to 800 ng/1. A comparison of Fig.6
with Fig. 4 which represents the gas chromatogram of fraction 5 of
the urine from case 1 shows that the principal steroids were
present in both fractions, but they differed in that 3β,16α-
dihydroxy-5-androsten-17-one (MU = 27.28) excreted as 1.60 mg/24 h

is present only in Fig. 4. The excretion of this steroid, although
not more commonly associated with low renin essential hypertension,
was indicative of mineralocorticoid potency which may be related
to the hypertensive state of this case 1. However up to now the
other C_{16} isomer, 3β, 16β-dihydroxy-5-androsten-17-one has been
found as the compound associated with such hypertension (6). Table
6 compares the amounts of the principal metabolites excreted by the
two patients.

TABLE 6

Gas chromatographic assessment of steroids excreted by two female
patients- Case 1 and Case 2, suffering from virilism of adrenal
origin as interpreted from Figs. 1,2 and 3 for the case report
Case 1 and from Fig. 6 for the case report Case 2.

	Case 1	Case 2
Plasma level of testosterone (ng/l)	400	800
Urinary 17-oxosteroids (Zimmerman method) (mg/24 h)	14.80	16.50
Androstane series		
. 3α-hydroxy-5-androsten-17 one	0.30	0.60
. 3α-hydroxy-5α-androstan-17-one	2.60	4.70
. 3α-hydroxy-5β-androstan-17-one	2.70	5.40
. 3β-hydroxy-5-androsten-17-one	1.60	2.55
. 3α-hydroxy-5α-androstan-11,17-dione	0.40	0.50
. 3α-hydroxy-5β-androstan-11,17-dione	1.00	0.10
.3α,11β-dihydroxy-5α-androstan-17-one	0.90	1.00
. 3α,11β-dihydroxy-5β-androstan-17-one	1.10	0.75
. 3β,16α-dihydroxy-5-androsten-17-one	1.60	0.20
Total amount	12.20	15.80
Pregnane series		
. 5β-pregnane-3α,20α-diol	0.80	3.10
. 5β-pregnane-3α,17,20α-triol	0.65	1.10
. 5-pregnene-3β,17,20α-triol	0.25	0.50
. 5α-pregnene-3β,16α,20α-triol	0.20	-

The application of the method to the investigation of the
steroids excreted during pregnancy is illustrated by the gas
chromatograms of urine samples obtained at 8 weeks (Fig.7) and 40
weeks (Fig.8) of gestation. These steroids analysed as TMS ether
derivatives by GLC were eluted in fraction 5 and identified as:
3α 16α-dihydroxy-5β-androstan-17-one (MU = 25.78), 3α,17-dihydroxy
-5β-pregnan -20-one (MU = 25.97), 3α,11β-dihydroxy-5β-androstan-17

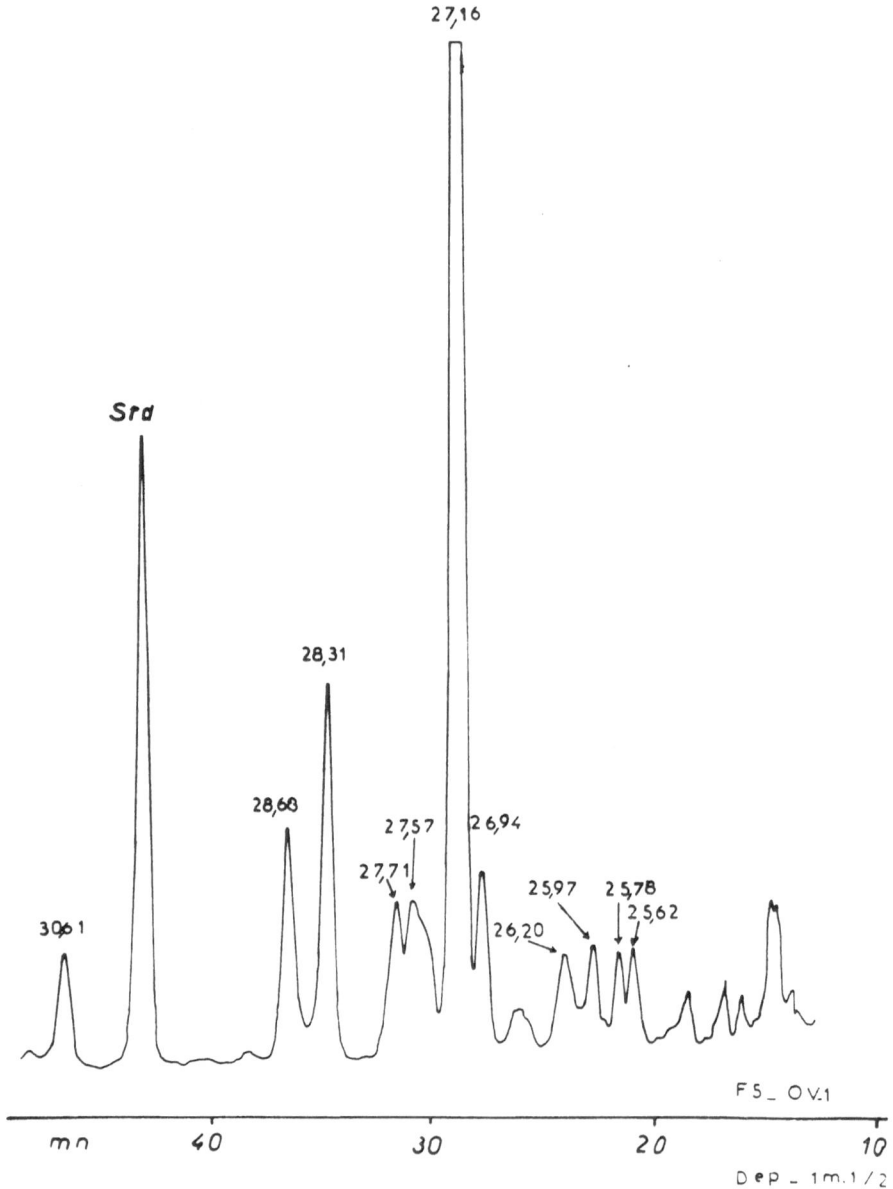

FIG. 7. Normal pregnancy at 8 weeks: gas chromatogram on
       OV-1 of TMS steroids eluted in Sephadex LH-20
       fraction 5 and numbered by their MU values. They
       were identified by GLC and GC-MS (see text for
       their nomenclature names).

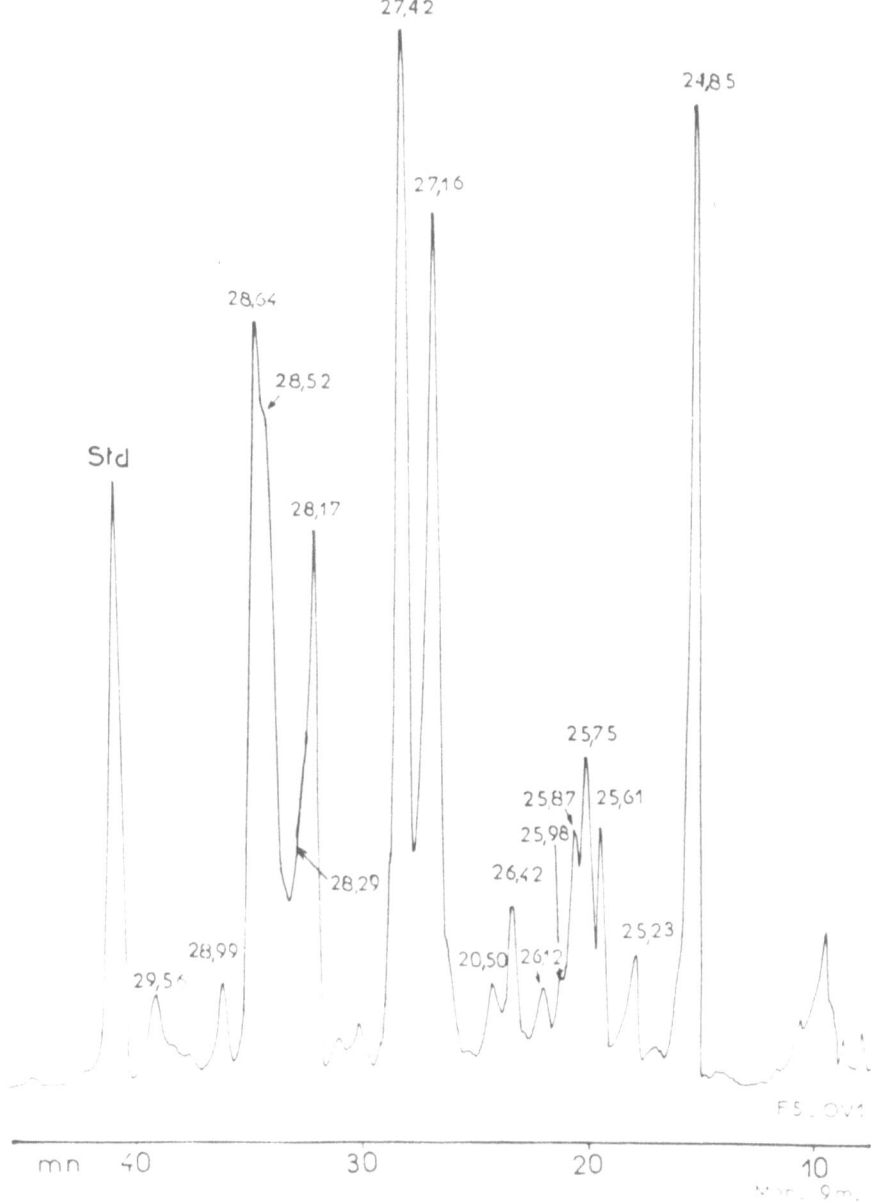

Fig. 8. Normal pregnancy of 40 weeks: gas chromatogram on OV-1 of TMS steroids eluted in Sephadex LH-20 fraction 5 and numbered by their MU values. They were identified by GLC and GC-MS (see text for their nomenclature names).

-17-one (MU = 26.20), 5α-pregnane-3α,20α-diol (MU = 26.94),5β-
pregnane-3α,20α-diol (MU = 27.16), 3α,16α-dihydroxy-5β-pregnan-20-
one (MU = 27.42), 5-pregnene-3β,20α-diol (MU = 27.71), 5-pregnene-
-3α,16α,20α-triol (MU = 28.31) and 5β-pregnane-3α,17,20α-triol
(MU = 28.68). Apart from the obvious quantitative differences the
40 week urine sample contained in fraction 5 the following additional steroids, as illustrated in Fig.8: estrone (MU = 25.23),
3α,16α-dihydroxy-5α-androstan-17-one (MU = 25.86), 3β,6α-dihydroxy
-5α-pregnan-20-one (MU = 28.17), 5β-pregnane-3α,16α,20α-triol
(MU = 28.64) and 5α-pregnane-3β,16α,20α-triol (MU = 29.51).

Fig. 9 shows the gas chromatogram of the TMS derivatives of
those steroids eluted in fraction 11 and which were contained in
a urine extract from a woman 38 weeks pregnant. It demonstrates
how the sequential elution neutral steroids - polar steroids can
be achieved with two different chromatographic solvents. In this
fraction, the following four estrogens were identified: 11-dehydro-
estradiol (MU = 25.26), estradiol 17β (MU = 25.96) (7), 16α-
hydroxy-estrone (MU = 27.05) and 16-oxoestradiol (MU = 27.37) (8)
and two components tentatively identified as 2ε, 3ε, 16ε-trihy-
droxy-5ε-pregnan-20-one (MU = 28.18) (9) and 1ε,3ε,16ε-trihydroxy-
5ε-pregnan-20-one (MU = 29.26).

Fig. 10 illustrates an application of the method in neonatal
pathology. Quantitative analysis of the 3β-hydroxy-5-ene steroids
which showed the characteristic steroids excreted by the new-born
and also those steroids associated with the 21-hydroxylase deficiency namely 3α,17-dihydroxy-5-pregnan-20-one, 5β-pregnane-3α,
17,20α-triol and 3α,17,20α-trihydroxy-5β-pregnan-20-one was
achieved by liquid-gel and gas-liquid chromatography. The gas
chromatogram of the three steroids quantitatively eluted in
fraction 6 was illustrated in Fig.10. The steroids analysed as
MO-TMS derivatives were identified as 3α,17-dihydroxy-5β-pregnan-
20-one (MU = 27.26), 3β,17β-dihydroxy-5-androsten-16-one (MU = .
28.00) and 3β,16α-dihydroxy-5-pregnen-20-one (MU = 29.19).

DISCUSSION

The few examples of clinical cases cited herein demonstrate
the usefulness of the liquid lipophilic gel and the gas-liquid
chromatographic tandem for steroid analysis. The method comprises
a combination of steps, each of which must be strictly standard-
ized and monitored. The volume of urine used when investigating
hormonal virilism was determined by the amount of 17-oxosteroid
excreted as evaluated colorimetrically and accordingly 1/20 to
1/100 of the total 24-hour urine sample is used. In pathological
pregnancies the volumes used were 1/10, 1/15 and 1/20 corresponding to the first, second and third trimesters of pregnancy
respectively. Neonatal analysis required 1/3 of the diurnal

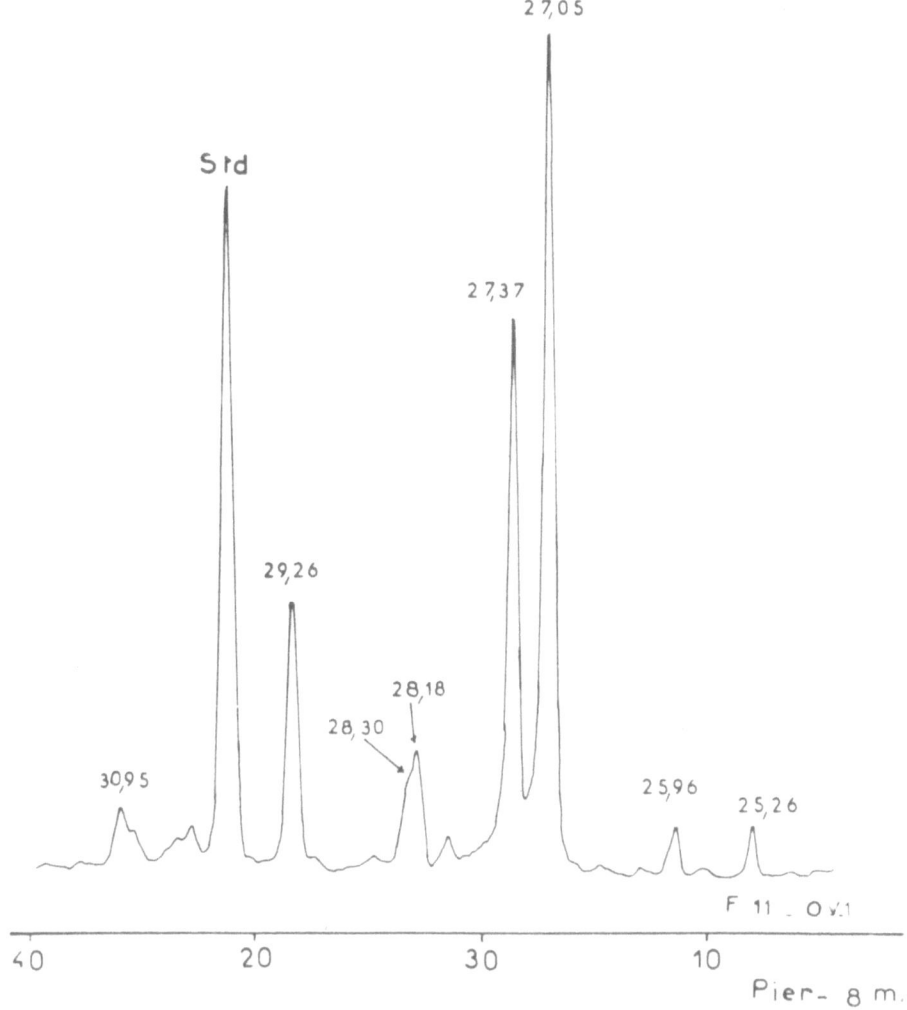

Fig. 9. Normal pregnancy of 38 weeks: gas chromatogram on OV-1 of TMS steroids eluted in Sephadex LH-20 fractio- 11 and numbered by their MU values. They were identified by GLC and GC-MS (see text for their nomenclature names).

volume. Sample pretreatment prior to chromatography consisted of four sequential processes: i. initial extraction of the conjugated steroids, ii. complete hydrolysis of the sulfoconjugates by solvolysis, iii. enzymatic hydrolysis of the glucuroconjugates by Helix Pomatia enzyme, iv. extraction of the free steroids into initially ethyl acetate and thence into diethyl ether. The

recovery percentage of the extraction process was determined
using the tritiated steroid appropriate to the steroid family
studied.   Chromatography on Sephadex LH-20 is both a means of
purifying the urine extract and of separating the constituent
steroids prior to quantitation by gas chromatography.   The order
of elution was reproducible and such that only the neutral
steroids (with the exception of estrone) are eluted by the first
solvent system, while the polar steroids were only eluted after
changing to a different solvent.   The differential elution of
neutral and polar steroids is of analytical importance when
investigating the excretion of C18, C19 and C21 steroids during

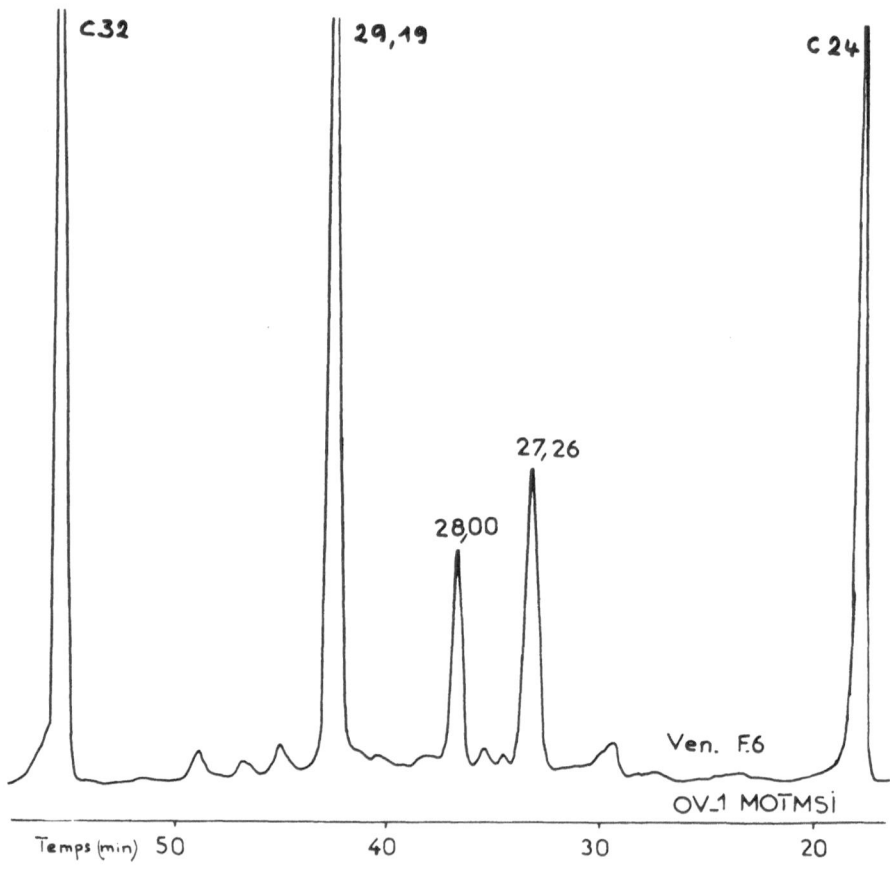

Fig. 10.   Congenital adrenal hyperphasia in a one day-old new-born:
gas chromatogram on OV-1 of MO, TMS steroids eluted in Sephadex LH
LH-20 fraction 6 and numbered by their MU values.   They were identified by GLC and GC-MS (see text for their nomenclature names).

normal and pathological pregnancies. The two chromatographic
procedures are complementary in that those compounds, such as: 3α-
hydroxy-5β-androstan-17-one and 5β-androstane-3α,17β-diol which
were not separated by GLC had different elution volumes which
permitted subsequent precise quantitation.

Temperature programming was employed for gas chromatographic
analysis. Comparison of retention times on OV-1 and OV-17 phases
of differing polarity with the values obtained for authentic
standards was a sufficient criterium for identification of commonly
occurring steroids. The 17-oxosteroids were analysed as MO-TMS
derivatives. However persilylation of 3α,11β-dihydroxyandrostan-
17-one required a minimum of 12 hours at 60°C in the following
reagent solution of : pyridine : BSTFA:TMCS (1:10:1 by vol.).
The other characteristic steroid group types, i.e.:androstane-3,17β
-diol, 3β-hydroxy-5-ene-steroids, 20-oxo-21-deoxysteroids, 20-
hydroxy-21-deoxy-steroids, 17-20-dihydroxy-21-deoxysteroids and
estrogens were analysed as TMS derivatives prepared with the
described silylating agent.

The excreted steroids were quantitated through relative
determinations to the internal standard 5β-cholestane-3α-ol which
had the advantage of being an exogenous steroid undergoing the same
derivatization process. Exhibiting a straight linear calibration
curve, it monitored the derivatization yield. It had a retention
time distinct from any other excreted steroids and yet was eluted
within the time scale used for steroid analyses.

Analyses by gas-chromatography-mass spectrometry were performed: i) on solutions of standards to verify the structure and
stability of the steroid derivatives; ii) on biological samples,
where it was necessary to establish or confirm the identity of a
compound showing an odd chromatographic behaviour, which might be
due to abnormal amount or to abnormal occurrence, after its initial
isolation by gas chromatography; iii) systematically in clinical
research to analyse biological samples that are representative of
well characterised endocrinological disorders. Effectively such
studies will permit the association of certain steroid metabolites
with known localized specific lesions be they of ovarian or
adrenal, fetal or placental origin.

CONCLUSION

The tandem chromatographic separative system consisting of
liquid-gel chromatography and gas-liquid chromatography and its
ancillary methods which are either different like colorimetry or
similar like mass spectrometry coupled to gas chromatography, has
acquired a high power of separativity to ensure a good resolution
of complex mixtures of steroids such as is found in urine. Such a

method therefore involves the expansion of well characterized steroids to include those which bear oxygen atoms at the C_2, C_6, C_7, C_{15}, C_6, C_{16}, and C_{19} positions of these steroids and which could be analysed by GLC and GC-MS. Investigations into the nature of new metabolites can only be achieved by a combination with mass spectrometry (10-13) having far reaching applications.

ACKNOWLEDGEMENTS

Grants from CNRS : ERA n° 267, INSERM : FRA n° 9 and DGRST: IAMOV were partly employed to support this work. We are very grateful to Mrs. Nicole Pitoizet for her assistance at the Mass Spectrometry Unit. Helen Losty was recipient of a study grant.

REFERENCES

1) R.A. Okerholm, C. Chattoraj, J.L. Pinkus, D. Charles and H.B. Wotiz, Steroids, 1970, 16, 66.
2) R. Vihko, Acta Endocrinol., 1966, suppl.109, 29.
3) R.J. Bègue, J. Desgrès, J.A. Gustafsson and P. Padieu, J. Chromatogr.Sci.,1974, 12, 763.
4) E.C. Horning, in "Gas Chromatography", K.B. Eik-Ness and E.C. Horning, Eds., Springer Verlag, Berlin, 1968, p.30.
5) J. Desgrès, R.J. Bègue and P. Padieu, Clin.Chim.Acta,1973, 46, 277.
6) G.W. Liddle and A.J. Sennett, J.Steroid Biochem., 1975, 5, 751.
7) T. Luukkainen and H. Aldercreutz, Biochim. Biophys. Acta, 1965, 107, 579.
8) H. Aldercreutz and T. Luukkainen, Acta Endocrinol., 1967, suppl. 124, 101.
9) H. Eriksson, J.A. Gustafsson and A. Poussette, Eur.J.Biochem., 1972, 27, 327.
10) J.A. Vollmin, Chromatographia, 1970, 3, 238.
11) B.F. Maume and J.A. Luyten, J. Chromatogr. Sci. 1973, 11, 607.
12) M. Axelson and J. Sjövall, J. Steroid Biochem.,1977, 8, 682.
13) J. Desgrès, D. Boisson and P. Padieu, this volume.

A MICROMETHOD FOR THE CLINICAL CHEMISTRY ROUTINE ANALYSIS OF AMINO
ACIDS IN BLOOD AND URINE BY CAPILLARY GAS LIQUID CHROMATOGRAPHY OF
ISOBUTYL ESTERS N(O)-HEPTAFLUOROBUTYRATE DERIVATIVES*

J.Desgres, D.Boisson, F.Veyrac, M.Susse and P.Padieu

Centre Hospitalier Régional Universitaire de Dijon,

Dijon , France

INTRODUCTION

For the last twenty years much work has been done on the
problem of gas liquid chromatography (GLC) of amino acids. Derivatization methods (1) since the first attempts (2,3) as well as
searches for separative liquid stationary phases (4-8) have been
widely investigated. Despite the many developments brought by
several workers since 1961 when Johnson et al.(9) succeeded in
separating thirty-three amino acids as their n-amyl ester,
N-acetate, and one year later when Zomzely et al. (10) used n-butyl
esters, N-trifluoroacetate to separate twenty-one amino acids, most
of the methods were proposed separately either for protein amino
acids or for non protein biological amino acids (11) except Siezen
and Mague's procedure (8). Most of the successful and considerable work originated from Gehrke's group (12) on the quantitative
assay of protein amino acids has been done with n-butyl ester,
N-trifluoroacetyl derivatives (13-17). Other investigations were
done on n-propyl ester, N-acetyl amino acids (18-20) but nowadays
alkyl ester, N-heptafluorobutyryl derivatives seem to be more often
used as n-propyl (1, 21-24), isoamyl (25,26) and isobutyl (7,8,27,
28) esters.

Therefore, the development of a method for the assay of amino
acids in human physiological fluids done as the routine task of a
clinical chemistry laboratory necessitated investigations of a
family of thirty amino acids which included the amino acids of
protein hydrolysate (eighteen amino acids including hydroxyproline
but without tryptophan destroyed in the acidic phase of the derivatization process),two plasma metabolites (ornithine and α-amino

377

butyric acid), and as described below, seven common urinary amino acids, two primary standards, one secondary standard and two odd amino acids, alloisoleucine and 5-hydroxylysine.

Results relevant to the research development of the methodology have been or will be published elsewhere : liquid phases (29), assay of histidine (30) according to Moodie (31), GC-MS study of derivative formation and elucidation of fragmentation patterns under various ionization modes (32-36).

MATERIAL

Amino acid standard. A 10^{-3}M standard solution was prepared in 0.1N HCl from standard solution or from pure compounds from Technicon Corporation (Tarrytown,N.Y.,USA), Sigma (Saint-Louis,Mo. USA), K. and K Labs. (Plainview,N.Y. USA) and Calbiochem (Los Angeles,Ca.USA).
Tables I gives the list of tested amino acids.
Reagents and solvents. Isobutanol, ethyl acetate, methylene dichloride and diethoxy-formic anhydride (EFA) were of analytical grade from Merck (Darmstadt, Germany). Heptafluorobutyric anhydride (HFBA) was purchased from Pierce (Rockford,Ill.USA). The ion exchange resin was Dowex 50Wx8, 100-200 mesh from Bio-Rad (Richmond Ca.USA). The capillary column from LKB (Bromma, Sweden) and purchased from Spiral (Dijon,France) was 25 m x 0.23 mm I.D. and coated with the OV-101 liquid stationary phase.
Gas chromatography and mass spectrometry. The gas chromatograph was a Becker 419 Research Gas Chromatograph (Becker,Delft, The Netherlands)equipped with a flame ionization detector and the above mentioned capillary column with an all glass solid injector according to Ros (37). Carrier gas was nitrogen with a flow rate of 2 ml/min. The mass spectrometric assays were done on an LKB 9000 (Bromma, Sweden) interfaced with a 1% SE-30 packed column and a Finnigan 3300-Computer 6100 (Sunnyvale,Ca. USA) or RIBERMAG GC-MS R10-10 with a computer (92502 Rueil-Malmaison,France) both with the capillary column mentioned above.

EXPERIMENTAL

Sample preparation and purification. Biological samples were cleaned up by deproteinization with sulfosalicylic acid at a concentration of 50 mg per ml of plasma and 25 mg per ml of urine. For purification by ion exchange, a glass column, 40 mm x 3 mm I.D. fitted with a glass-wool plug was filled up to a height of 1 cm with Dowex resin (H$^+$) in water. A volume of 20 to 500 µl of plasma or 100 to 1000 µl of urine with the primary standard such as Nε-mono-methyllysine (MML) or homoarginine (hARG): 4 to 100 mole in the case of plasma or 20 to 200 nmole in the case of urine and adjusted to pH 2.5 by 6M Hcl was layered on top of the resin. After absorp-

tion of the liquid layer, the column was washed with 2 ml of
distilled water. Then the amino acids were eluted with 2 ml of
4M NH4OH at a flow-rate of 1 drop every 5-10 sec. To the amino
acid fraction collected in a screw-cap tube, cycloleucine (cLEU),
the secondary standard, was added in the same amount as the MML or
hARG standard. Ammonia was evaporated to dryness by heating at 90°C
in a sand bath under a dry nitrogen stream, care was taken to
achieve a complete dessication of the tube and the screw-cap avoid-
ing any trace of moisture. The sample of the standard solution of
amino acids underwent the same preparation with the same primary
and secondary standards in order to calculate the relative molar
ratio (RMR).

    Sample derivatization into isobutyl ester,(N(10)-heptafluoro-
butyrate and into isobutyl ester,N(10)-heptafluorobutyrate,N-EF in
the case of histidine. After centrifugation, the dried sample was
taken up with 500 µl of anhydrous isobutanol in which dry gaseous
HCl had been dissolved to about 4M and then heated 45 min at 110°C
in the sand-bath. After cooling down cap isobutanol-HCl was
evaporated to dryness under a nitrogen stream at 40°C. The 80 µl
of ethyl acetate and 20 µl of heptafluorobutyric anhydride were
added. The acylation reaction was conducted in the sand-bath at
110°C for 20 min. The solution ready for GLC can be diluted at
will with ethyl acetate. All amino acids were completely derivatized
as isobutyl ester,N(O)heptafluorobutyrate (IBU,N(OH)-HFB), except
histidine. For this amino acid the acylation mixture was completely
evaporated at room temperature. To the dry tube, 400 µl of methylene
dichloride and 10 µl of diethoxiformic anhydride were added. The
derivatization of the imidazol amino group was carried out by heating in the sand-bath at 110°C for 15 min (30,31) to give a solution
of isobutyl ester, N(O)-heptafluorobutyrate, N-ethoxyformate,(IBU,
N(O)-HFB,N-EF), histidine.

    GLC analytical parameters. The most favourable chromatographic
parameters were : temperature 250°C for injection port and 270°C for
detector; gas rate flow, 250 ml/min for air and 25 ml/min for
hydrogen; temperature programming: 2°C/min in general and 3°C/min
starting from 90°C for isobutyl ester, N_α-heptafluorobutyryl, N_τ-
ethoxyformyl histidine.

    Quantitation. Quantitative analyses were performed using the
method of internal standardisation with cLEU. Foe each amino acid,
the relative molar response (RMR) to the internal standard amino
acid (i.s.) was established (38) since the absolute molar response
factor cannot be accurately calculated (39) from the electricity
produced by the detector (molar response in Coulomb/mole (40)),
therefore:

$$RMR = \frac{\text{peak height of amino acid}}{\text{peak height of cLEU}}$$

From the nmole amount, qi.s., of cLEU added to the biological
sample, the amount of each amino acid in nanomole was calculated
by:

$$Q = \frac{1}{RMRstd} \cdot RMR_{sample} \cdot q_{i.s.}$$

RMR_{std} = relative molar response for the standard amino acid solution

RMR_{sample} = relative molar response for amino acid sample

$q_{i.s.}$ = nanomoles of cLeu added to the biological sample.

To prevent loss of peak height recording due to the electro-
mechanical inertia of the pen recorder, it is advisable to use a
reporting integrator such as the HP 3380 or 3885 (Hewlett-Packard,
Orsay, France).

RESULTS

Studies of amino acid standards. Owing to the high complexity
of amino acid composition of physiological fluids especially in
metabolic diseases, a method has been developed for the complete
separation of a standard solution of thirty two amino acids (see
Table I).

Gas chromatographic derivatization of compounds. From our
own experience the isobutyl esters, N(O)-heptafluorobutyrate
derivatives have been chosen since they are completely resolved on
common stationary phases such as SE-30, OV-1 and OV-101. They are
very stable and reproducible in packed or capillary colunns and in
addition, widely used in GC-MS coupling. The most critical steps
during derivative formation are: reagents devoid of any trace of
water moisture, complete elimination by evaporation of preceding
reagent (tubes and caps) before undertaking the next derivatization
step. Side-reactions occurred during the overall procedure: complete
cysteine oxidation into cystine, hydrolysis of glutamine and
asparagine into glutamic and aspartic acids; full conversion
of methionine sulfoxide into methionine; paryial reaction of cystine
and homocystine to produce together cysteinyl, homocysteinyl
disulfide.

GLC analysis characteristics. The salient feature of the proposed method are: complete resolution of the most common physiological compounds, sensitivity and rapidity, quality control of amino acid quantitation.

Chromatographic separation. Fig. 1 demonstrates a complete resolution of twenty-nine amino acids achieved on OV-101 capillary. They correspond to the presence of : i) the seventeen protein amino acids, ii) two common plasma metabolites: α-aminobutyric acid (αABA) and ornithine, iii) seven urinary compounds : β-alanine (βALA), β-aminoisobutyric acid (βAIBA), γ-aminobutyric acid (γABA),methionine sulfone (MSO_2), lanthionine (Lan), cystathionine (CTT) and homo-cystine $(hCys)_2$; iv) cycloleucine (cLeu) used as a secondary standard for quantitative determinations, v) two alternative primary standards, N_ϵ -monomethyllysine (MML) or homoarginine (hArg).

Histidine which did not give any suitable derivative with this method should undergo specific derivatization with EFA. A special chromatographic setting as specified in the caption of Fig.2 was devised for this compound.

Furthermore, the capillary column allowed the separation of alloisoleucine (aIle) in between leucine and isoleucine (Fig.3a) bringing a clear-cut biochemical diagnosis of maple syrup urine disease which was not possible up to now with packed column (Fig.3b).

Practicability: analysis duration and sensitivity. The time needed to complete the whole assay of the twentynine amino acids was between 60-75 min depending on chromatographic settings (see Experimental). Chromatographic conditions for shorter elutions can be adjusted to restrict the quantitation to few related amino acids: Pro and Hyp; Phe and Tyr; Orn, Lys, Arg and $(Cys)_2$. The practicable sensitivity limits range between 10 pmoles and 150 pmoles. The lower limit makes it possible to use an initial sample of 20 µl of plasma in order to obtain a peak height five times the noise from the injection of a 5 hundredth aliquot of the derivatized sample.

Quality control during the different steps of the analysis.
1. Reproducibility: Table I shows the relative molar responses (RMR) to cLeu for each of the thirty one amino acids of the standard solution with ten different preparations of the derivatized standard sample. The coefficient of variation did not exceed 8% with a mean of 5%.
2. Purification steps : compounds such as lipids, carbohydrates and pigments are eliminated by ion exchange purification prior to derivatization. The most favorable conditions have been established from the standard solution to determine: i) the column loading capacity of the column which comes to 500 nmoles of every amino acid together; ii) no losses after 3 ml distillated water washing; iii) safe elution by 2 ml of 4M NH_4OH without hampering the perform-ance of a series of assays by excessive time for aqueous evapora-tions; iv) flow-rate of elution as already mentioned. Table I shows

TABLE I

GLC parameters of standard amino acids on OV-101 coated glass
capillary column. Retention time (RT) and temperature (T°C) are
given for two temperature programs : 3°C/min and 2°C/min from
90°C to 275°C. The relative molar response (RMR) to cycloleucine
(cLeu) in crude and cleaned up sample for each amino acid corresponds
to the mean ± SD from twelve samples routinely made during a six
month period.

§Pure alloisoleucine was not available. Its RMR to cLeu was
obtained from molar mixture of isoleucine and alloisoleucine
(approximately   1:1).

§§The observed retention time and temperature of histidine
corresponded to its peculiar chromatographic settings : 4°C/min
from 150°C to 260°C. Its RMR was related to N-monomethyllysine.

§§§Cysteinyl-homocysteinyl disulfide not being available, no
RMR could be calculated. Its formation seems (see text) to be
a derivatization artifact.

TABLE 1 Amino Acid	3°C/min		2°C/min		RMR Std. Solution	
	RT	T°C	RT	T°C	Crude Sample	Cleaned up sample
α-Alanine (Ala)	12.7	128	12.8	115.6	0.99+0.06	1.00+0.05
Glycocolle(Gly)	13.2	129.6	13.3	116.6	0.88+0.04	0.83+0.05
α-Aminobutyric acid(αABA)	15.2	136.6	15.7	121.4	0.99+0.04	0.96+0.04
β-Alanine(βAla)	16.1	138.3	16.7	123.4	0.83+0.04	0.84+0.04
Valine (Val)	16.7	140.1	17.5	125	1.14+0.07	1.08+0.07
β-Aminoisobutyric acid (βAIBA)	17	141	17.8	125.6	1.05+0.06	1.03+0.05
Threonine(Thr)	17.4	142.2	18.4	126.8	0.91+0.08	0.89+0.08
Serine (Ser)	18.1	144.3	19.2	128.4	0.84+0.07	0.83+0.07
Leucine(Leu)	19.3	147.9	20.8	131.6	0.91+0.04	0.90+0.04
Alloisoleucine(al (alle)§	19.6	148.8	-	-	1.02 -	1.02 -
Isoleucine (Ile)	19.8	149.4	21.5	133	1.09+0.08	1.06+0.07
γ-Aminobutyric acid (γABA)	21.3	153.9	23.4	136.8	0.70+0.05	0.70+0.06
Cycloleucine(cLeu)	22.2	156.6	24.6	139.2	1.00 IS	1.00 IS
Proline(Pro)	23.1	159.3	25.8	141.6	0.90+0.06	0.88+0.06

TABLE I (cont'd)

Amino Acid	3°C/min		2°C/min		RMR Std. Solution	
	RT	T°C	RT	T°C	Crude Sample	Cleaned up sample
4-Hydroxyproline (HPR)	26.8	170.4	31.1	152.2	0.85+0.04	0.84+0.05
Methionine(Met)	28.5	175.5	33.4	156.8	0.59+0.05	0.56+0.05
Aspartic acid(Asp)	30.7	182.1	36.7	163.4	0.90+0.06	0.91+0.06
Phenylalanine(Phe)	32.4	187.2	39.1	168.2	1.10+0.08	1.10+0.08
Ornithine(Orn)	33.6	190.8	41.1	172.2	0.72+0.04	0.68+0.04
Glutamic acid(Glu)	35.3	195.9	43.4	176.8	0.98+0.06	0.96+0.05
Lysine (Lys)	37.8	203.4	47.1	184.2	0.74+0.05	0.73+0.06
Tyrosine (Tyr)	38.4	205.2	47.9	185.8	1.00+0.07	0.96+0.04
Methionine Sulfone (MSO$_2$)	39.0	207	48.7	187.4	0.45+0.05	0.44+0.05
N$_\epsilon$-monomethyllysine (MML)	39.6	208.8	49.7	189.4	0.86+0.04	0.87+0.04
Arginine (Arg)	41.4	214.2	52.7	195.4	0.66+0.05	0.62+0.06
Histidine(His)[§§]	15.2	210.8	15.2	210.8	0.52+0.05	0.47+0.06
Homoarginine(hArg)	45.3	225.9	58.4	206.8	0.66+0.05	0.61+0.06
Lanthionine(Lan)	48.8	236.4	63.5	217.0	0.73+0.05	0.70+0.05
Cystathionine(CTT)	52.3	246.9	68.5	227.0	0.81+0.06	0.80+0.07
Cystine(Cys)$_2$	54.2	252.6	71.3	232.6	0.50+0.06	0.48+0.06
Cys-S-S-Homocys (CyshCys)	57.5	262.5	76.1	242.2	§§§	§§§
Homocystine(hCys)$_2$	60.6	271.8	80.8	251.6	0.88+0.07	0.84+0.07

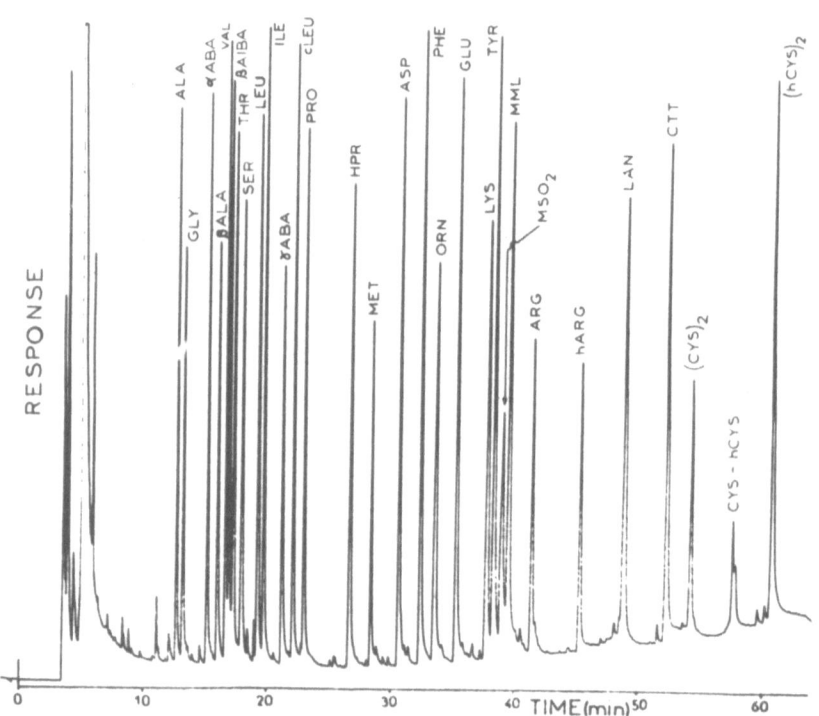

Fig. 1 : GLC of 29 standard amino acids (50 pmole each) as IBU-N(O)-HFB derivatives. See text for GLC conditions : temperature programming was 3°C/min.

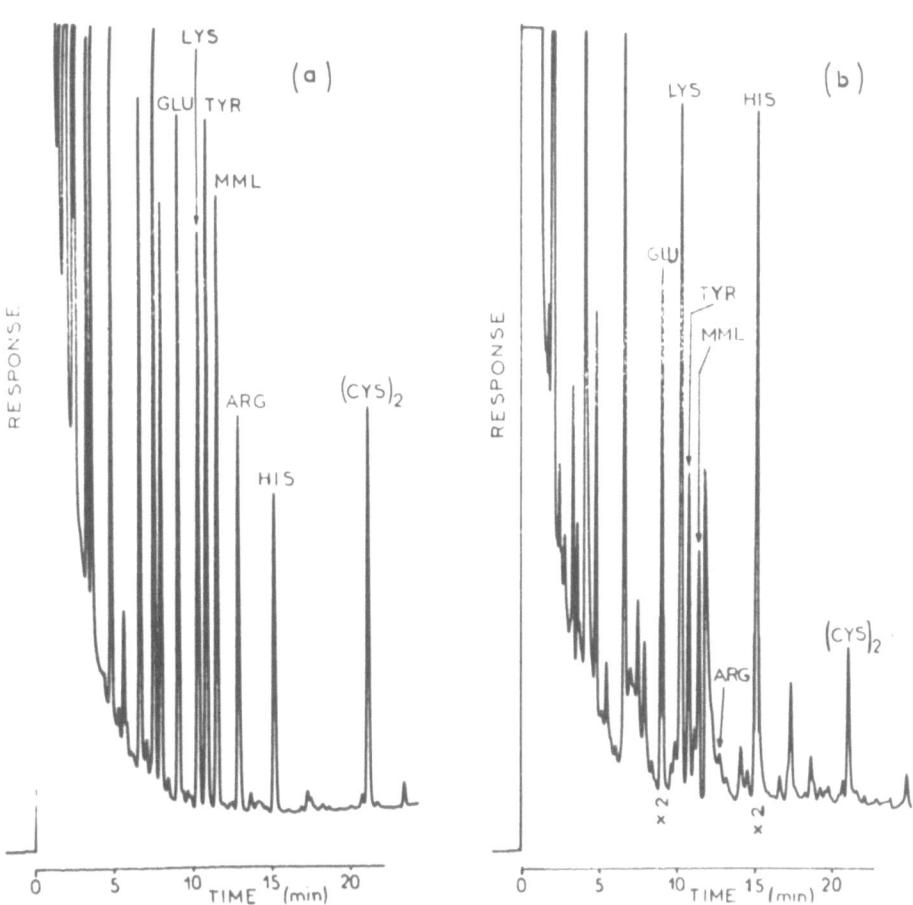

Fig. 2   Gas chromatograms of the isobutyl ester, Nα-heptafluorobutyrate, Nτ-ethoxyformate derivative of histidine: 2a : in standard mixture, 2b : in urine of normal patient. TP : 4°C/min, 150°C–240°C.

that under these conditions the RMR of each amino acid of a
standard solution cleaned up or not with Dowex is the same.
3. Choice of standards : MML of hArg was added to the sample before
purification, while cLeu, the secondary standard, was then added
before derivatization. The RMR of MML to cLeu in the amino acid
standard solution and the biological samples handled together through
the same purification procedure were compared to the RMR calculated
from an amino acid standard solution derivatized without going
through the ion exchange resin step (see Table II). Altered RMR
values would have indicated losses during this crucial stage. In
addition MML is used as a secondary standard in the case of histidine
assay since cLeu is eluted in the solvent front. During the
evaporation of the sample deposited on the Ros injector needle losses
of the more volatile derivatives may occur. An increased RMR value
of MML to cLeu would have been a warning of such a situation.

TABLE II.
Quality control assay of the biological sample subjected to the
whole procedure carried out by comparison of RMR values of MML
as the internal standard to cLeu as the secondary one in cleaned
up urine and plasma to crude or cleaned up standard solution. The
value of the mean $\pm$ SD was obtained from twelve preparations.

	Crude sample	Cleaned up Sample		
RMR to cLeu	Amino acid Standard solution	Amino acid Standard solution	Plasma sample	Urine sample
MML	0.86 $\pm$ 0.04	0.87 $\pm$ 0.04	0.85 $\pm$ 0.06	0.93 $\pm$ 0.07

4. Derivative stability : at room temperature, most of the derivatives remained stable for three weeks, but Lys and $(Cys)_2$ responses
decreased by about 15% in 12 days while Met and Arg derivatives were
rapidly lost. Therefore derivatized samples should be analysed
within three days, a period of time sufficient to check again, but
a new standard solution sample should be prepared for each series
of analyses.
General Procedure.   The flow diagrams shown on Scheme 1 and 2
summarize the various criteria that have been selected after several
years of continuous application of the method in clinical biochemistry.

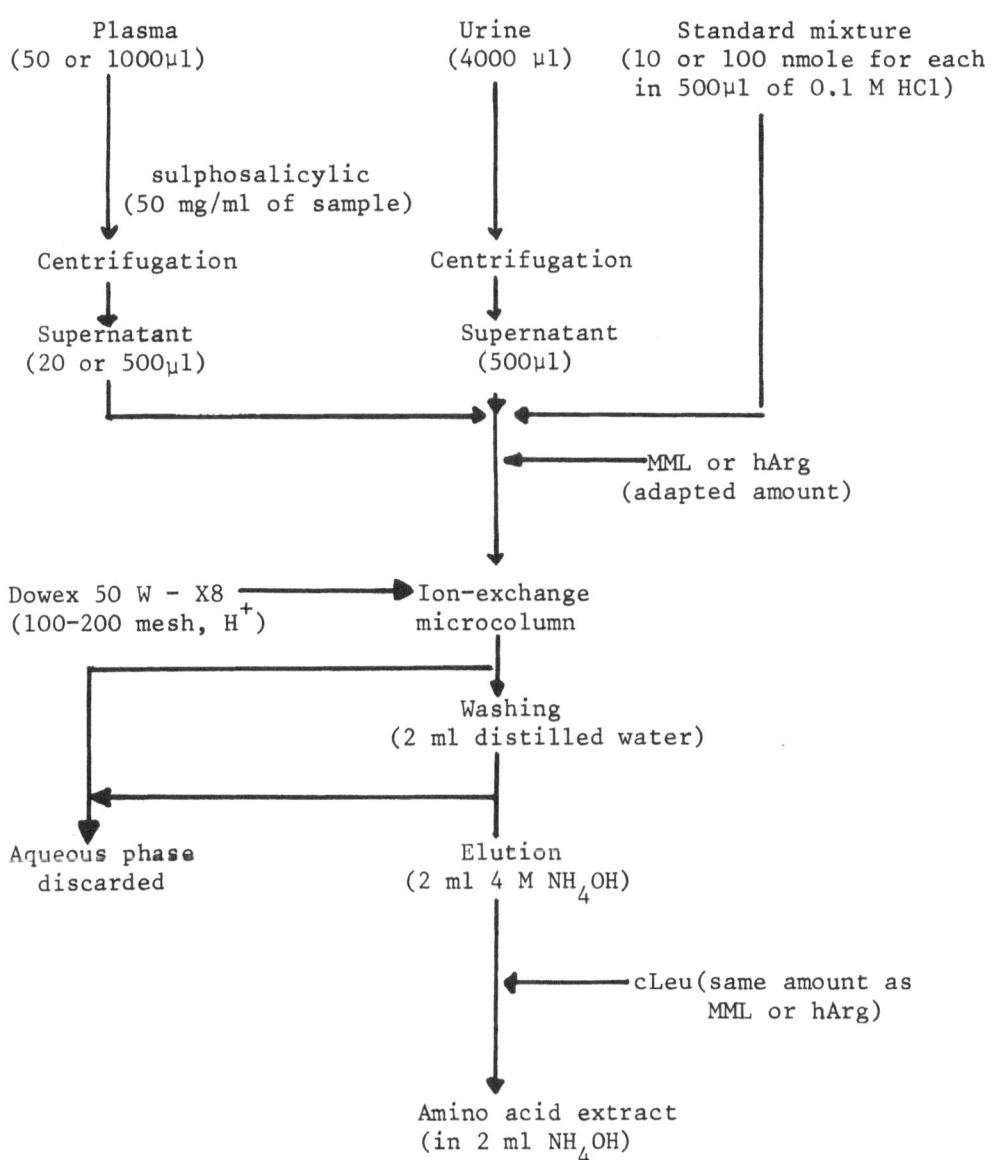

Scheme 1

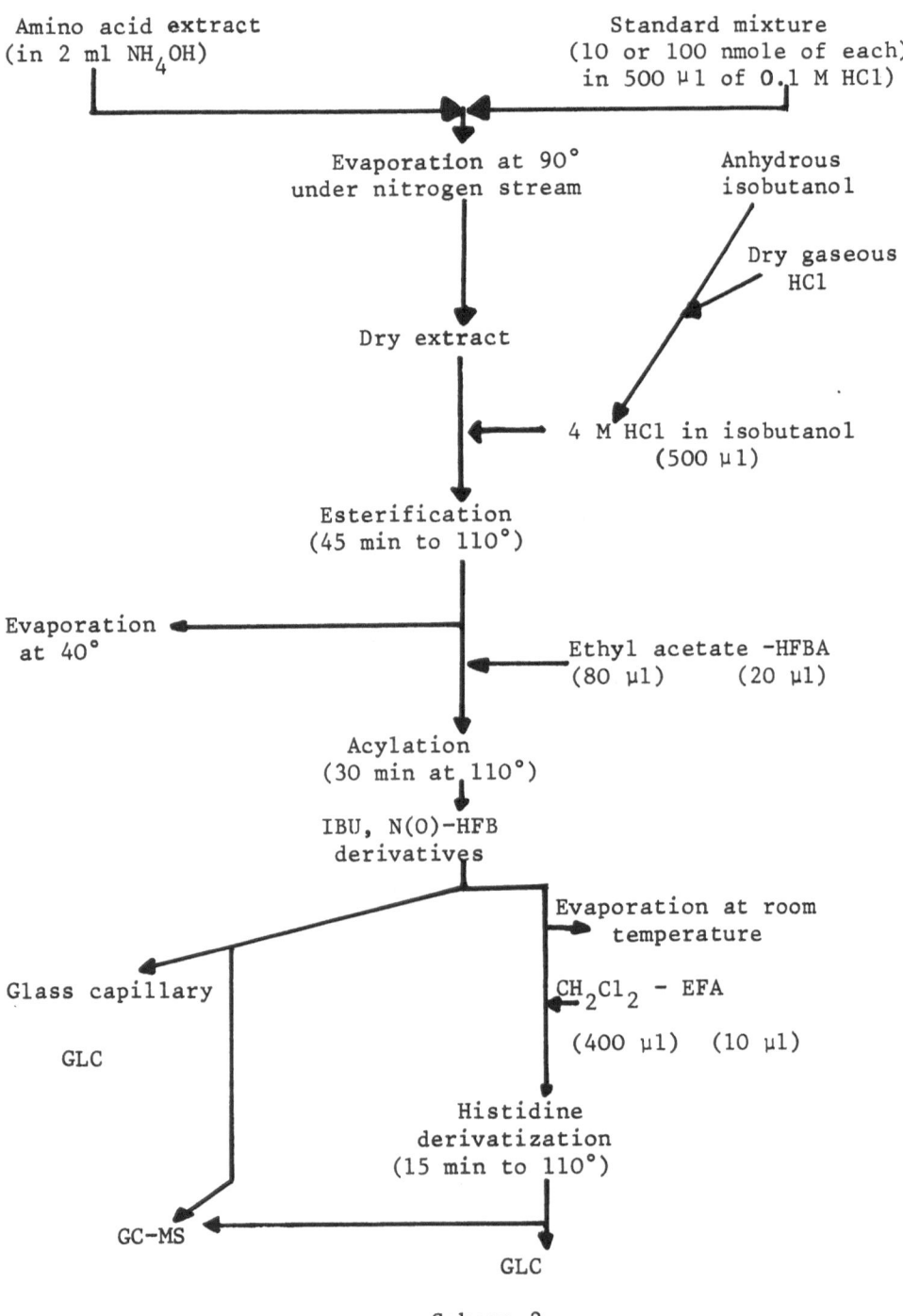

Scheme 2

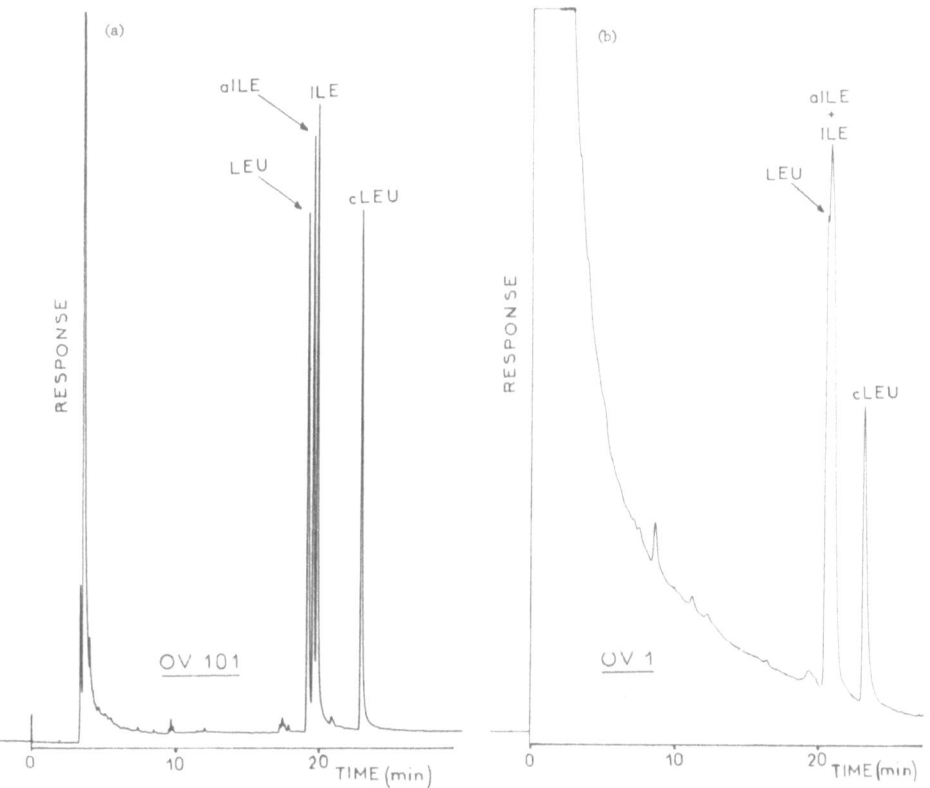

Fig. 3.   GLC of Leu, Ileu, aIle and cLeu on OV-101 capillary(a) and OV-1 packed(b) column.

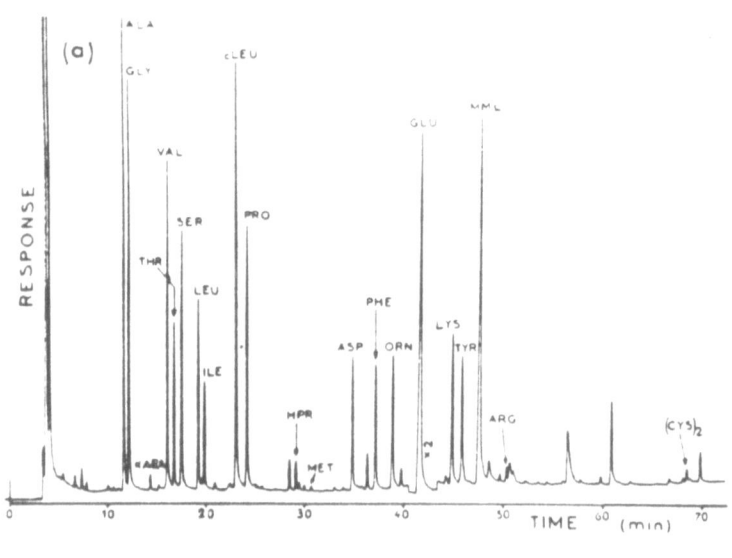

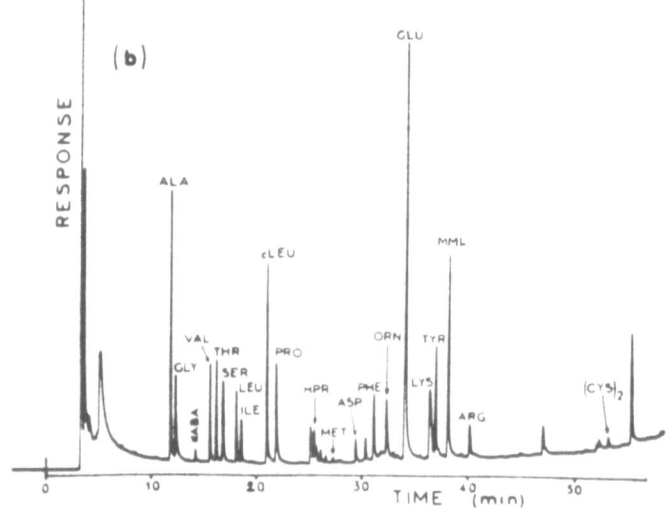

Fig. 4.   Gas chromatogram on OV-101 coated glass capillary column
of two normal plasma samples subjected to the complete procedure.
The TP:2°C/min (5a) and 3°C/min (5b), 90°C - 270°C.

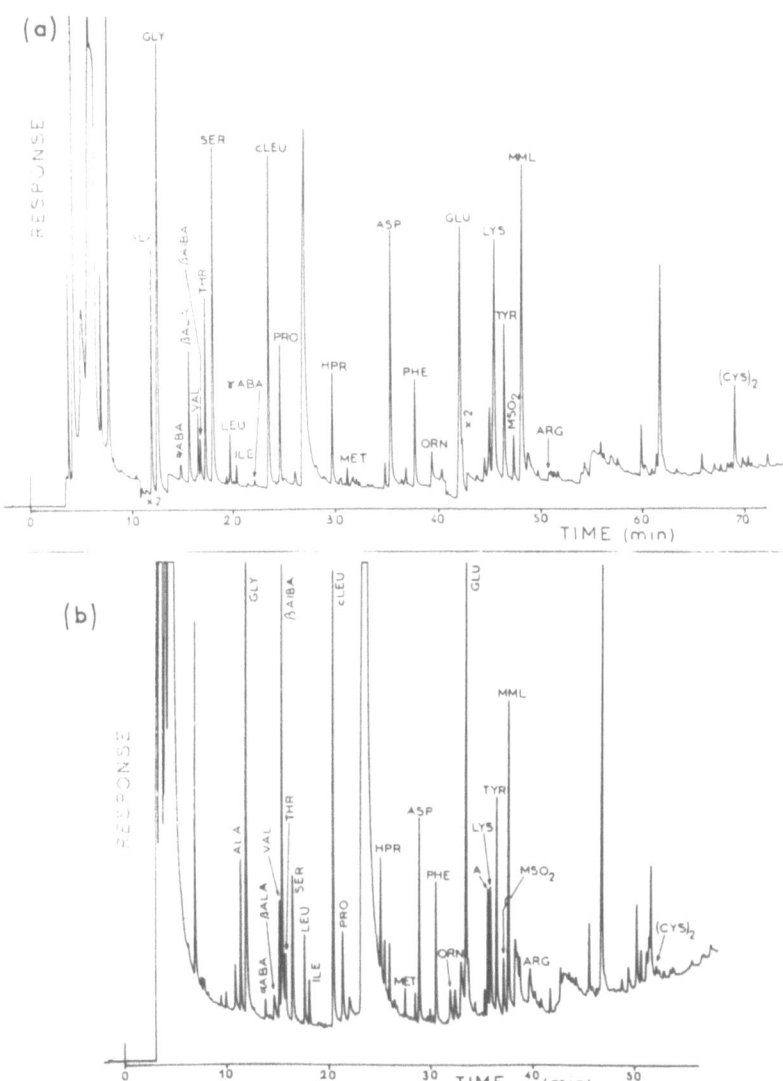

Fig. 5.   Gas chromatograms on OV-101 coated glass capillary column
of two normal urine samples subjected to the complete
procedure.
TP : 2°C/min (5a) and 3°C/min (5b), 90°C - 270°C.

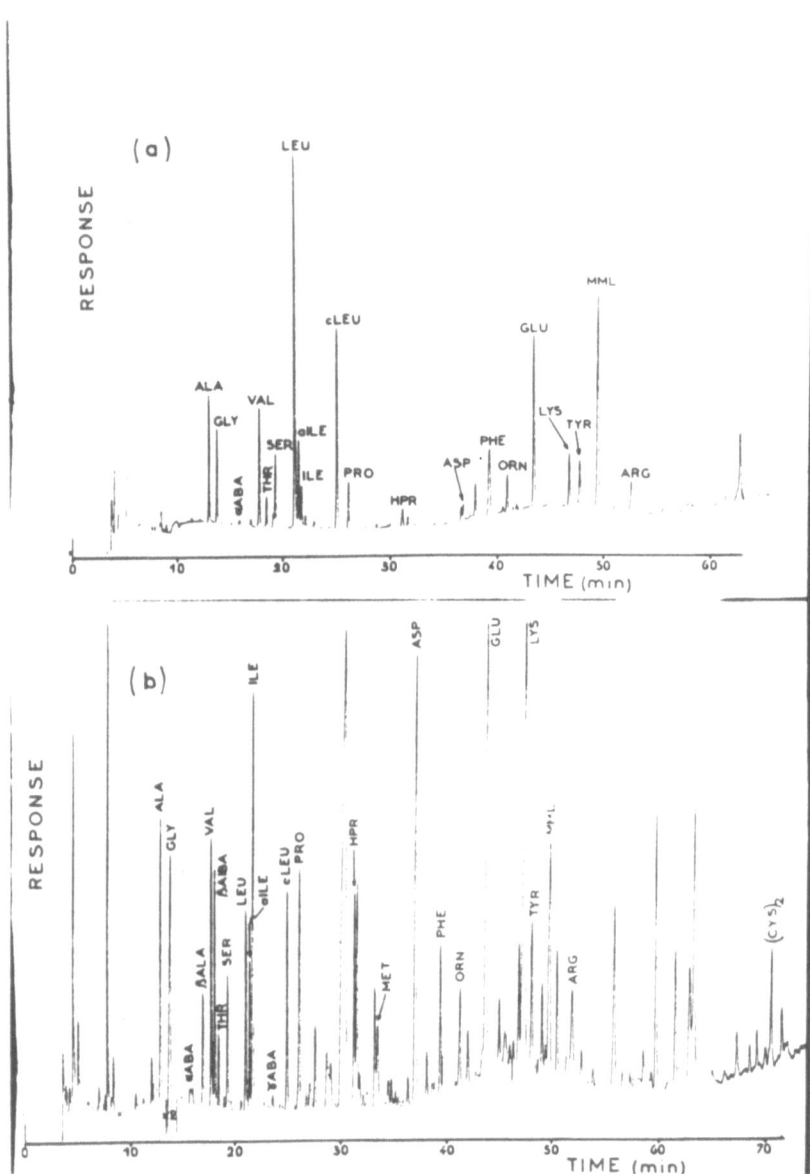

Fig. 6.   Gas chromatograms on OV-101 coated glass capillary column
          of two samples obtained from plasma (6a) and urine (6b)
          of a new-born infant with maple syrup disease.
          TP : 2°C/min,   90°C – 270°C.

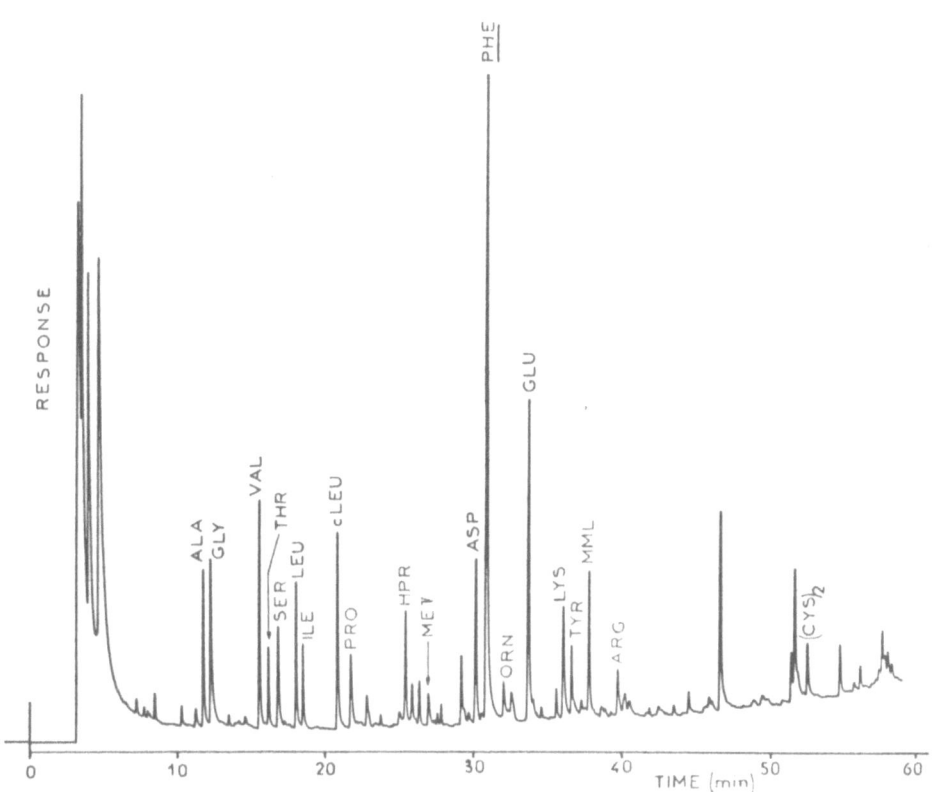

Fig. 7. Gas chromatogram on OV-101 coated glass capillary column of a sample obtained from plasma of a phenylketonuria. TP : 3°C/min, 90°C – 270°C.

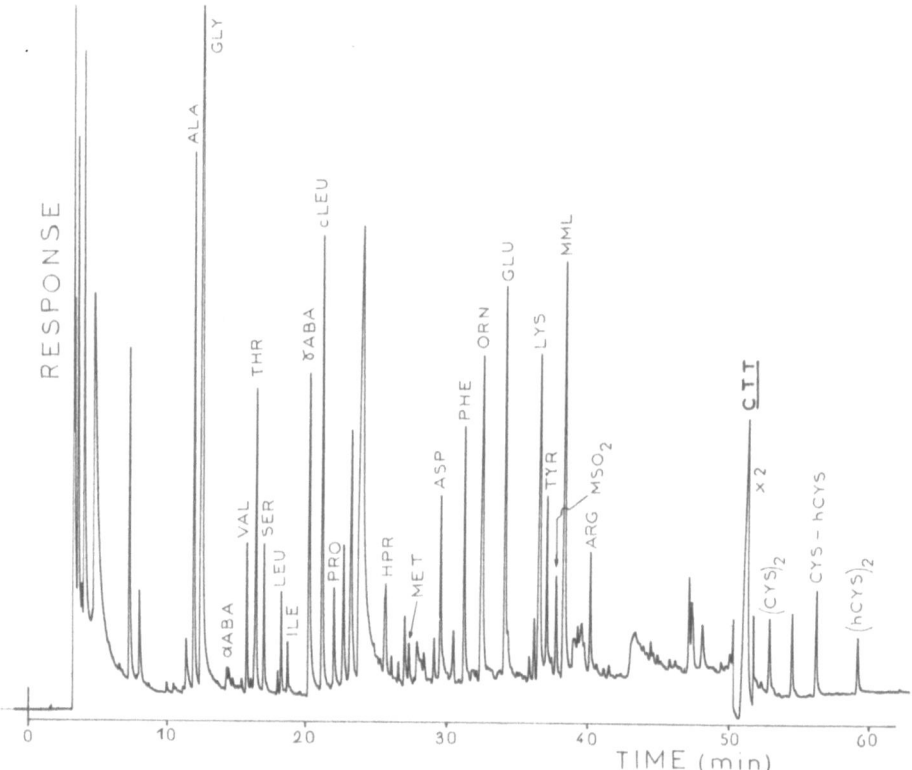

Fig. 8. Gas chromatogram on OV-101 coated glass capillary column of urine amino acids with hypercystationurie disease. The homocystine is also identified in the sample. TP : 3°C/min, 90°C – 270°C.

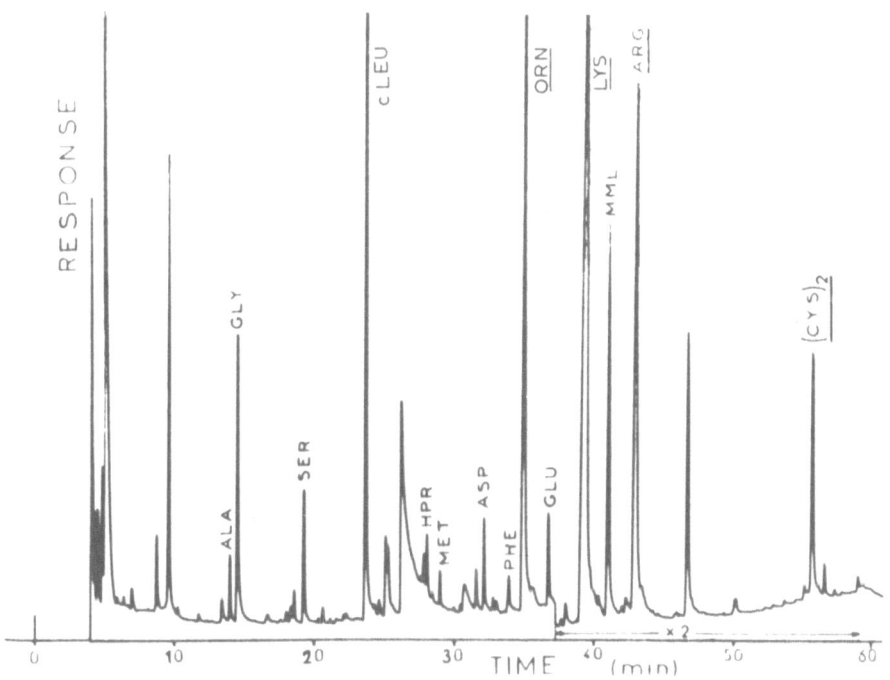

Fig. 9.   Gas chromatogram on OV-101 coated glass capillary column of urine amino acids from a patient suffering from ornithine-lysine-arginine-cystinuria.   TP : 3°C/min, 90°C - 270°C.

APPLICATION TO PHYSIOLOGICAL FLUIDS IN CLINICAL BIOCHEMISTRY

Identification of chromatographic peaks from biological samples.
Chromatographic peaks of biological samples were identified by
comparison of their retention time to those of the standard amino
acid solution. Table 1 shows the data for two different analytical
conditions. The nature of the presumed amino acid was checked by
individual spiking off. Finally exact identifications were obtained
by comparison of the mass spectrum of the biological compound and of
the authentic reference standards. The detailed results of this
study have been partly reported in articles (32, 33) which will be
completed by forthcoming publications (29,30,34).
GLC analysis of normal physiological fluid from human. Figures 4a
and 4b show the separation of amino acids as IBU,N(O)-HFB from two
normal plasmas. The chromatographic settings were different :
Fig.4a was obtained with a temperature program of 2°C/min instead of
3°C min as in Fig.4b. The 15 min increase of the elution time
brought a better separation. The advantage of this improvement
appears more conspicuous in the case of amino acids of normal urine
(Fig. 5a and 5b), especially when considering the group composed
of βAla, Val, βAIBA and Thr, the HPR and also the group of Lys, Tyr
MSO$_2$ and MML. Such charts clearly demonstrate the extent of the
physiological variations of amino-acid concentrations or in the
case of βAla, βAIBA, Ser,Lys. In Fig. 5b, the peak preceding Lys
has been identified by GC-MS to 5-hydroxylysine. This amino acid
eluted with Lys on packed columns has been found in variable
amounts in most urine samples from normal patients assayed so far.
Its physiological occurrence and significance remain to be seen.
Interest of the capillary column. The chromatographic response
of a capillary column compared to a packed one was found to be
thirty times more sensitive. In the case of plasma amino acids,
this feature is of great interest since the plasma sample size can
be scaled down to 20-50 μl for blood specimens from premature or
new-born infants.
Pathological plasma and urine.   Fig. 6 to 9 show four examples of
amino acid separationa and assays in inborn disorders of amino
acid metabolism.

     In maple syrup disease, plasma (Fig.6a) and urine (Fig.6b)
amino acid levels and excretions in Val, Leu and Ile were increased
while allo-isoleucine (aIle) was very well separated on the capillary column and was identified by GC-MS.

     Fig. 7 from the plasma of an infant affected with an inborn
phenylketonuria shows the high increase of Phe.

     A congenital defect of cystathionase is illustrated by Fig. 8
with an excretion peak of CTT. Cystinuria (Fig. 9) which is known
to result from the alteration of specific reabsorption receptors

in kidney tubules leads to increased levels of $(Cys)_2$ but also
Orn, Lys and Arg.

DISCUSSION

In clinical chemistry, gas liquid chromatography has acquired a
widespread diffusion in the field of steroid hormones for many
years (see reference in 41) since the pioneering work of
Luukkainen et al.(42). In consideration of the prominent work of
Gehrke and his associates (12-17,38), the same outcome was not found
for amino acid assay and no such appreciation of this chromatographic
technique can be found in clinical chemistry.   Since the beginning
of our interest in this field (43-47) re-investigations of already
published methods and in the course of the development of specific
techniques considerable benefits accrued from the coupling of the
gas chromatograph to the mass spectrometer (34,48) for the determination of amino acids eluted and the elucidation of new compounds
belonging to this family (32,49-52). The use of GLC for analysis
of aminoaacids of protein and human plasma is well documented (1,7,
17,25,38,49) but many of the authors if not all, except Adams (19),
had not reached the level of achievment which would open competi-
tivity with the well established ion exchange liquid chromatography
used in reliable automatic analysers.

Benefiting from the improved separation due to sample purification and capillary column, up to thirty two amino acids could be
easily analysed on a routine basis for human plasma sample, in
clinical chemistry, although histidine has still to be assessed by
the method of Moodie (30,31). However the development of a similar
method for the analysis of all the urinary amino acids revealed more
compelling difficulties to be overcome as regards satisfactory uses
for patients. Difficulties in urine amino acid assay arise from the
fact that the number of amino acids is greater than in plasma and
from the presence of many metabolites having similar gas chromatographic properties of amino acids and which are detected by flame
ionization. Purification of the urine sample by ion exchange resin
and the use of glass OV-101 coated capillary column were the two
improvements that made it possible to separate and quantitate these
amino acids in a standard mixture as well as in the biological
samples.

A complete analysis can be performed in 60 or 75 min. Even with
a duration of 75 min when more than seventeen protein amino acids
have to be separated and quantitated, ion exchange liquid chromatography cannot compete with GLC in the case of urine sample. Of
course one must take into account the time spent for sample purification and derivatization. But six samples can be handled at once.
The sensitivity can reach 10 pmoles for a signal noise ratio of five

Such sensitivity made it possible to analyse very small samples down to 20 µl of plasma as in the case of infants who cannot sustain large withdrawals of blood. Cerebrospinal fluid and dialyzing fluids from extra-renal clearance of toxic metabolites in the case of inborn amino acid metabolism error can be easily analysed as well as food to control the diet from which toxic amino acids must be restricted. Finally the possibility of identifying peaks by mass spectrometry using the coupling of the same column is of great help and is in many respects obligatory in the field of metabolic disorders. Since OV-101 or SE-30 glass coated capillary columns used in this method are very reproducible they can easily be interfaced directly on low resolution mass spectrometers without any separator.

CONCLUSION

The glass OV-101 coated capillary column achieved a complete resolution of the protein amino acids by GLC in such a manner that up to twenty-eight amino acids commonly found in urine could be separated in about the same time and then quantitated. The time for a chromatographic run can be adjusted between 60 to 75 min or less in case of analysis of a small group of amino acids. The sensitivity of the glass capillary column with the solid injector made it possible to work on plasma specimens as small as 20 µl. The use of the same column and liquid phases that are commonly found in GC-MS ideally procured a safe control of the method and the possibility of trying to identify any new peak as shown by the clear-cut identification of allo isoleucine and by repeated finding of 5-hydroxylysine in normal urine samples. These analytical features as well as the increase of interest in GLC and GC-MS in biomedical sciences should stimulate further research in the field of amino acid metabolism and chemical pathology and therefore have far reaching application.

ACKNOWLEDGEMENTS

Grants from CNRS : ERA 267, INSERM : FRA 9, DGRST : ACC IAMOV and ACC BFM were partly employed to support this work.

We are extremely grateful to Dr. B.F. Maume for much help during the studies on the Finnigan GC-MS of the Université de Dijon and to the Societé RIBERMAG for the generous utilization by us of the GC-MS R10-10 with the direct glass capillary column to the dual ionization source without separator in their Application Laboratory during the course of this work. We thanks Mrs. Nicole Pitoizet for her assistance at our Mass Spectrometry Unit.

*This paper was presented at the 9th International Symposium on Chromatography and Electrophoresis, held at Riva del Garda, Lake Garda, Italy, 15-17 May 1978.

REFERENCES

1) J.F. March, Anal.Biochem., 1975, 69, 420.
2) C.G. Youngs, Anal Chem., 1959, 31, 1019.
3) E. Bayer, in "Gas Chromatography (1958)", D.H.Desty,Ed., Butter Butterworth, London, 1958, p. 333.
4) A. Darbre and K. Blau, J.Chromatogr., 1967, 29, 49.
5) A. Darbre and K. Blau, Biochim.Biophys.Acta, 1966, 126, 591.
6) P. Husek and K. Macek, J.Chromatogr., 1975, 113, 139.
7) S.L. MacKenzie and D. Tenaschuk, J.Chromatogr., 1974, 97, 19.
8) R.J. Siezen and T.H. Mague, J.Chromatogr., 1977, 130, 151.
9) D.E. Johnson, S.J. Scott and A. Meister, Anal. Chem., 1961, 33, 669.
10) C. Zomzely, G. Marco and E.Emery, Anal. Chem.,1962, 34, 1414.
11) V. Amico, G. Oriente and C. Tringali, J. Chromatogr., 1977, 131, 235.
12) C.W. Gehrke, H. Nakamoto and R.W. Zumwalt, J. Chromatogr, 1969, 45, 24.
13) C.W. Gehrke and D.L. Stalling, Separ. Sci., 1967, 2, 101.
14) C.W. Gehrke, R.W. Zumwalt and K.Kuo, J. Agr.Food Chem., 1971, 19, 605.
15) R.W. Zumwalt, K. Kuo and C.W. Gehrke, J. Chromatogr., 1971, 57, 193.
16) C.W. Gehrke and H. Takeda, J. Chromatogr., 1973, 76, 63.
17) F.E. Kaiser, C.W. Gehrke, R.W. Zumwalt and K.C. Kuo, J. Chromatogr., 1974, 94, 113.
18) J. Graff, J.P. Wein and M. Winitz, Fed.Proc., 1963, 22, 244.
19) R.F. Adams, J. Chromatogr., 1974, 95, 189.
20) I.M. Moodie and R.K. George, J. Chromatogr., 1976, 124, 315.
21) C.W. Moss, M.A. Lambert and F.J. Diaz, J. Chromatogr., 1971, 60, 134.
22) C.W. Moss and M.A.Lambert, Anal. Biochem., 1974, 59, 259.
23) F.J. Diaz, C.W. Moss and M.A. Lambert, Rev.Ass.Biog.Argentine, 1971, 36, 67.
24) J. Jönsson, J. Eyem and J. Sjöqvist, Anal.Biochem., 1973, 51, 204.
25) J.P. Zanetta and G. Vincendon, J. Chromatogr., 1973, 76, 91.
26) P. Felker and R.S. Bandurski, Anal. Biochem., 1975, 67, 245.
27) S.L. MacKenzie and D. Tenaschuk, J. Chromatogr., 1975, 104, 176.
28) S.L. MacKenzie and D. Tenaschuk, J. Chromatogr., 1975, 111, 413.
29) F. Barbier-Chapuis, G. Lavoué and P. Padieu, submitted for publication.
30) D.H. Jo, J. Desgrès and P. Padieu, J. Chromatogr., 1978, 146, 413.
31) J.M. Moodie, J. Chromatogr., 1974, 99, 495.
32) F. Barbier-Chapuis, B.F. Maume and P. Padieu, in "Mass Spectrometry in Biochemistry and Medicine", A. Frigerio and N. Castagnoli,Eds., Raven Press, New York, 1974, p.119.

33) D.H. Jo, Diplôme d'Etudes Approfondies, Université de Dijon, 1975.

34) P. Padieu, J. Desgrès, B.F. Maume, G. VanderVelde and R.S. Skinner, in "Proc. VIIth International Mass Spectrometry Conference, Heyden and Sons, London, 1978, p.1604.

35) P. Padieu, J. Desgès and B.F. Maume, Finnigan Spectra, 1978, 7, 1.

36) D.H. Jo, J.Desgrès, B.F. Maume and P. Padieu, in "5th.Internat. Symposium on Mass Spectrometry in Biochemistry and Medicine", Rimini, Italy 19-21 June 1978, Abstract p.34.

37) A. Ros, J. Gas Chromatogr. 1965, 3, 252.

38) C.W. Gehrke, D. Roach, R.W. Zumwalt, D.L. Stalling and L.L. Wall, Analytical Biochemistry Laboratories Inc. Publisher, Colombia(Mo), 1968, p.27.

39) A. Delfavero, A. Darbre and M. Waterfield, J. Chromatogr., 1969, 40, 213.

40) D.H. Desty, C.J. Geach, and A. Goldup, in "Gas Chromatography 1960", R.P.W. Scott,Ed., Butterworth, London, 1960, p.46.

41) J. Desgrès, R.J. Bègue and P. Padieu, Clin.Chim.Acta, 1973, 46, 277 and 1974, 52, 381.

42) T. Luukkainen, W.J.A. VanderHeuvel, E.O.A. Haahti and E.C. Horning, Biochim.Biophys.Acta, 1961, 52, 599.

43) P. Padieu, N. Maleknia and A.M. Thireau, Bull.Soc.Chim. de France, 1963, p.2690.

44) P. Padieu, N. Maleknia, and A.M. Thireau, Bull.Soc.Chim. Biol., 1963, 47, 493.

45) G. Schapira, J. Rosa, P. Padieu and N. Maleknia, in "Methods in Enzymology", L. Grossman and K. Moldave, Eds., Vol.XIIb,"Nucleic Acids", Academic Press, New York, 1968, p.747.

46) P. Padieu, I. Harary and F. Barbier, Pathol.Biol., 1970, 18, 1041.

47) F. Barbier, G. Lavoué and P. Padieu, in " Les Aminoacidopathies", Journées Internat.Pharmacie., Grenoble, France, 1972, p.157.

48) E. Gelpi,W.A. Koenig, J. Gibert and J. Oro, J. Chromatogr.Sci., 1969, 7, 604.

49) K.M. Williams and B. Halpern, Aust. J.Biol. Sci., 1974, 26, 831.

50) F.P. Abramson, M.W. MacCaman and R.E. MacCaman, Anal.Biochem., 1974, 51, 482.

51) M.F. Schulman and F.P. Abramson, Biomed.Mass Spectrom., 1975, 2, 9.

52) S.L. MacKenzie and L.R. Hoogge, J. Chromatogr., 1977, 132, 485.

METABOLIC PROFILE OF VENTRICULAR CEREBROSPINAL FLUIDS

I.Matsumoto, N.Hisanaga, T.Shinka, T. Kuhara, M. Yoshida,

N.Kusano and S.Kuramoto; Kurume University School of

Medicine, Kurume, Fukuoka, Japan

INTRODUCTION

The complexity of biochemical reaction in the central nervous system is easily suspected from the fact that the central nervous system in the human and also in animals plays a very important role in integration of the whole function of the individual. Biochemical reaction in the central nervous system has a specificity because this system is separated and protected by the blood-brain-barrier from other organs. Various scientific approaches have been performed to clarify mechanism of the central nervous system underlying delicate and nearly mysterious integrative function, but so far this mechanism seems to reject any sophisticated scientific approach (1-3). The investigation of biochemical reaction in the central nervous system has contributed much to the understanding of the mechanism related to neuropsychiatric disease and also its diagnoses and treatment. For the investigation of metabolism of the central nervous system in its broad sense we have analyzed organic acid and amino acid in cerebrospinal fluids utilizing gas chromatography and mass spectrometry. The cerebrospinal fluid was obtained from the intraventricular and lumber subarachinoid space of patients who are suffering from various kinds of neurological disorders. The result will be briefly reported.

MATERIALS

Lumbar cerebrospinal fluids were obtained by lumbar puncture from 2 Parkinsonians and one Huntington's chorea. The ventricular CSF was obtained during stereotaxic surgery from 2 Parkinsonisms and

one dystonia musculorum deformans. The ventricular CSF from a patient
with thalamic tumor was obtained during stereotoxic biopsy of the
tumor.

EXTRACTION

Two ml of cerebrospinal fluids, after addition of 100 μg of
n-eicosane and 300 μg of α-aminooctanoic acid as internal standards,
were deproteinized by adding three volumes of cold ethanol. After
ethanol in supernatant was evaporated under reduced pressure, the
remaining aqueous layer was adjusted to pH 1.0 with 2N-HCl and
extracted twice with diethyl ether and twice with ethyl acetate.
The organic solvent was combined and concentrated to dryness under
reduced pressure. The residue was trimethylsilylated with 150 μl
of N,O-bis-trimethylsilylacetamide and 2 μl of the reaction mixture
was injected onto gas chromatography and GC-MS. The remaining aqueous
layer was applied to cation exchange column chromatography. After
amino acids were eluted with 3N-ammonia, the eluate was concentrated
to dryness under vacuum and trimethylsilylated with 200 ml of N,O-
-bis-trimethylsilyltrifluoroacetamide in acetonitrile (1 : 1) and
subjected to gas chromatography and GC-MS.

GAS CHROMATOGRAPHY-MASS SPECTROMETRY

A JEOL JMS D-100 GC-MS on line to JMA 2000 data acquisition
system was used throughout the experiments. For gas chromatographic
analysis, the temperature of injection port was maintained at 300°C,
and separator and ion source were kept at 280°C. Column temperature
was programmed at 6°C/min from 80°C to 300°C for the analysis of
organic acids, and was programmed at 6°C from 80°C to 200°C and at
4°C/min from 180°C to 300°C for amino acid analysis respectively.
An ionization voltage and an ion current were 75eV and 300 μA, and
scanned every 5 seconds.

RESULTS AND DISCUSSION

Fig. 1 shows gas chromatographic profile of amino acids of the
CSF within the ventricle. Urea and glutamine are noted in large
amounts. Ornitine was also detected. Valine, leucine, glycine proline,
serine, threonine, creatinine, phenylalanine and lysine are also
detected. Glutamic acid and glutamine are known to play a very
important role in the mechanism which handles ammonia in the brain.
The detection of urea, arginine and ornithine in the intraventricular
CSF as well as a large amount of glutamine and glutamic acid is
very interesting in the sense that this fact suggests presence of
urea cycle in the central nervous system. When ventricular CSF was
incubated with 5 units of urease at 30°C for 1 hr, urea peak

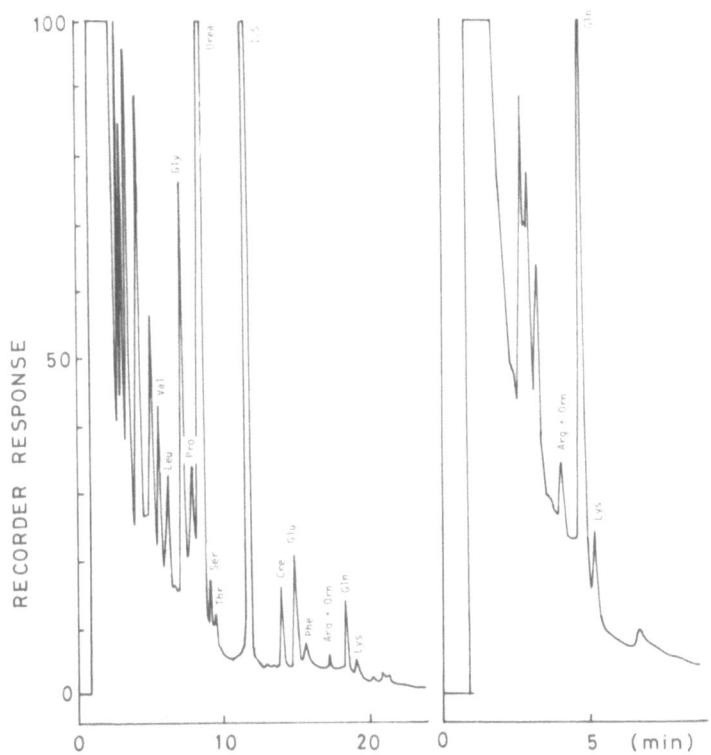

FIG. 1. Gas chromatogram of amino acids in V-CSF.
Column temperatures were programmed at 6°C/min
from 80°C to 200°C (left) and 4°C/min from 180°C
to 300°C (right).

disappeared almost completely. Thus the existence of urea in the
ventricular CSF was firmly established. In 1959, Sporn et al. (4)
observed the formation of ^{14}C-labeled urea from ^{14}C-arginine after
intracisternal injection to rat brain. As they also found that the
rat brain can metabolize proline to ornithine and arginine in vivo,
they proposed that brain was capable of urea synthesis in vivo.
However, the quantitative significance of urea cycle as a method of
ammonia detoxication in the brain was not presented. Our results
support the presence of urea cycle in human brain and are indicative
of an important role of urea synthesis in the metabolism of the brain,
because of the highest urea peak in the TIM chromatogram. Thus this
is the first paper to propose a significant role of urea synthesis
in the central nervous system.

   TIM chromatogram of organic acids of ventricular cerebrospinal
fluid was shown in Fig. 2. Most of the identified peaks represent

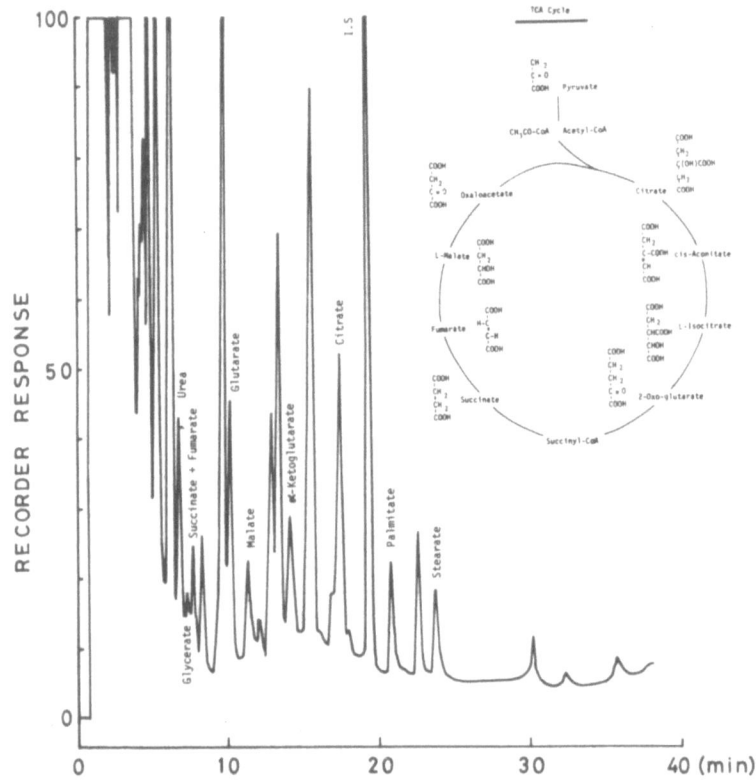

FIG. 2. Gas chromatogram of volatile organic acids in
V-CSF. Column temperature was programmed at
6°C/min from 80°C to 300°C.

participants in TCA cycle which is well known to exist in the brain
in high activity and this TIM chromatographic profile is quite
compatible with this fact.

Glutarate normally occurs in human urine only in trace amounts.
Increased excretion was described inconsistently in human ketosis
after the administration of hypoglycin A, the toxic agent in Jamaican
vomiting sickness. Hypoglycin A has been proposed to inhibit
isovaleryl CoA dehydrogenase and glutaryl CoA dehydrogenase. More
recently, elevated excretion of glutaric acid was reported in a
patient with glutaric aciduria 'type 11' by Przyrembel et al. (5).
This dicarboxylic acid is formed as the Coenzyme A derivative in
the degradation of L-tryptophan, L-lysine and hydroxy-L-lysine in
mammalian tissues. However, nothing has been known about the presence
of glutarate in the mammalian brain nor the significance of this
carboxylic acid. Further investigations are still in progress.

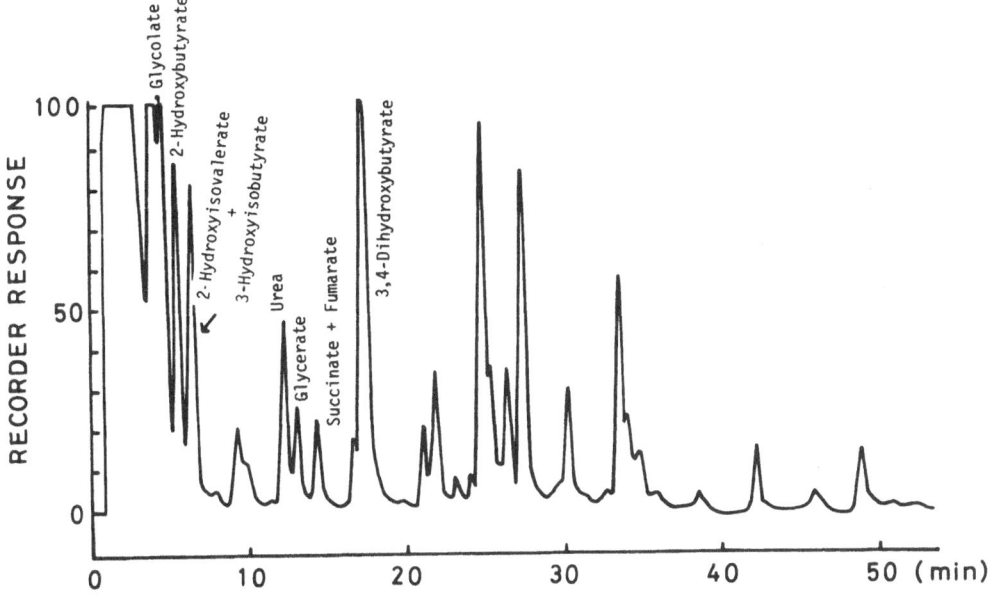

FIG. 3. TIM chromatogram of organic acids in V-CSF.
Column temperature was programmed at 4°C/min
from 70°C.

Fig. 3 shows the TIM chromatogram of organic acids in ventricular
cerebrospinal fluids. In this case, column temperature was programmed
at 4°C/min from 70°C in order to analyze highly volatile compounds
in detail. The following compounds are identified by GC-MS analysis;
glycolate, 2-hydroxybutyrate, 3-hydroxyisobutyrate, 2-hydroxyiso-
valerate, urea, glycerate, succinate plus fumarate and 3,4-dihydroxy-
butyrate. Interestingly enough, very similar profiles to those of
urinary organic acids were obtained (6). The mass spectrum of
glyceric acid identified in CSF is shown in Fig. 4. The ion at m/e
307 is $/ M - 15 /^+$. The abundant ion at m/e 292 is produced by
McLafferty rearrangement of trimethylsilyl group (7). This ion is
characteristic for the 3,4-dihydroxy acid TMS derivatives. Fig. 5
shows the mass spectrum of this peak which is identified as 3,4-
-dihydroxybutyrate by comparison with a mass spectrum of authentic
sample reported by Petersson (8). The ion at m/e 321 $/ M - 15 /^+$,
246 $/ M - 90 /^+$ and 233 $/ M - 15 - 103 /^+$ are present in high
abundance. The ion at m/e 189 is due to $/ M - 15 - HCH_2COOTMS /^+$.
The presence of 3,4-dihydroxybutyrate in urine was first reported
by Lee and Jollitt (9). It was found that 3,4-dihydroxybutyrate,
malate and tartarate in urine fluctuated in catatonic patients in
accordance with the clinical manifestation and decreased steeply at

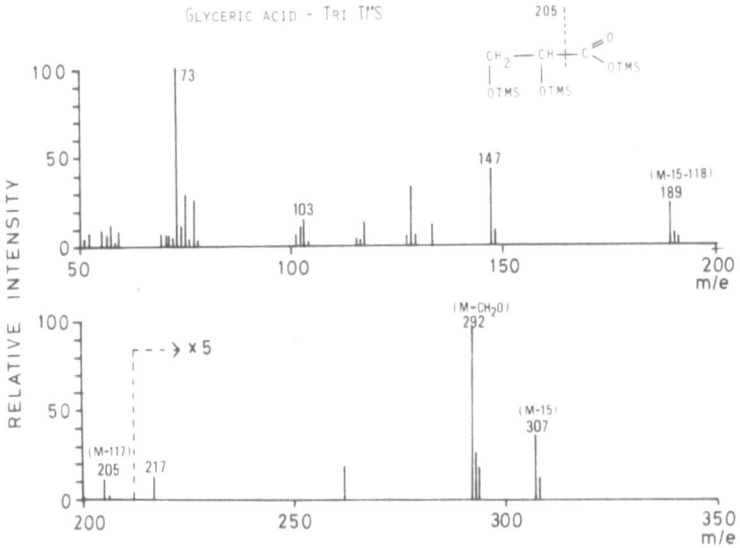

FIG. 4. Mass spectrum of glyceric acid-triTMS.

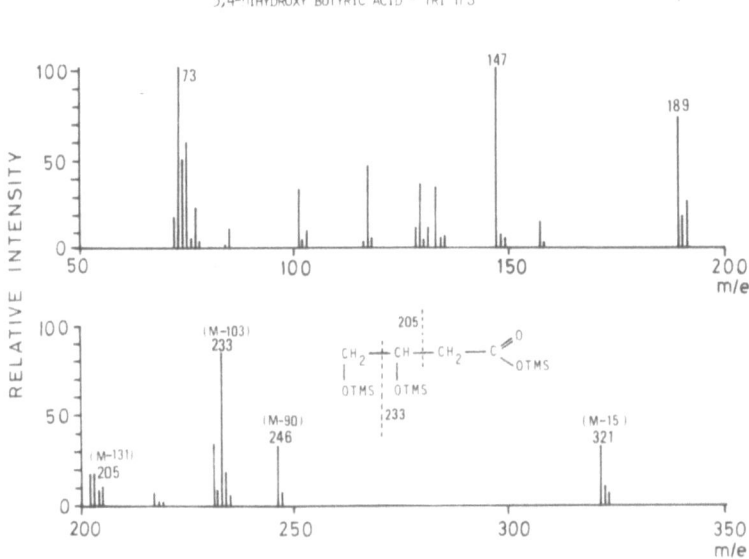

FIG. 5. Mass spectrum of 3,4-dihydroxybutyric acid-triTMS.

the initation of the stuporous period. They suggested fluctuation of
3,4-dihydroxybutyrate in urine reflected metabolism in the brain
and this compound was considered to be a metabolic product of β-
-oxidation of GABA, although not varified, in the central nervous
system in vivo. The presence of large amount of 3,4-dihydroxybutyrate
in cerebrospinal fluids in this study further supports the importance
of this material in the central nervous system. 3-Hydroxyisobutyrate,
a catabolic intermediate of branched chain amino acid, valine, was
also identified in cerebrospinal fluids (Fig. 6).

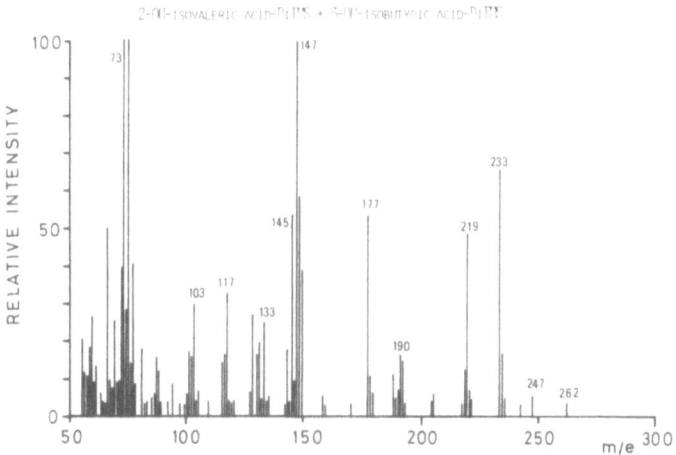

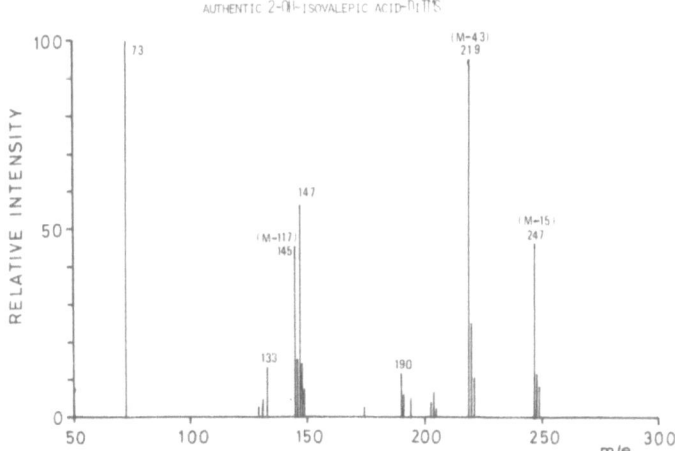

FIG. 6. Mass spectra of the peak which is identified as the
mixture of 3-hydroxyisobutyric acid and 2-hydroxyiso-
valeric acid (upper) and authentic 2-hydroxyisovaleric
acid-diTMS (lower).

This compound is normally found in human urine. In CSF, however, the peak of 3-hydroxyisobutyrate contained another component abundantly. This substance was estimated as 2-hydroxyisovaleric acid by the presence of peak at m/e 145 due to α-cleavage. Estimation of chemical structure was confirmed by comparison of its retention time and mass spectrum   with those of authentic sample. This compound is also a catabolic intermediate of valine.However, 2--hydroxyisovalerate is not a normal constituent of human urine. As for other catabolic intermediates of branched chain amino acid, 3-hydroxyisovalerate, derived from leucine is detected in human urine abundantly. However, in CSF we could not detected this compound in the present study.

To summarize, we obtained metabolic profiles of both amino acids and organic acids in the cerebrospinal fluid from patients with various kinds of neurological disorders. We confirmed the presence of the urea cycle in the brain and its significant role in the detoxication of ammonia by identification of urea in large amounts and a part of ornitine is derived from arginine during the sample preparation, especially during trimethylsilylation procedure. In the organic acid fraction, the highly active metabolism in TCA cycle was confirmed by the presence of all the intermediates abundantly. We also identified glyceric acid and 3,4-dihydroxybutyric acid for the first time in cerebrospinal fluids. 2-Hydroxyisovaleric acid, 3-hydroxyisobutyric acid and glutaric acid were also identified, all of them are biochemically interesting catabolic intermediates of valine, lysine or tryptophane. The present study proved that GC-MS analysis is the most suitable for organic acid profiling in cerebro-spinal fluids as this has already been accepted in urinary organic acid profiling and has enabled new findings of inborn errors of organic acid metabolisms. Although in the present study we studied metabolic profiles of cerebrospinal fluids as a whole without concern for the individual neurological disorders, this approach may enable us to clarify a part of the complex biochemical mechanisms which underly in various neurological disorders and general metabolism in the central nervous system.

REFERENCES

1) E.H.F. McGale, I.F. Pye, C. Stonier, E.C. Hutchinson and G.M. Aber, J. Neurochem., 1977, 29, 291.
2) E.M. Wright, Proceedings of the Physiological Society, 1977, p.31.
3) R.D. Malcolm and R. Leonards, Clin. Chem., 1976, 22, 623.
4) M.B. Sporn, W. Dingman, A. Defalco and R.K. Davies, Nature, 1959, 183, 1521.
5) H. Przyrembel, V. Wendel, K. Becker, H.J. Bremer, L. Bruinvis, D. Ketting and S.K. Wadman, Clin. Chim. Acta, 1976, 66, 227.

6) J.A. Thompson, S.P. Markey and P.V. Fennessey, Advances in Mass Spectrometry in Biochemistry and Medicine, 1976, 2, 1.

7) G. Petersson, Org. Mass Spect., 1972, 6, 577.

8) G. Petersson, Tetrahedron, 1970, 26, 3413.

9) C.R. Lee and R.J. Pollitt, in "Mass Spectrometry in Biochemistry and Medicine", A. Frigerio and N. Castagnoli, Eds., Raven Press, New York, 1974, p. 365.

ENERGY AND METASTABLE CHARACTERISTICS IN PEPTIDES. ELECTRON IMPACT APPEARANCE POTENTIALS OF SEQUENCE IONS IN TRIPEPTIDES

Z.V.I. Zaretskii and P. Dan

Isotope Department, The Weizmann Institute of Science

Rehovot, Israel

INTRODUCTION

Mass spectrometry has been demonstrated to be a sensitive method for sequence determination in peptides (1).

While it is observed that in many cases the identification of so called sequence ions can be made without difficulties, there are many noncharacteristic peaks in the mass spectra of peptides, formed as a result of both nonspecific cleavages in the molecular ions and after secondary decompositions of sequence ions.

Such a complication, which makes interpretation of mass spectrum more difficult, is due to the fact that electron energy routinely used in mass spectrometry (70 eV) substantially exceeds the appearance potentials (AP) of sequence ions. It is possible to decrease the complexity of the spectra by using low energy electrons in order to decrease the probability of secondary ruptures in characteristic fragments which yield sequence information.

It is known that while below 20 eV the sensitivity of mass spectrometric devices decreases rapidly there are electron energies between 11 and 20 eV that can be used to obtain less complex spectra but without considerable losses in sensitivity.

To facilitate the choice of the optimum electron energy producing less complex mass spectra of peptides it was decided to measure appearance potentials (AP) of both sequence ions and the ions formed as a result of further decompositions of the latter.

N-Trifluoroacetyl - derivatives of tripeptide methyl esters

411

($\underline{1}$ - $\underline{9}$) were chosen for the study (Scheme 1).

TFA-Ala-Gly-Val-OMe ($\underline{1}$)
TFA-Ala-Gly-Phe-OMe ($\underline{2}$)
TFA-Val-Gly-Val-OMe ($\underline{3}$)
TFA-Val-Gly-Leu-OMe ($\underline{4}$)
TFA-Val-Gly-Phe-OMe ($\underline{5}$)
TFA-Val-Gly-Pro-OMe ($\underline{6}$)
TFA-Phe-Gly-Val-OMe ($\underline{7}$)
TFA-Phe-Gly-Phe-OMe ($\underline{8}$)
AC-Ala-Ala-Ala-OMe ($\underline{9}$)

SCHEME 1.

We also used energetics of formation of sequence ions, as well as metastable transitions (MST) in the first field-free region (f.f.r.), to elucidate in detail the mechanisms of their fragmentation. Indeed, one should bear in mind that unambigous conclusions as to the structure of the peptide could be drawn only on the basis of the analysis of all types of fragmentation of its molecular ion.

EXPERIMENTAL

Low resolution mass spectra of peptides were determined at 70 and 12 eV and an emission 0.8 and 0.1 mA respectively, on a Varian MAT-731 double focusing Mass Spectrometer using a direct inlet system and the standard sample probe (± 0.3°) temperature stabilization. The ion source temperature was maintained at 250-255°. The spectra were taken after reaching the minimum temperature necessary to volatilize the sample while producing an adequate ion current.

High resolution mass spectra were obtained at a resolution of 10.000 and at 70 eV, using Varian MAT-731 Mass Spectrometer in conjunction with Spectro-System SS-100.

The metastable ions were discovered in the first field-free region (f.f.r.) or the double-focusing Varian MAT-731 Mass Spectrometer scanning the accelerating voltage with fixed electric sector voltage and fixed magnetic field ("high voltage scan") (2).

Appearance potentials (AP) were measured with Varian MAT-731 Mass Spectrometer according to the procedure described previously (3). The peptide was admitted to the ion source using direct insertion probe (± 0.3° temperature stabilisation) simultaneously with diphenyl ether (admitted from Refernce Inlet System), as calibrating compound, its ionization potential being 8.82 eV (4). At least three sets of replicate AP determination were carried out on different days with each compound. The AP's were reproducible to ± 0.05 eV.

RESULTS AND DISCUSSION

Ions A_1 and B_1 (Scheme 2). The intensities of the ions A_1 are generally very low, and in the case of peptide 4, 7, and 8 they are even absent from the spectra.

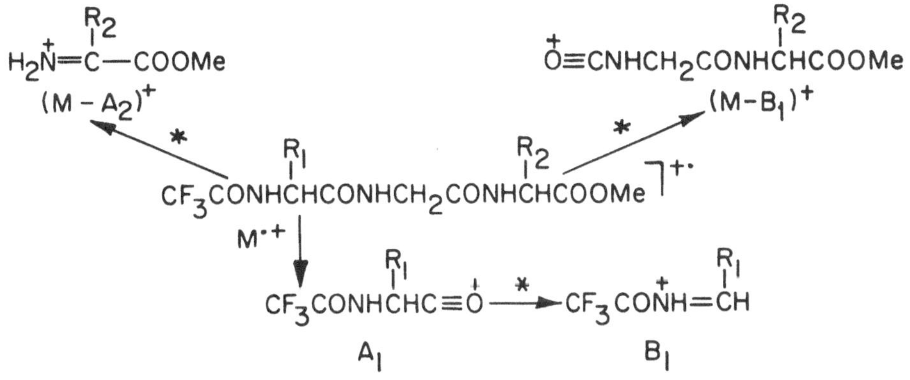

SCHEME 2.

Nevertheless the formation of ions B_1, as indicated by the appropriate MST, in about all cases proceeds via ions A_1 even when the latter could not be detected in the spectrum (Table 1).

The inspection of Table 1 also shows that formation of ions B_1 in TFA peptides (1-8) requires relatively high energy. Indeed in all, but one, cases AP values were found to exceed 11 eV, and in

TABLE 1.
Appearance potentials (AP, eV)[a] and origins of
B_1 ions in tripeptides ($\underline{1}$ - $\underline{9}$).

$$\begin{array}{c} R_1 \\ \text{CF}_3\text{CONHCH} \cdots \text{CO} \cdots \text{NHCH}_2\text{CO} \text{---} \text{NH} \text{---} \overset{R_2}{\text{CH}} \text{---} \text{COOMe} \\ B_1 \qquad A_1 \end{array}$$

Compound			Ion B_1	
No.	Sequence	m/e	AP	Origin
($\underline{1}$)	Ala-Gly-Val	140	12.34[b]	$A_1 \to B_1$[c]
($\underline{2}$)	Ala-Gly-Phe	140	13.08	$A_1 \to B_1$
($\underline{3}$)	Val-Gly-Val	168	11.50	$A_1 \to B_1$
($\underline{4}$)	Val-Gly-Leu	168	11.04	$A_1 \to B_1$
($\underline{5}$)	Val-Gly-Phe	168	12.24	$A_1 \to B_1$
($\underline{6}$)	Val-Gly-Pro	168	11.55[d]	$A_1 \to B_1$
($\underline{7}$)	Phe-Gly-Val	216	10.84[d]	$A_1 \to B_1$ $M \cdot^+ \to B_1$
($\underline{8}$)	Phe-Gly-Phe	216	12.67	$A_1 \to B_1$
($\underline{9}$)	(Ac)-Ala-Ala-Ala	86	10.46[e]	$A_1^f \to B_1$
($\underline{9}$)	(Ac)-Ala-Ala-Ala	87[g]	9.84[g]	no m̂

[a] Mean errors (S.D.) ± 0.05 to ± 0.1 eV unless otherwise stated;
[b] S.D. ± 0.11; [c] Also $B_2 + H \to B_1$; [d] S.D. ± 0.12; [e] S.D. ± 0.13;
[f] m/e 114, AP 10.38; [g] $B_1 + H$.

peptides $\underline{1}$, $\underline{2}$, $\underline{5}$ and $\underline{8}$ the energies required for the formation of
ions B_1 exceed 12 eV.

This fact is, probably, another indication of a two-step
mechanism of formation of ions B_1 from their molecular ions (via
ions A_1).

Ions a + 2H, a +H and A_2 (Schemes 3 and 4). The comparison of
appearance potentials and MST showed that ions A_2 originate from
a + 2H (in peptides $\underline{1}$, $\underline{2}$, $\underline{4}$, $\underline{6}$ and $\underline{7}$), or from a + H ions - in
case of peptides ($\underline{2}$ and $\underline{8}$), containing C-terminal Phe (Schemes 3
and 4, Tables 2 and 3).

It was reported earlier (5) that ion a + 2H are formed in
peptides with C-terminal Val and Ileu but not in peptides
containing C-terminal Leu residues. Our results indicate that the

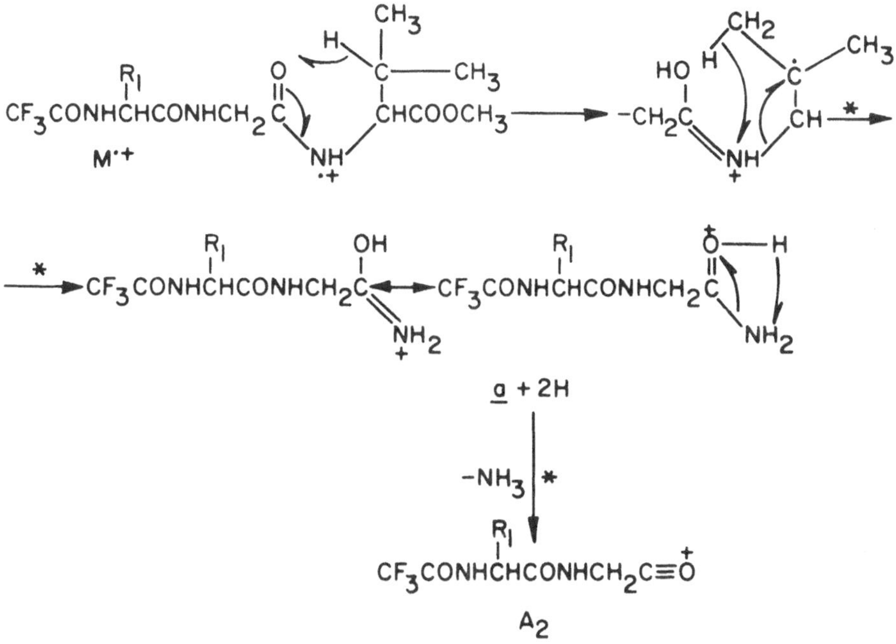

CH_3

H

C — CH_3

O

$CF_3CONHCHCONHCH_2$ C $CHCOOCH_3$ ——→

R_1

$M^{\cdot+}$

NH

$\cdot+$

CH_2

HO H C — CH_3

$-CH_2C$ CH —*→

NH

$+$

OH

*——→ $CF_3CONHCHCONHCH_2C$ ←—— $CF_3CONHCHCONHCH_2C$

R_1 NH_2 $+$ R_1 NH_2

O—H

$\underline{a} + 2H$

$-NH_3$ | *

R_1

$CF_3CONHCHCONHCH_2C\equiv\overset{+}{O}$

A_2

SCHEME 3.

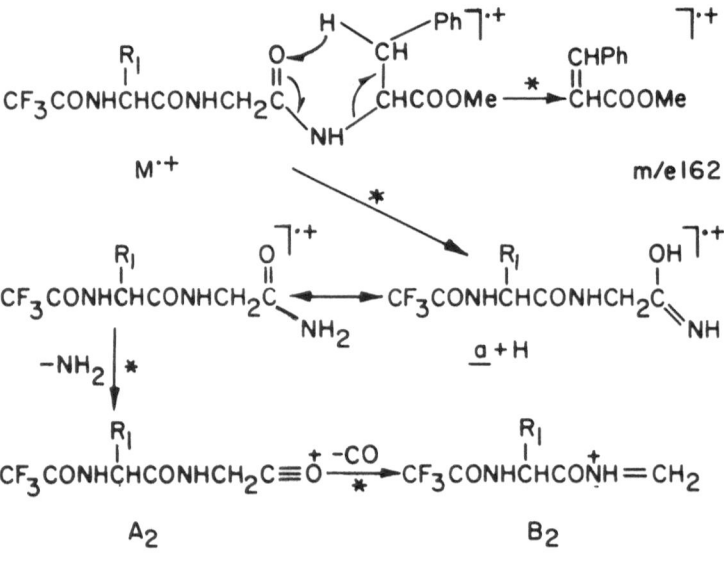

SCHEME 4.

TABLE 2.
Appearance potentials (AP, eV)[a] and origins of A_2
ions in tripeptides ($\underline{1}$ - $\underline{9}$).

$$CF_3CONH\overset{R_1}{\underset{|}{C}}HCONHCH_2CO \overset{}{\underset{A_2}{\curvearrowleft}} NH \overset{R_2}{\underset{\underline{a}}{\curvearrowleft}} CHCOOMe$$

Compound			Ion A_2	
No.	Sequence	m/e	AP	Origin[b]
($\underline{1}$)	Ala-Gly-Val	225	10.00	$\underline{a}$ + 2H → A_2
($\underline{2}$)	Ala-Gly-Phe	225	10.08	$\underline{a}$ + H → A_2
($\underline{3}$)	Val-Gly-Val	253	9.96	$\underline{a}$ + 2H → A_{2c}
($\underline{4}$)	Val-Gly-Leu	253	10.20	$\underline{a}$ + 2H$_d$ → A_2[c]
($\underline{5}$)	Val-Gly-Phe	253	10.11	-
($\underline{6}$)	Val-Gly-Pro	253	10.17	$\underline{a}$ + 2H → A_2
($\underline{7}$)	Phe-Gly-Val	301	9.93	$\underline{a}$ + 2H → A_{2e}
($\underline{8}$)	Phe-Gly-Phe	301	9.95	$\underline{a}$ + H → A_2
($\underline{9}$)	(Ac)-Ala-Ala-Ala	185	10.05	$(M\cdot^+-CO_2Me)$ → A_2

[a] Mean error: ± 0.05 up to ± 0.1 eV; [b] In $\underline{3}$, $\underline{7}$, $\underline{8}$ also $M\cdot^+$ → A_2;
[c] Also: $(M-C_4H_8)\cdot^+$ → A_2; [d] Not measurable owing to low intensity
of m^ peak; [e] Also: $(M-CO_2Me)^+$ → A_2.

TABLE 3
Appearance potentials (AP, eV)[a] and origins of
$\underline{a}$ + 2H ions in tripeptides ($\underline{1}$ - $\underline{8}$).

$$CF_3CONH\overset{R_1}{\underset{|}{C}}HCONHCH_2CONH \overset{R_2}{\underset{\underline{a}}{\curvearrowleft}} CHCOOMe$$

Compound			Ion $\underline{a}$ + 2H	
No.	Sequence	m/e	AP	Origin
($\underline{1}$)	Ala-Gly-Val	242	9.38	$M\cdot^+$ → $\underline{a}$ + 2H
($\underline{3}$)	Val-Gly-Val	270	9.35	$M\cdot^+$ → $\underline{a}$ + 2H
($\underline{4}$)	Val-Gly-Leu	270	-[b]	$(M^+-CO_2Me)^+$ → $\underline{a}$ + 2H
($\underline{6}$)	Val-Gly-Pro	270	-[b]	$(M-MeOH)\cdot^+$ → $\underline{a}$ + 2H
($\underline{7}$)	Phe-Gly-Val	318	9.22	$M\cdot^+$ → $\underline{a}$ + 2H
($\underline{8}$)	Phe-Gly-Phe	317[c]	8.96[c]	$M\cdot^+$ → $\underline{a}$ + H

[a] Mean error: ± 0.05 up to ± 0.1; [b] Not measurable owing to low
intensity of the peak; [c] $\underline{a}$ + H.

formation of ions $\underline{a}$ + 2H takes place also in case of tripeptide (4) with C-terminal Leu, but the respective peak is about 15 times less intense as compared with that of peptides $\underline{1}$, $\underline{3}$ and $\underline{7}$, containing Val residues at the C-terminal position.

Moreover, two step processes $M^{.+} \rightarrow \underline{a}$ + 2H (or $\underline{a}$ + H) $\rightarrow A_2$ (Schemes 3 and 4) seem to be the general way of formation of the ions A_2 in all acylderivatives of tripeptide methyl esters with the exception of tripeptide (9) containing Ala residue at C-terminal position (Table 2). So the appearance of ions $\underline{a}$ + 2H or $\underline{a}$ + H are of great importance in peptide sequencing by mass spectrometry. The $\underline{a}$ + 2H and $\underline{a}$ + H ions have relatively low APs, (differed by only about 0.2 eV from IP of tripeptide molecule, cf. Tables 3 and 7) and, hence, they are very abundant in low evergy (12 eV) mass spectra, along with their daughter A_2 ions, the mass differences being 17 and 18 mass units respectively. The driving force for the formation of ion $\underline{a}$ + 2H is a double hydrogen transfer of the side chain H-atoms to the charged fragment (Scheme 3).

Indeed, in peptides with C-terminal Phe (e.g. $\underline{8}$), where only one active (benzylic) hydrogen is available for such a migration, the formation of $\underline{a}$ + 2H was suppressed, and, instead of the latter, the ion $\underline{a}$ + H was formed.

On the other hand the critical step to begin with this fragmentation seems to be the migration of the first hydrogen from secondary C-atom of the side chain of the C-terminal residue.

Indeed, in tripeptide (9), with C-terminal Ala-residue the formation of $\underline{a}$ + 2H ions did not occur, the respective A_2 ion having been formed from $(M-CO_2Me)^+$ ion (Table 2).

It is to be noted that in spite of the fact that APs of ions $\underline{a}$ + 2H ($\underline{a}$ + H) are lower than that of ions A_2, in some tripeptides ($\underline{3}$, $\underline{7}$ and $\underline{8}$), the MST corresponding to the formation of ions A_2 from the respective molecular ions have also been found. This finding does not contradict the proposed mechanism (Scheme 3) as, in this case too, the observed MST may result from two stepwise processes occurring within the first field – free region (6).

It is worthwhile noting that energy requirements for the formations of ions A_2 respectively from $\underline{a}$ + 2H and $\underline{a}$ + H ions are different: while 0.6-0.7 eV is required for the reaction $\underline{a}$ + 2H $\rightarrow A_2$ in tripeptides ($\underline{1}$, $\underline{3}$ and $\underline{7}$), the formation of A_2 ion from $\underline{a}$ + H ion in tripeptide ($\underline{8}$) required the energy of about 1 eV (the differences between the APs of the A_2 ions and APs of the respective $\underline{a}$ + 2H and $\underline{a}$ + H ions have been used as a measure of the energies required for the above reactions (see Tables 2 and 3).

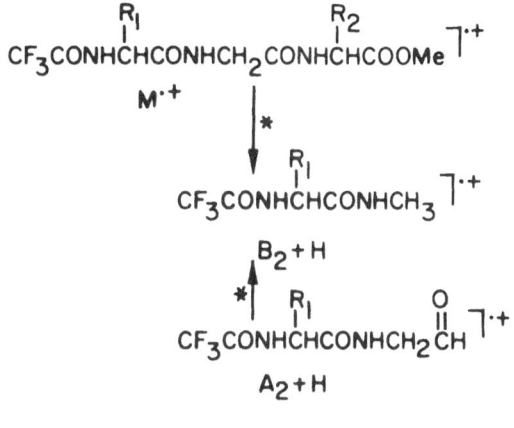

SCHEME 5.

SCHEME 6.

Ions B_2 + H and B_2 (Schemes 5 and 6). The formation of ions B_2 + H is accompanied by the migration of amido hydrogen to the charged fragment and required the energies (AP-IP) of about 0.30-0.55 eV (Tables 4 and 7).

TABLE 4

Appearance potentials (AP, eV)[a] and origins of B_2 + H ions in tripeptides ($\underline{1}$ - $\underline{9}$).

$$\begin{array}{ccc} & R_1 & & R_2 \\ & | & & | \\ CF_3CONHCHCONHCH_2 & \overbrace{}^{} CO \overbrace{}^{} & NHCHCOOMe \\ & B_2 \quad\quad A_2 & \end{array}$$

No.	Compound		Ion B_2 + H	
	Sequence	m/e	AP	Origin
($\underline{1}$)	Ala-Gly-Val	198[b]	9.73[b]	A_2+H → B_2+H
($\underline{2}$)	Ala-Gly-Phe	197[b]	11.09[b]	A_2 → B_2
($\underline{3}$)	Val-Gly-Val	226	9.53	A_2+H → B_2+H
				$M^{\cdot+}$ → B_2+H
($\underline{4}$)	Val-Gly-Leu	226	9.58	A_2+H → B_2+H
				$M^{\cdot+}$ → B_2+H
($\underline{7}$)	Phe-Gly-Val	274	9.35	A_2+H → B_2+H
				$M^{\cdot+}$ → B_2+H
($\underline{9}$)	(Ac) Ala-Ala-Ala	158[b]	9.22[b]	A_2+H → B_2+H
($\underline{9}$)	(Ac)Ala-Ala-Ala	157[b]	9.24[b]	A_2 → B_2

[a] Mean error: ± 0.05 up to ± 0.1 eV;   [b] B_2.

The MST showed that ions B_2 + H are formed from both molecular and A_2 + H ions, in spite of the fact that the latter are very low intensities.

Anyway, low energy requirements for the formation of ions B_2 + H, as well as the respective MST, suggest that these fragments may originate directly from the molecular ions. On the other hand, in tripeptide ($\underline{2}$) the ion B_2 only is formed, the energy required for the respective reaction A_2 → B_2 having been found to be about 2 eV (AP_{B_2} - AP_{A_2}, see tables 2 and 4).

Ions $(M-B_1)^+$ (Scheme 2). While in f.f.r. no MST were found in tripeptides ($\underline{2-5}$ and $\underline{9}$), the occurrence of the respective MST ($M^{\cdot+}$ → $(M-B_1)^{+}$) in peptides ($\underline{6}$ and $\underline{7}$) as well as low energy requirements found (AP-IP = 0.6-0.65 eV, see Tables 5 and 7) for the

TABLE 5
Appearance potentials (AP, eV)^a and origins
of $(M - B_1)^+$ ions in tripeptides ($\underline{1}$ - $\underline{9}$).

$$
\begin{array}{cc}
R_1 & R_2 \\
| & | \\
CF_3CONHCH & CONHCH_2CONHCHCOOMe
\end{array}
$$
$$\longrightarrow (M - B_1)^+$$

Compound			Ion $(M - B_1)^+$	
No.	Sequence	m/e	AP	Origin
($\underline{1}$)	Ala-Gly-Val	215	9.67	$(M-CF_3CONH)^+\rightarrow(M - B_1)^+$
($\underline{2}$)	Ala-Gly-Phe	263	9.62	no m⌢
($\underline{3}$)	Val-Gly-Val	215	9.74	no m⌢
($\underline{4}$)	Val-Gly-Leu	229	9.76	no m⌢
($\underline{5}$)	Val-Gly-Phe	263	9.78	no m⌢
($\underline{6}$)	Val-Gly-Pro	213	9.85	M·$^+$ → $(M - B_1)^+$
($\underline{7}$)	Phe-Gly-Val	215	9.94	M·$^+$ → $(M - B_1)^+$
($\underline{9}$)	(Ac)Ala-Ala-Ala	201	9.30	no m⌢

^{a}Mean error: ± 0.05 up to ±0.1 eV.

formation of $(M-B_1)^+$ ions makes it possible to suggest that these
fragments, in general, originate from the molecular ions (Scheme 2).

It is remarkable that appearance potentials of $(M-B_1)^+$ are
1.5-2.5 eV lower than that of B_1 ions. There is definitely a
competition for retaining the charge, between two parts of tripeptide
molecules (B_1 versus $(M-B_1)^+$).

Ions $(M - A_2)^+$ (Scheme 2). The appearance potentials of these
ions in tripeptides ($\underline{1}$), ($\underline{3}$) and ($\underline{5}$) are close to 11 eV - e.g.
substantially higher as compared to that of A_2 ions (Table 6 and 2).
Owing to this fact the low energy (12 eV) spectra reveal practically
no $(M - A_2)^+$ peaks. At the same time, the peaks of ions A_2 are
relatively abundant under these conditions.

On the other hand, in tripeptide ($\underline{6}$) the energy required for
the formation of ion $(M - A_2)^+$ is lower than that of the respective
ion A_2. In this case, while the ion $(M - A_2)^+$ is a base peak in
both 70 eV and 12 eV spectra, the ion A_2 is small in 70 eV and is
absent from 12 eV spectra.

While the MST's M·$^+$ → $(M -A)^+$ were found in the f.f.r. (Table
6), the relatively high energy requirements for the formation of

TABLE 6
Appearance potentials (AP, eV)[a] and origins
of $(M - A_2)^+$ ions in tripeptides ($\underline{1} - \underline{6}$).

$$R_1 \qquad\qquad R_2$$
$$CF_3CONHCHCONHCH_2CO \overbrace{} NHCHCOOMe$$
$$\longrightarrow (M - A_2)^+$$

| Compound | | | Ion $(M - A_2)^+$ | |
No.	Sequence	m/e	AP	Origin
($\underline{1}$)	Ala-Gly-Val	130	10.92	
($\underline{3}$)	Val-Gly-Val	130	10.90	$M^{\cdot +} \to (M - A_2)^+$ [b]
($\underline{5}$)	Val-Gly-Phe	178	10.47	$M^{\cdot +} \to (M - A_2)^+$ [b]
($\underline{6}$)	Val-Gly-Pro	128	9.80	$M^{\cdot +} \to (M - A_2)^+$ [b]

[a] Mean error: ± 0.05 up to ± 0.1 eV; [b] Also other very weak m^peaks.

ions $(M - A_2)^+$ in tripeptides ($\underline{1}$, $\underline{3}$ and $\underline{5}$) suggest that these
fragments may originate from the stepwise process rather than from
direct decomposition of the respective molecular ions.

Ions m/e 162 and 131 (Scheme 4). The formation of m/e 162 ions
is characteristic of peptides containing C-terminal Phe residue
(Scheme 4) (1). These ions are formed by the same fragmentation,
the McLafferty rearrangement, as the ions a + H, but with the
retention of the charge by the cinnamic ester, originated from the
C-terminal part of the peptide molecule. The energies required for
this reaction are very low (AP - IP $\sim 0.2 - 0.35$ eV, see Tables 7
and 8), and, owing to this fact, the m/e 162 peaks are the base
peaks in both 70 eV and 12 eV spectra.

CONCLUSION

The results of this study may have an important analytical
application. Indeed, for N-trifluoro-acetyl derivatives of tripeptide
methyl esters ($\underline{7} - \underline{8}$) the appearance potentials of all ions,
bearing sequence information, with only one exception (ions B_1)
do not exceed 11 eV.

This finding makes it possible, when working at 11 - 12 eV,
to produce considerably simplified mass spectra, containing very

TABLE 7

Appearance potentials (AP, eV)[a] of some diagnostically important ions in tripeptides (1 - 9).

R_1 $(M-B_1)^+$ R_2 $(M-MeOH)^{.+}$

$$CF_3CONHCH \longrightarrow CONHCH_2CONHCH \longrightarrow CO \longrightarrow OMe$$

$(M-CO_2Me)^+$ $(M-CO_2Me)^+$ ·H

No.	Compound Sequence	$M^{.+}$ m/e	$M^{.+}$ IP	$(M-CO_2Me)^+$ m/e	$(M-CO_2Me)^+$ AP	$(M-B_1-HCO_2Me)^+$ m/e	$(M-B_1-HCO_2Me)^+$ AP
(1)	Ala-Gly-Val	355	-[b]	296	9.41	155	11.12
(2)	Ala-Gly-Phe	403	8.96	344	-[b]	203	11.49
(3)	Val-Gly-Val	383	-[b]	324	9.37	155	11.55
(4)	Val-Gly-Leu[c]	397	9.14	338	9.35	169	-
(5)	Val-Gly-Phe	431	9.20	372	-[b]	203	11.52
(6)	Val-Gly-Pro	381	9.22	322	-[b]	153	11.61
(7)	Phe-Gly-Val	431	9.04	372	9.26	155	11.84
(9)	(Ac)-Ala-Ala-Ala	287	-[b]	228	9.19	141	-[b]
(8)	Phe-Gly-Phe	479	9.00	420	-[b]	203	-[b]

[a] Mean error: ± 0.05 up to ± 0.1 eV; [b] Not measureable owing to low intensity of the peak; [c] $(M-C_4H_8)^{.+}$, m/e 341, AP 9.36.

TABLE 8
Appearance potentials (AP, eV)[a] of m/e 162 and
131 ions characteristic of tripeptides containing
C- terminal phenylalanyl residue (2, 5 and 8).

$$CF_3CONHCHCONHCH_2CONH \longrightarrow CHCOOMe$$

with R_1 above the first CH and CH_2Ph above CHCOOMe, and $-H \longrightarrow 162$

Compound		AP	
No.	Sequence	m/e 162[b]	m/e 131[c]
(2)	Ala-Gly-Phe	9.33	11.74
(5)	Val-Gly-Phe	9.40	11.54
(8)	Phe-Gly-Phe	9.35	11.82

[a] Mean error: 0.05 up to   0.1 eV;   [b] M.+ ^162;   [c] 162 ^131.

few numbers of sequence ions together with relatively abundant
molecular ion peaks (7).

Such low energy mass spectra do not contain the non-specific
ions and provide all necessary information on the amino acid
sequence in peptide molecules. We believe that studying the energy
characteristics for sequence ions in various peptides could open
the way to the "regulative" fragmentation of peptides under electron
bombardment. Indeed, the AP data might perhaps be used as an aid
in proper selection of the electron energies needed for the
selective cleavages of peptide linkages.

REFERENCES

1) P.J. Arpino and F.W. McLafferty, in "Determination of Organic
   Structure by Physical Methods", F.C. Nachod, J.J. Zuckerman and
   E.W. Randall, Eds., Vol. 6, Academic Press, New York, 1976, p. 1.
2) M. Barber and R.M. Elliott, in "American Society for Mass
   Spectrometry , 12th Annual Conference on Mass Spectrometry and
   Allied Topics", Montreal, Canada, 1964.
3) Z.V.I. Zaretskii, V.L. Sadovskaya, N.S. Wulfson, V.S. Sizoy
   and V.G. Merimson, Org. Mass Spectrom., 1971, 5, 1179.
4) L.J. Franclin, J.G. Dillard, H.M. Rosenstock, J.T. Herron, K.
   Draxl and F.M. Field, in "Ionization Potentials, Appearance

Potentials and Heats of Formation of Gaseous Positive Ions",
National Bureau of Standards, Washington, 1969, p. 133.

5) A. Prox and F. Weygand, in "Peptides", H.C. Beyerman, A. Van de
Linde and W. Maasen Van den Brink, Eds., North Holland,
Amsterdam, 1967, p. 158.

6) D.M. Smith, A.M. Duffield and C. Djesassi, Org. Mass Spectrom.,
1973, 7, 367; b. K.R. Jennings, Chem. Commun., 1967, 283;
c. E. Caspi, J. Wicha and A. Mandelbaum, Chem. Commun., 1967,
1161.

7) Low energy mass spectra of peptides will be published elsewhere.

THE CONTINUOUS RECORDING OF EXPIRED $C^{16}O^{18}O$ BY RESPIRATORY MASS

SPECTROMETRY AFTER INJECTION OF $H_2^{18}O$ INTO THE VASCULAR SYSTEM

K.-P. Pflug, K.-D. Schuster, H. Förstel° and J.P. Pichotka

University of Bonn, Nussallee, Bonn; °Nuclear Research

Center (KFA), Jülich, West Germany

INTRODUCTION

The radioactive isotopes of oxygen are in general not suited for measurements on biological and medicine objects, because of their very short half-lives. The use of stable oxygen isotopes as tracers has been limited so far to analyzing concrete, individual samples, not knowing their fractionations. The distribution of the oxygen isotopes ^{16}O and ^{18}O between body fluids and carbon dioxide is not the same in both phases (1).

The determination of these and other fractionations were performed under stationary conditions of natural abundance (1). We have compared samples of body fluids from different compartments and samples of expired carbon dioxide in different physiological situations. These measurements were performed with a mass spectrometer which was designed for the determination of isotopic ratios.

The body fluids of the different compartments are identical in their ^{18}O compositions (1). The comparison of the $^{18}O/^{16}O$ ratio in body fluids and carbon dioxide shows that after equilibration the ^{18}O concentration in carbon dioxide is about 40%₀ higher than in body fluids.

Because of the constancy of these values, the $^{18}O/^{16}O$ ratio of expired carbon dioxide gives the $^{18}O/^{16}O$ ratio of body fluids under stationary conditions. If $H_2^{18}O$ is given into the venous blood, a part of the ^{18}O which was introduced appears immediately within the expired carbon dioxide (2). This effect is the base of a new method to determine the lung perfusion which is demonstrated in the second part of this paper. A small amount of $H_2^{18}O$ is injected into the circulation and its dilution within the flowing blood is determined by analysing the expired CO_2.

The measurements of the $^{18}O/^{16}O$ ratio of the expired carbon dioxide were performed with a respiratory mass spectrometer. For the continous measurement of the isotope ratio it is particularly useful to have a norm of measurement similar to the δ-scale used in isotope ratio mass spectrometry. Therefore a norm of measurement has been deduced which is described in the first part of the paper.

METHOD OF MEASUREMENT

In the high precision technique of isotopic abundance measurements a single sample is analysed by comparing it several times with a laboratory standard of known composition. The isotopic composition of the sample is given as δ-value relative to the standard defined as (3):

$$\delta = \frac{\left(\dfrac{C^{16}O^{18}O}{C^{16}O_2}\right)_{sample} - \left(\dfrac{C^{16}O^{18}O}{C^{16}O_2}\right)_{reference}}{\left(\dfrac{C^{16}O^{18}O}{C^{16}O_2}\right)_{reference}} \qquad (\underline{1})$$

For the continuous determination of the deviation of the $^{18}O/^{16}O$ ratio with the respiratory mass spectrometer, it is necessary to have an equivalent norm of measurement to the δ-value. The isotopes ^{16}O and ^{18}O together with hydrogen constitute the molecules $H_2^{16}O$ and $H_2^{18}O$; in carbon dioxide they occur as $C^{16}O_2$, $C^{16}O^{18}O$ and $C^{18}O_2$ with the masses 44, 46 and 48. Due to the low abundance $C^{18}O_2$ is neglected. The masses 44 and 46 were continuously measured with a double collector. At the beginning of the experiment the expired air has a composition of X. The voltage of the mass peak 46 X_{46} is compensated to zero with a fraction P of the voltage X_{44}, which comes from the mass peak 44:

$$X_{46} - PX_{44} = 0 \qquad\qquad (\underline{2})$$

After the injection of $H_2^{18}O$ into the circulation, the oxygen isotopes change reciprocally between water and the carbon dioxide dissolved in water. The change can be formulated by the following equation:

$$H_2^{18}O + C^{16}O_2 \rightleftarrows H_2^{16}O + C^{16}O^{18}O \qquad (\underline{3})$$

The expired air is then said to have the altered composition Y. By unchanged amplifications in both channels a deviation D of the zero point is registrated

$$Y_{46} - PY_{44} = D \qquad\qquad (4)$$

If we multiply the equation (4) by the factor F, so that

$$FY_{44} = X_{44} \qquad\qquad (5)$$

results, and if equation (2) is substracted from the relation (4), we come to the equation (6):

$$FY_{46} - X_{46} = F \cdot D \qquad\qquad (6)$$

After division of the equation (6) by X_{44} and X_{46}/X_{44} we get the result:

$$\cfrac{\cfrac{F \cdot Y_{46}}{X_{44}} - \cfrac{X_{46}}{X_{44}}}{\cfrac{X_{46}}{X_{44}}} = \cfrac{F \cdot D}{X_{46}} \qquad\qquad (7)$$

Under consideration of the relation (5), the left side of the equation (7) is equivalent to the δ-deviation between the alveolar air before the injection of $H_2^{18}O$ into the circulation and the alveolar air after the injection:

$$\delta = \cfrac{F \cdot D}{X_{46}} \qquad\qquad (8)$$

The deviation D can be taken as the difference to the zero point of the registrated curve of measurement from the recorder.

F is found equivalent to (5) from the voltages of the mass peaks 44 in the zero point situation and the situation of measurement Y. The voltage X_{46} is determined after having balanced the zero point when the difference formation has been switched off.

The norm of measurement supplies the correct values only if the whole range of measurement for both kinds of molecules is linear.

For proof of the linearity, samples of expired air were collected.

They were measured, both with the respiratory mass spectrometer and
also after separation of carbon dioxide from the other ingredients
in a vacuum distillation apparatus with the mass spectrometer
designed for the measurement of isotopic ratios (Micromass 602 C;
Vacuum Generators).

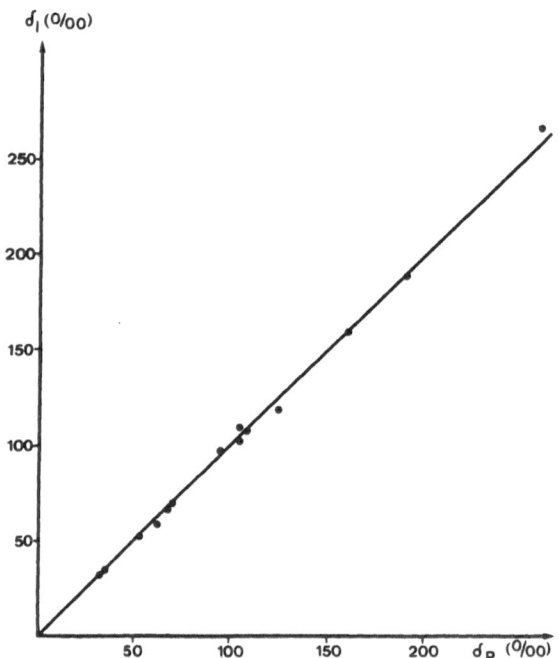

FIG. 1. Comparison between the measurements of the
respiratory mass spectrometer and those of
the isotope ratio mass spectrometer. The
δ_R-values have been measured with the
respiratory MS the δ_I-values with the
Micromass 602 C.

In Fig. 1 we present a comparison between the measurement of
the respiratory mass spectrometer and those of the isotope ratio
mass spectrometer. On the abscissa are shown the δ_R-values which
have been measured with the respiratory mass spectrometer. On the
ordinate are shown the δ_I-values which have been measured with the
Micromass 602 C. The resulting curve describes a linear function
over the whole range of measurement.

EXAMPLE OF APPLICATION AND RESULTS

As an example of application for this method of measurement we determined the perfusion of lungs after venous injection of $H_2^{18}O$.

These experiments were performed on rabbits anaesthetized with urethane. The surgical procedure was performed as described in a previous paper (Fig. 2) (2).

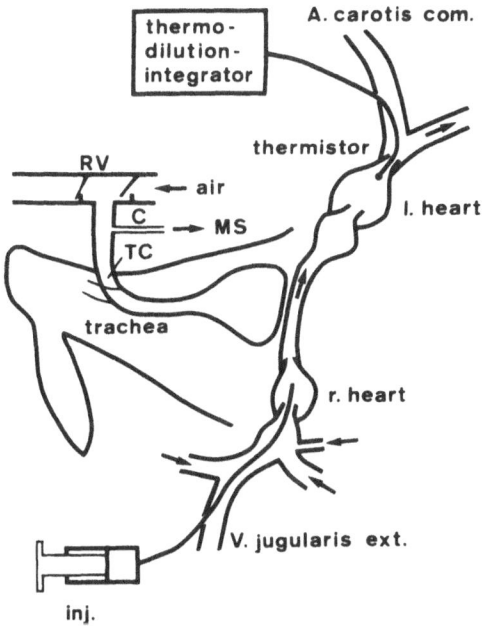

FIG. 2. Scheme for the experimental set-up on rabbits. TC: tracheal catheter; C: inlet capillary; MS: mass spectrometer; RV: respiratory valve.

The rabbits were breathing spontaneously. The catheter was introduced into the jugular vein and advanced through the right heart into the pulmonary artery. Tracer solutions were given in instantaneous injections of 1 ml. The capillary inlet of the respiratory mass spectrometer was connected to the tracheal catheter. By the respiratory valve inspiratory and expiratory air were separated. Besides the $^{18}O/^{16}O$ ratio several other values were determined. Gas concentrations of inspiratory and expiratory air were recorded continuously by mass spectrometer. The average concentrations of O_2 and CO_2 in expired air and respiratory minute volume were also

recorded. Simultaneously with the tracer injection the perfusion
of the lungs was determined by thermodilution (4). Fig. 3 gives the
time course of the $^{18}O/^{16}O$ ratio in expired carbon dioxide after
instantaneous injection of $H_2^{18}O$ for one experiment.

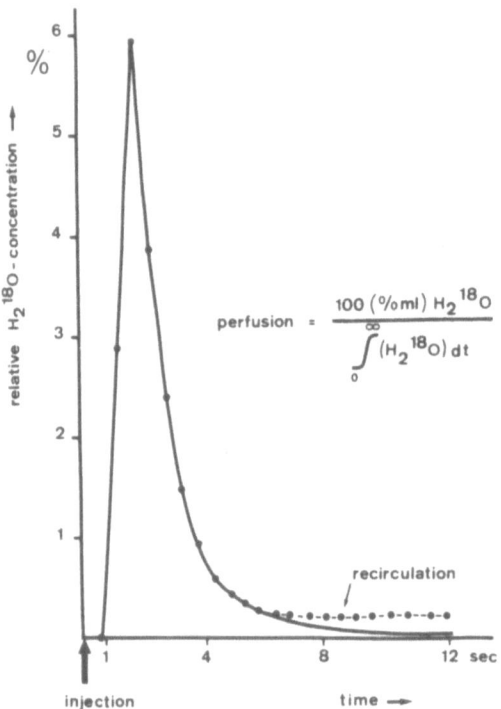

FIG. 3. The time course of the $H_2^{18}O$ concentration
          calculated from the $^{18}O/^{16}O$ ratios in
          expired carbon dioxide after instantaneous
          injection of 1 ml Ringer's solution
          containing 5% $H_2^{18}O$.

The points on the curve are the end-expiratory values of CO_2 with
respect to the fraction of ^{18}O between CO_2 and blood (1). The units
on the ordinate refer to the concentration of the tracer solution;
100% on this scale would be the ^{18}O concentration in carbon dioxide
equilibrated with the injected solution. Hence a hundred divided
by the ordinate values gives the actual dilution of the tracer
within the blood. In these experiments $H_2^{18}O$ was given with the
cold injection for the determination of the perfusion of the lungs.

We may therefore assume that the change of $H_2^{18}O$ and as well of the temperature of this cold injection have the same distribution volume in the blood.

The perfusion of lungs can be calculated from the distribution volume according to Hamilton (5). The areas were intergrated after subtracting the effect of recirculation from the curve. Therefore the points on the curve between the maximum and the 8th second after the injection had been fitted to an exponential function.

The perfusion values measured by the $H_2^{18}O$ method are given in Table 1, column 3. Column 2 contains the values measured with the thermodilution method. The mean values amount 490 ml/min and 462 ml/min respectively.

TABLE 1.
Comparison between values of lung perfusion measured by the thermodilution method and those measured by the $H_2^{18}O$ method.

experi- ment	thermo- dilution $ml \cdot min^{-1}$	$H_2^{18}O$ $ml \cdot min^{-1}$	difference c3-c2 $ml \cdot min^{-1}$
5	511	517	6
7	461	498	37
9.1	575	604	29
9.2	366	401	35
10	398	432	34
mean	462	490	28
SD	85	79	13

DISCUSSION

The ordinates in Fig. 3 have been given in a relative form to indicate directly the situation of dilution. With respect to the procedure of measurement the absolute values in the δ-notation should be considered. The maximum values of the recorded curves are in the order 1000 to 1500‰. The lowest values which we had to use for the integration of the areas are in the order of 70‰. The limit of accuracy of the mass spectrometer has been improved up to 1‰. Hence the recorded isotope ratios can be considered of sufficient accuracy.

The difference of the perfusion values in Table 1, column 4, showed that the values measured by the thermodilution method in the mean are 28 ml lower than those values measured by the $H_2^{18}O$ method. The difference is constantly positive; the deviation is systematic.

The $H_2^{18}O$ method turns out to be a real dye dilution method from the measure-principle as well as from the degree of accuracy (6).

$H_2^{18}O$ given into the venous system and the measurement of the tracer in expired air stands out against the present conventional clinical methods to determine the perfusion of the lungs as a noninvasive method of measurement. The cost of this method is limited by the presence of a suitable mass spectrometer to ca. 1 ml $H_2^{18}O$ per one measurement. One ml $H_2^{18}O$ in the concentration of the applicated step of enrichment costs at present ca. 5.- DM. Another advantage of the $H_2^{18}O$ method in comparison with other dilution methods is that the preparation of the patient for this measurement is less difficult, as it is not necessary to introduce a catheter in the heart. The time of measurement is in the range of seconds.

REFERENCES

1) K.P. Pflug, K.D. Schuster, H. Förstel and J.P. Pichotka, in "Proceedings Third International Conference on Stable Isotopes in Chemistry, Biology and Medicine", Argonne, 1978, in press.
2) K.D. Schuster, K.P. Pflug, H. Förstel and J.P. Pichotka, in "Proceedings Third International Conference on Stable Isotopes in Chemistry, Biology and Medicine", Argonne, 1978, in press.
3) H.C. Urey, J. Chem. Soc., 1947, 1, 562.
4) H. Slama and J. Piiper, Z. Kreislauf-forschg., 1964, 53, 322.
5) W.F. Hamilton, J.W. Moore, J.M. Kinsman and R.G. Spurling, Am. J. Physiol., 1928, 85, 377.
6) P.G. Spieckermann and H.J. Bretschneider, Arch. Kreislaufforschg, 1968, 55, 211.

A CONTINUOUS MASS SPECTROMETRIC ANALYSIS OF EXPIRED PHYSIOLOGICAL

GASES: THE PULMONARY GAS EXCHANGE

S. Damato

Istituto di Tisiologia e Malattie dell'Apparato Respiratorio dell'Università degli Studi di Milano, Milano, Italy

INTRODUCTION

Recently a new method was presented of continuous analysis of gases and flow rate in the expirate, breath by breath, by computing the data from a mass spectrometer and flow meter (1).

This technique was applied to the continuous analysis of oxygen and carbon dioxide during air breathing in normal non-smoker subjects; the results will be presented in a more appropriate form in a paper in preparation (2). Here a report is given on it in a concise form.

In the present paper the analysis has been extended to nitrogen during air breathing and first breath of a multi breath nitrogen wash out with an oxygen-argon mixture. These results especially concern pulmonary gas exchange.

EXPERIMENTAL

A mass spectrometer (Centronic Limited Croydon) with a response time of 80 msec. and a lag time of 100 msec. was used continuously to analyze oxygen, carbon dioxide and nitrogen expired, in connection with a flow meter, whose pneumotachograph was placed at a suitable opening of a "box-bag" used for inspired and expired gas collection. Each sitting subject wore a nose clip and mouthpiece and breathed through a solenoid valve box, according to Lee and Crisp (3), having a low resistance and a small dead space, which allows a change, at the end of a normal expiration, from an

inspiratory mixture to another without needing the subject's co-
-operation.

The sample inlet system of the mass spectrometer, connected with
the valve box, was a non-heated disposable one, so that the water
response time was extremely prolonged. At the same time an automatic
sensitivity control was employed to remove the influence of the change
in water vapour concentration upon the gases under analysis (4).

As a result, a dry gas was always measured in the analysis of
gases and flow rate.

The analog signals were recorded upon a magnetic tape recorder
(Tandberg 100) and subsequently fed into the analog to digital
converter of a digital computer (Varian 73). The computer programme
syncronised the signals, multiplied the flow rate by the concentration and cumulated the result. At the same time the flow rate was
cumulated to yield the expired volume at any time during the breath.
These two variables were plotted as evolved gas volume versus expired
volume.

In this way, it is possible to compute the tidal volume (TV)
for each breath, and, for each analyzed gas, a linear regression
which is simply depicted in Fig. 1. Therefore it is also possible
to obtain, for each breath and gas, the series dead space (VDS),
the alveolar ventilation (AV), the volume of gas exchanged and/or
evolved (V), the mean alveolar gas concentration ($F_{\bar{A}}$) and the alveolar
efficiency.

The results are referred to normal non-smoker subjects, both
males and females, whose routine functioning lung tests were normal,
and are reported and discussed better to point out their significance
in the pulmonary gas exchange.

RESULTS AND DISCUSSION

A). The series dead space (VDS) data in 22 non smoker subjects
regarding 627 different sized breaths, both for oxygen and carbon
dioxide, are concisely reported in Fig. 2.

The values of the dead space for oxygen and carbon dioxide
differ significantly, whilst the anatomy is identical for both
gases, the anatomical dead space being defined as the volume
between the lips and the area of gas mixing and gas exchange. Since
the flow rate is also the same for both gases, the difference in
behaviour is probably attributable to their different diffusivity.

We observed that carbon dioxide, with its molecular weight of

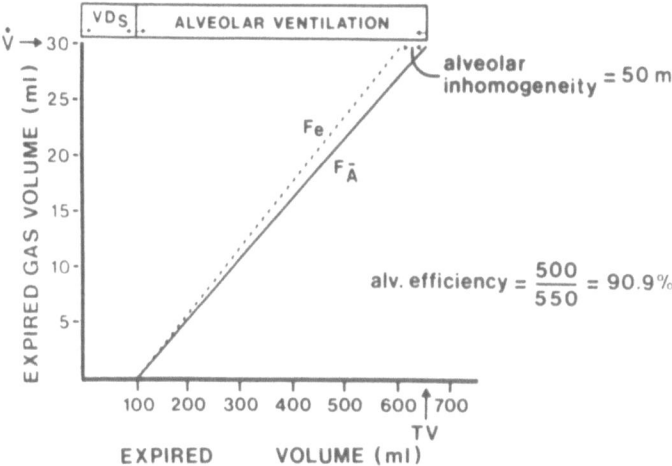

FIG. 1. The evolution of gas volume (ordinate) in
expired volume (absissa). The intercept of the
regression line (continuous) upon the volume
axis indicates the volume of the conducting airways,
whilst its slope indicates the mean alveolar gas
concentration ($F_{\bar{A}}$). Fe is the end tidal gas
concentration, VDS the series dead space, VT the
tidal volume and V the volume of gas exchanged
and/or evolved.

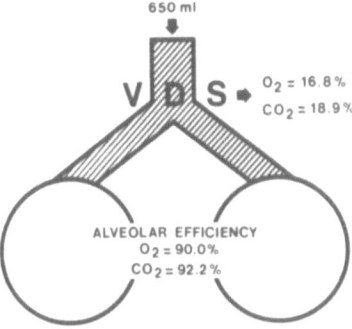

FIG. 2. Experimental evidence of series dead space (VDS),
expressed as fraction of the tidal volume, and alveolar
efficiency from 627 different sized breaths in 22 subjects.

44, produces a larger dead space than oxygen, with its molecular
weight of 32. Therefore the dead space cannot be considered as a
single value for all the exchanged gases, also as alveolar
ventilation level is different for each gas.

The usual way of measuring alveolar ventilation is to compute
the physiological dead space, or wasted ventilation. The Bohr
equation, which is used for this calculation, takes into account
the mixed expired concentration and the arterial carbon dioxide
value, in equilibrium with alveolar carbon dioxide. The result
includes in a single figure all causes of the high ventilation
perfusion ratio, necessarily assuming that the anatomical dead
space is a part of the gas exchange area. The latter assumption,
which was acceptable because of difficulties in defining and analysing the alveolar ventilation, can no more be accepted.

The application of the mass spectrometry proposed by Langley
and Cumming (1) is helpful in studying the gas inhomogeneity of
the alveolar air, which takes part in pulmonary gas exchange.

Fig. 1 depicts a continuous line, whose slope indicates the
mean alveolar gas concentration, and a dotted one, which expresses
the end tidal gas concentration; the last one could be replaced
by the arterial value.

The difference between the mean and the end gas concentrations
is a function of the alveolar gas inhomogeneity and could be
expressed as usual by calculating an "alveolar dead space". This is
an intraalveolar compartment which, if subtracted, gives a complete
homogeneous alveolar ventilation. I believe this procedure is
incorrect because of the following two considerations: i) it divides
the alveolar volume into inexistent compartments in agreement with
a "probable" parallel lung model; ii) it requires a steady-state
condition (multi-breath) which assumes that the difference between
oxygen uptake and carbon dioxide output is balanced by a small
continous flux to the ambient air from the body. The procedure is
also invasive because of the arterial blood sample.

The application of a compartmental analysis to the above mass
spectrometric technique is impossible at the present time because
of difficulties in analysing oxygen and carbon dioxide in arterial
blood, in vivo, breath by breath.

As a consequence, I propose that at the present stage the best
way of expressing the alveolar inhomogeneity is to compute the
"alveolar efficiency", as depicted in Fig. 1; this value is 100
per cent when there are no differences between mean alveolar and end
tidal gas concentration during the expired volume.

The alveolar efficiency, as shown simply in Fig. 2, is greater

for oxygen than for carbon dioxide (the opposite of what occurs in
the dead space). This difference is in agreement with the regional
distribution of ventilation perfusion ratios and the oxygen
dissociation curve, from which is expected that the alveolar inhomogeneity is at least three times greater for oxygen than for
carbon dioxide (5). Even if the difference is in the same direction
of the regional distribution of gas exchange, it is never of the
same amount as expected. This phenomenon suggests that another
mechanism is also involved in the alveolar space inhomogeneity,
and therefore in the pulmonary gas exchange, which mixes the gases
in this area, smoothing their differences (in this case between oxygen and carbon dioxide). The mechanism could be the diffusive gas
mixing, whose action in the alveolar space on SF_6 and neon has been
experimentally demonstrated (6).

Regional distribution and diffusive mixing are both present in
the alveolar space and then both of them are important in the
pulmonary gas exchange. This is simply shown by studying the
evolution of oxygen and carbon dioxide in the expired volume; they
both have an alveolar plateau with an upward direction (the oxygen
signal has to be electronically inverted and the situation becomes
clearer).

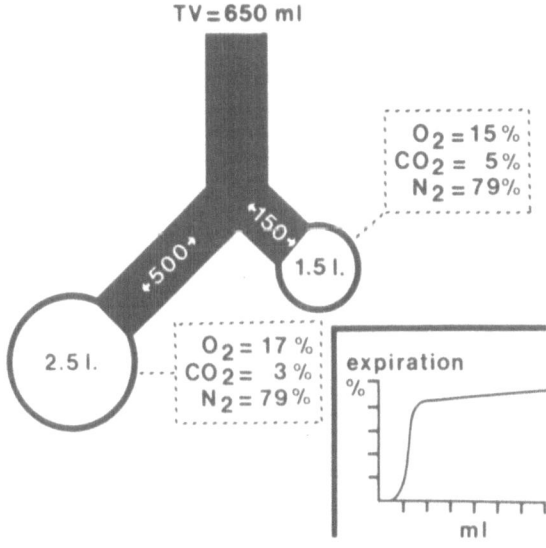

FIG. 3. A two compartment example of a parallel lung
        model of alveolar ventilation distribution. The
        better ventilated area has lower carbon dioxide
        and higher oxygen concentration values. In the
        lower part is depicted the expected alveolar plateau
        for both gases, the oxygen signal having been inverted.

Fig. 3 depicts a parallel lung model, based upon the principle
that the dependent part of the lung is better ventilated (7), to
test the hypothesis of the parallel distribution influence on the
alveolar inhomogeneity for oxygen and carbon dioxide. Since the
dependent part of the lung also has the lower time constant (8),
it empties first and therefore the alveolar plateau for both gases
is the same as that experimentally observed.

This simple parallel model of the lung has to be corrected,
in the case of the exchangeable gases, for the regional distribution
of ventilation perfusion ratios. I have made this correction on
the previous parallel lung model in Fig. 4, where the gas concentrations within each compartment have been modified in agreement
with a defined value of the respiratory quotient (R).

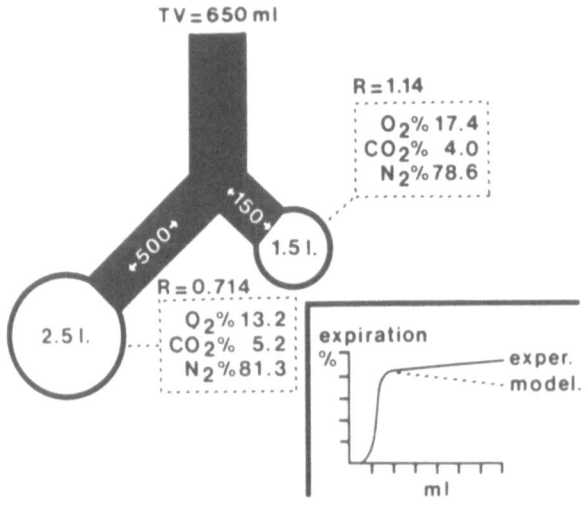

FIG. 4. A two compartment example of a parallel lung
        model of the alveolar ventilation distribution,
        corrected for the regional distribution of
        ventilation perfusion ratios. R is the respiratory
        quotient. The better ventilated area has higher
        carbon dioxide and lower oxygen concentration
        values. In the lower part is depicted the expected
        (dotted line) and the experimental alveolar
        plateau (continuous line) for the exchangeable
        gases.

In this case the expected alveolar plateau for the three exchangeable gases has to be in the opposite direction to the experimental one.

Since the exchangeable gases are evolved as volumes at increasing concentration values, the pattern of their evolution is in agreement with a stratified lung model; on the contrary the different inhomogeneity between oxygen and carbon dioxide is in agreement with a parallel lung model.

B). It is now clear that nitrogen could be quite a useful gas to be analysed in the attempt to better evaluate the contribution of the regional distribution of ventilation perfusion ratios and the alveolar diffusive mixing to the pulmonary gas exchange.

Since the difference in concentration between inspiration and expiration for nitrogen during air breathing was very small, the precision of dead space measurement was too low to yield useful information.

Table 1 shows the mean values of tidal volume (TV), oxygen uptake (V O_2), carbon dioxide output (V CO_2), respiratory quotient (R) and nitrogen exchanged (V N_2), all measured breath by breath in 14 normal non-smoker subjects, 9 males and 5 females, breathing air in the sitting position, using the above mentioned technique and procedure.

The difference between the expected nitrogen to be exchanged (V O_2 - V CO_2) and the experimental nitrogen exchanged is computed as

$$\Delta = V\ N_2 - (V\ O_2 - V\ CO_2)$$

This value, reported in the same Table, is not statistically significant in one subject, is negative in 11 subjects and positive in two.

Since no gas exchange occurs in the conducting airways, their contribution in determining the expired nitrogen volume during air breathing is negligible also because the nitrogen series dead space is expected to be very low.

As the alveolar equation has to be met within each alveolar (intra-lobular) phenomenon, the experimental has to be ascribed to an inter-lobular phenomenon. It could be expected that the presence of a negative balance (Δ negative) of nitrogen has to be ascribed to a nitrogen uptake (high VA/Q areas) and the converse (low VA/Q areas) for a positive balance (Δ positive).

An excess of nitrogen uptake or output versus a mean value expected by the difference between oxygen uptake and carbon dioxide

TABLE 1

Experimental results from expired gas analysis during air breathing — TV is the tidal volume, $V O_2$ is the oxygen uptake, $V CO_2$ is the carbon dioxide output, R is the respiratory quotient, $V O_2 - V CO_2$ is the nitrogen expected to be evolved, $V N_2$ is the nitrogen uptake(−) or output(+) and Δ is $V N_2 - (VO_2 - VCO_2)$.

Subject	Breaths n.	TV ml	$V O_2$ ml	$V CO_2$ ml	R	VO_2-VCO_2 ml	$V N_2$ ml	Δ ml	"t" std. p	Δ/TV %
A.C.	30	1072	38.27	42.30	1.091	− 4.03	−12.66	− 8.63	0.0005	0.80
F.L.	47	345	15.79	11.87	0.751	+ 3.92	+ 0.55	− 3.37	0.0005	0.97
J.M.	11	834	37.27	35.45	0.951	+ 1.82	− 8.55	−10.36	0.0005	1.24
J.B.	21	956	39.62	33.14	0.836	+ 6.48	− 0.66	− 7.14	0.0005	0.74
R.A.	27	466	29.56	22.56	0.763	+ 7.00	+ 6.70	− 0.30	NS	−
F.L.	23	514	32.00	22.45	0.701	+ 9.55	+ 6.22	− 3.33	0.0005	0.64
V.S.	39	279	16.72	11.33	0.677	+ 5.39	+ 4.61	− 0.78	0.0005	0.28
M.C.	23	627	39.57	28.52	0.720	+11.05	+ 1.30	− 9.75	0.0005	1.55
S.E.	21	793	46.43	37.52	0.808	+ 8.91	− 0.81	− 9.72	0.0005	1.22
A.G.	9	1091	61.44	55.11	0.897	+ 6.33	− 3.55	− 9.89	0.0005	0.90
A.W.	27	412	19.70	16.81	0.853	+ 2.89	− 1.03	− 3.92	0.0005	0.95
J.S.	20	470	27.95	18.45	0.660	+ 9.50	+ 6.85	− 2.65	0.0005	0.56
B.M.	31	534	33.35	22.42	0.673	+10.93	+14.45	+ 3.52	0.0025	0.66
J.D.	24	636	29.72	24.92	0.837	+ 4.83	+ 7.25	+ 2.42	0.025	0.38

output, has to be imputed to the proportional excess of some areas
at very low or very high VA/Q over a mean value, which means inhomogeneity of the regional (parallel) distribution.

Therefore, the value of Δ is the expression of the alveolar
inhomogeneity produced by the parallel distribution, and its
consequence on the pulmonary gas exchange could be expressed as a
fraction of the tidal volume. This value is reported in Table 1
for each subject. Since only two have a parallel distribution inhomogeneity for presence of alveolar areas at very low VA/Q, one has
a virtually homogeneous parallel distribution and the others have
an inhomogeneity for presence of alveoli at very high VA/Q.

C). We are still facing the problem of estimating the stratified
component of alveolar inhomogeneity, irrespective of the parallel
one. This could again be possible by analysing the expired nitrogen
in a non-exchangeable state, which the first breath of a multi-
-breath nitrogen washes out with an oxygen-argon inspired mixture.
Indeed the amount of nitrogen exchanged in this situation is quite
negligible in relation to the gas washed out from the alveolar
space.

In 10 normal non-smoker subjects, 5 males and 5 females, a
nitrogen wash out was performed applying the above mentioned
procedure and technique for expired nitrogen, oxygen and carbon
dioxide. The first breath and the complete nitrogen wash out were
recorded, the last being performed to calculate the ventilated
lung volume (LV), as shown by Prowse and Cumming (9).

In Table 2 the data are reported, in the first breath, of the
series dead space for the three gases, as in this state it is also
possible to measure the nitrogen dead space. This latter shows the
lowest value in every subject, in agreement with its molecular
weight of 28.

Since the ventilated lung volume was computed, it is possible
for the first breath to calculate the expected volume of nitrogen
which has to be evolved by an ideal lung, having a value of
nitrogen dead space experimentally measured. This procedure is
outlined in Fig. 5. The consequence of the dead space can be evaluated by preliminarily increasing the nitrogen concentration in
the inspired tidal volume by a proportional amount.

Therefore, for a breath of 650 ml and a nitrogen dead space
of 100 ml the inspired nitrogen concentration for the alveolar
volume rises to 10.5 per cent; as the functional residual capacity,
having a nitrogen concentration of 79 per cent, is 4 liters, the
expected nitrogen to be evolved in the expired tidal volume, again
correcting for the dead space interference during the expiration,
must be 400 ml.

TABLE 2
Experimental results from first breath of a multi breath
nitrogen wash-out - TV is the tidal volume, VDS is the
series dead space, $V O_2$ is the oxygen uptake, $V CO_2$ is the
carbon dioxide output and R is the respiratory quotient.

Subject	TV	VDS-N_2	VDS-O_2	VDS-CO_2	$V O_2$	$V CO_2$	R
	ml	ml	ml	ml	ml	ml	
S.J.	492	81	96	103	28	20	0.714
P.S.	406	81	90	98	20	17	0.850
F.L.	504	65	71	78	26	21	0.807
V.S.	308	48	51	61	15	12	0.800
B.M.	467	82	97	107	19	17	0.894
A.S.	603	90	99	110	28	22	0.785
W.B.	585	71	90	116	26	21	0.807
S.N.	752	76	90	115	30	26	0.866
M.F.	697	74	98	120	32	23	0.718
D.S.	808	93	105	130	29	23	0.793

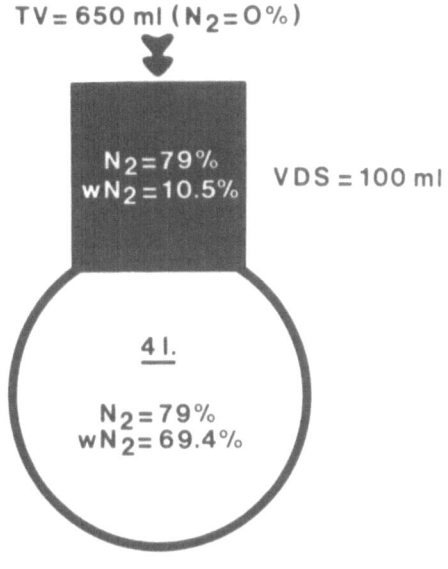

TV = 650 ml ($N_2 = 0\%$)

$N_2 = 79\%$
$wN_2 = 10.5\%$ VDS = 100 ml

4 l.

$N_2 = 79\%$
$wN_2 = 69.4\%$

expired $N_2 = 400$ ml (61.57%)

FIG. 5. A model of the ideal lung with two compartments: the series dead space and the homogeneous alveolar volume. N_2 is the concentration of nitrogen during air breathing and wN_2 is the concentration of nitrogen during the first breath of a multi breath nitrogen wash out.

The same calculation can be easily made for each subject and the result can be related to the measured nitrogen volume evolved in the first breath. The difference between these two values is expressed as a Δ in Table 3, also as a fraction of the tidal volume.

It is necessary to further discuss the physiological meaning of such a measurement. The first hypothesis is that the nitrogen inhomogeneity in the alveolar space, which produces the above measured phenomenon, is promoted by the parallel distribution of the alveolar ventilation, based upon the regional differences in time constant. I tested this hypothesis with respect to the oxygen carbon dioxide alveolar plateau during air breathing. The same model can now be tested for nitrogen during the first breath of a multi breath nitrogen wash out, as in Figs. 6 A and B. In this case

TABLE 3
Experimental results from first breath of a multi breath nitrogen
wash-out - LV is the ventilated lung volume, V N_2 is the nitrogen
evolved during first breath, Δ = V N_2 ideal - V N_2 experimental.

Subject	TV	VDS-N_2	LV	V N_2	Δ	Δ/TV
	ml	ml	l	ml	ml	%
S.J.	492	81	3.20	252	- 51	10.4
P.S.	406	81	2.40	192	- 52	12.8
F.L.	504	65	2.60	267	- 40	7.9
V.S.	308	48	2.00	158	- 32	10.4
B.M.	467	82	2.50	217	- 57	12.2
A.S.	603	90	3.80	306	- 67	11.1
W.B.	585	71	4.00	312	- 58	9.9
S.N.	752	76	4.30	438	- 70	9.3
M.F.	697	74	3.90	408	- 56	8.0
D.S.	808	93	3.80	407	- 80	9.9

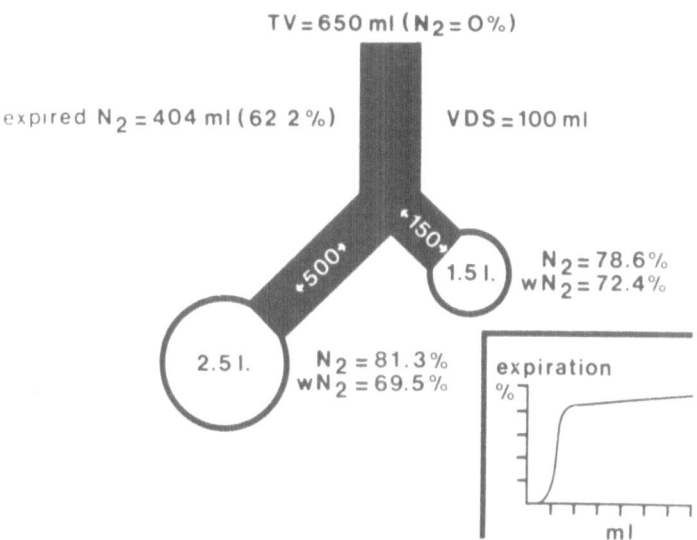

FIG. 6A. The same model as Fig. 3 during the first breath
of a multi breath nitrogen wash out. In the lower
part the expected nitrogen alveolar plateau is
depicted and on the top the amount of nitrogen to be
evolved in the expired tidal volume.

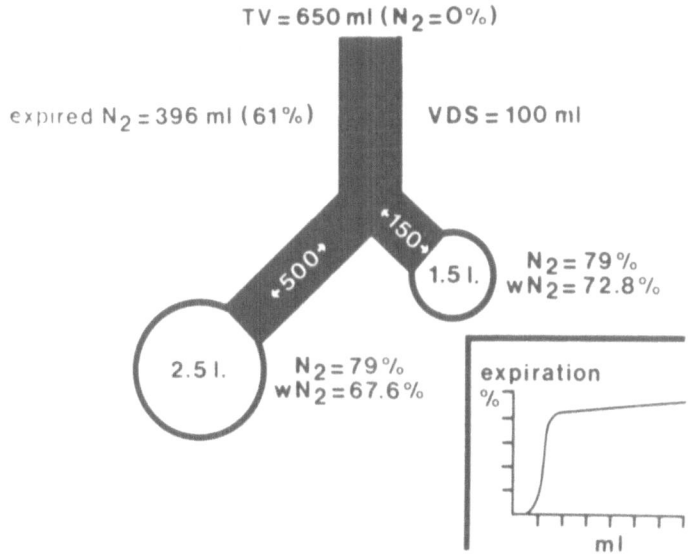

FIG. 6B. The same model as Fig. 4 during the first
breath of a multi breath nitrogen wash out.

the expected nitrogen alveolar plateau has the same direction as the
experimental one, also if the regional distribution of ventilation
perfusion ratios is taken into account. This figure could suggest
that the parallel distribution is the main determinant of the
nitrogen inhomogeneity in such a condition.

However it must be stressed that also for a stratified lung
model the evolution rate has to be the same: evolution of volumes
at increasing nitrogen concentration. Therefore neither of the
hypotheses can be disproved.

A problem arises in the attempt to explain the experimentally
measured defect of the nitrogen evolution by means of a lung parallel
model distribution, as in Fig. 6A. In this case the nitrogen expected
to be evolved in 396 ml, which is only 4 ml less than the same
ideal lung depicted in Fig. 5.

The situation is slightly different when the regional
distribution of ventilation perfusion ratios is also considered,
such that the 2.5 liters area has a nitrogen concentration of
81.3 per cent and the other is 78.6 per cent (see Fig. 6B). In this
case the expected nitrogen to be evolved is 404 ml, which is 4 ml
over the same ideal lung model.

On the contrary, when I have tested the hypothesis of a
stratified model of the same lung, as in Fig. 7, the expected
nitrogen to be evolved is 365 ml, which is a value smaller than the
same ideal lung in the size of 35 ml; this latter value is of the
same size as the experimental one shown in Table 3.

D). The importance of the new method of analysis of the exhaled
gases, which was suggested by Cumming, in the evaluation of the
gas exchange may be evidenced in that it enables an experimental
measurement of the mean alveolar oxygen and carbon dioxide
concentration, thus solving the old problem of the analysis and definition of alveolar air.

The application of this method to the measurement of the dead
space of oxygen and carbon dioxide and of the alveolar efficiency
of both, for the pulmonary gas exchange, shows two important facts:
i) the dead space, induced by the presence of the airways in series
with the alveolar area, is of such functional size and importance
as to be considered in the evaluation of the gas exchange also in
connection with the diffusivity of the gas under study; ii) alveolar
air takes part in the gas exchange as a single and indivisible
compartment, with features of gas inhomogeneity induced by the
regional distribution of the ventilation perfuaion ratios and by
diffusive mixing.

The use of distributional pulmonary models of parallel or

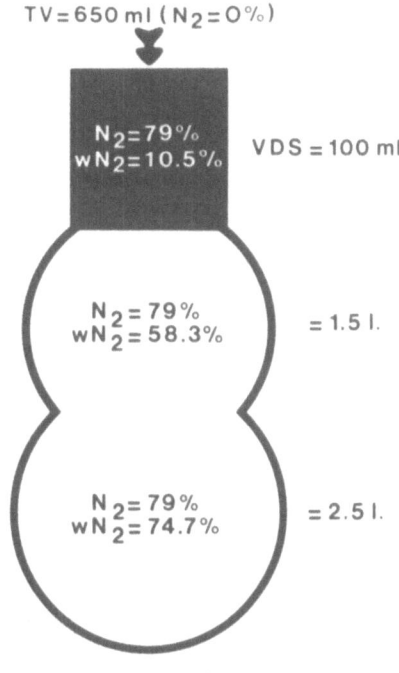

$TV = 650\ ml\ (N_2 = 0\%)$

$N_2 = 79\%$
$wN_2 = 10.5\%$ $VDS = 100\ ml$

$N_2 = 79\%$
$wN_2 = 58.3\%$ $= 1.5\ l.$

$N_2 = 79\%$
$wN_2 = 74.7\%$ $= 2.5\ l.$

expired $N_2 = 365\ ml\ (56.23\%)$

FIG. 7. A stratified lung model of the alveolar
ventilation distribution during the first
breath of a multi breath nitrogen wash out.

stratified type is an attempt to analyse the two components which
exert simultaneous actions in making the gases available for the
pulmonary exchange in the terminal units. Incidental comments
about the parallel or stratified model to corroborate assumption
on the pulmonary gas exchange have no valid experimental substrate,
but the use of either of them may be helpful for a separate study
of two simultaneous phenomena.

In this sense, I only propose to evaluate the parallel
component of alveolar inhomogeneity by measuring the discrepancy
between exchanged and to be exchanged nitrogen volume during air
breathing (ΔN_2 air breathing). The same discrepancy measured
between ideal pulmonary model and experimental state during the
first breath of a nitrogen wash out test with an oxygen-argon
mixture (ΔN_2 wash out) is proposed as evaluation of the component
of alveolar inhomogeneity depending on diffusive mixing.

REFERENCES

1) F. Langley, G. Cumming, P. Even, P. Duroux and R.L. Nicolas,
   Inserm, 1975, 51, 209.
2) S. Damato and G. Cumming, paper in preparation.
3) K.D. Lee and H.A. Crisp, J. Appl. Physiol., 1973, 17, 93.
4) G. Cumming, Progressi in Medicina Respiratoria, Ed. Scientifiche
   Terme di Sirmione, 1974, p. 35.
5) J.B. West, J. Appl. Physiol., 1962, 17, 893.
6) G. Cumming, K. Horsfield, J.G. Jones and D.C.F. Muir, Respiration
   Physiol., 1967, 2, 386.
7) J. Milic Emili, J.A.M. Henderson, M.B. Dolovich, D. Trop and
   K. Kaneko, J. Appl. Physiol., 1966, 21, 749.
8) A.B. Otis, C.B. McKerraw, R.A. Bartlett, J. Mead, M.B. McIllory,
   N.J. Silverstone and E.P. Radford, J. Appl. Physiol., 1956,
   8, 427.
9) K. Prowse and G. Cumming, J. Appl. Physiol., 1973, 34, 23.

ADAPTATION OF RESPIRATORY MASS SPECTROMETER TO CONTINUOUS RECORDING OF ABUNDANCE RATIOS OF STABLE OXYGEN ISOTOPES

K.-D. Schuster, K.-P. Pflug, H. Förstel° and J.P. Pichotka

University of Bonn, Nussallee, Bonn; °Nuclear Research

Center (KFA), Jülich, West Germany

INTRODUCTION

Respiratory mass spectrometers have been developed for a continuous determination of several gas components in inspiratory and expiratory air. The first generation of this kind of mass spectrometers is represented by the type M3 of Varian MAT (Fig. 1). It is a mass spectrometer with a 180° magnetic sector having four collectors to determine four different components of a gas mixture simultaneously. Its sensitivity is 10 ppm and its resolution about 25. For some years a second generation of respiratory mass spectrometers with magnetic deflection has been available. They represent an improvement in their stability of measurement and operating conditions but are of decreased sensitivity. Some mass spectrometers of the quadrupole type have a sensitivity up to 0.2 ppm but there may arise problems in measuring two components with very different abundances simultaneously. Respiratory mass spectrometers are well suitable to analyse the major components of air but they are not suitable for the measurement of isotopic abundance.

On the other hand powerful diagnostic methods have been developed using stable isotopes as tracers. The stable isotope of carbon, ^{13}C, is used to study the function or disfunction of metabolism (1) particularly of liver, (2, 3), bile (4) and ileum (5). The stable isotope of oxygen, ^{18}O, can be used to determine the perfusion of the lungs (6) or to study the distribution of water within the body (7). The breath tests with ^{13}C are nowadays performed using the technique of isotopic abundance measurements, that means that single samples of expiratory air are processed in a complicated procedure for analysis by a mass spectrometer

451

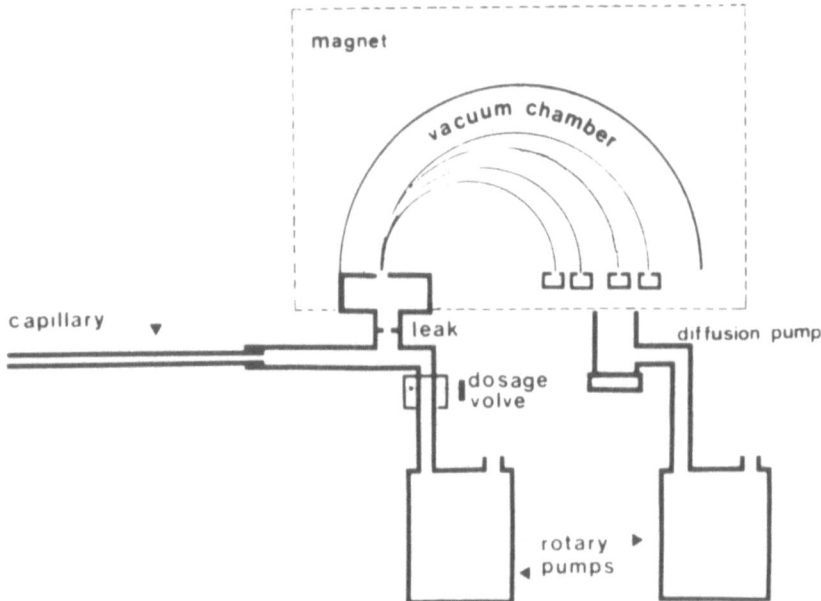

FIG. 1. Respiratory mass spectrometer M3 of Varian MAT.
The dosage valve is inserted afterwards to
increase the inlet pressure.

designed especially for the detection of isotopic ratios. The
investigations using ^{18}O are involved with the continuous determination of the $^{18}O/^{16}O$ ratio in expired carbon dioxide by the respiratory mass spectrometer. The instrument type M3 of Varian MAT was
adapted to the particular problem to analyse isotope ratios within
respiratory air continuously. In the following the improvements and
modifications of the M3 are described which have been performed for
this purpose.

RESOLUTION PROBLEMS

An increase in resolution is connected with a decrease in
sensitivity. Therefore it is useful to find a compromise between
the two essentials.

Increase of resolving power. The resolving power A of a mass
spectrometer is characterized by the ability to separate two
substances of the mass numbers M and M+ΔM being present simultaneously. A has been defined generally as

$$A = \frac{M}{\Delta M}$$

It is necessary to specify the conditions at which two substances
can be considered separated. This has been done in various ways.
A definition often used states that two adjacent peaks are resolved
when the height ΔH of the valley between them is less than a factor
F. F may be taken as 0.001, 0.01, 0.05 or 0.1 (8). In this paper
we will take a factor F of 0.05. The resolution of a single focussing
mass spectrometer with 180° deflection depends on the radius R of
the ion orbit and on the width of the ion source slit SI and
collector slit SC (9):

$$\frac{M}{\Delta M} = \frac{R}{SI + SC + B \cdot R}$$

B is called the aberration constant. The largest chooseable ion
orbit has a radius R of 50 mm in the M3. The slits SI and SC had
originally the width 0.5 mm and 0.8 mm respectively, the resolving
power was 25. Reducing the width of the slits to SI = 0.15 mm and
SC = 0.6 mm, the resolving power has been increased to 50.

The partial pressure of $C^{16}O_2$ and $C^{16}O^{18}O$ changes between
inspiratory and expiratory air.

These changes are much larger than those of their ratio to one
another to be measured. It is necessary to eliminate the respiratory
alternations by compensating the signal of $C^{16}O^{18}O$ with a fraction
of $C^{16}O_2$. Therefore both signals must be measured simultaneously. We
have built a double collector shown in Fig. 2 for this purpose.
The electrode surface of the trap collecting the ion of m/e = 46 is
arranged acute-angled to the ion beams to reduce the escape of the
cage of secondary emissions (10).

The additional collector has been connected to a reserve pin.
The mass spectrometer is equipped with three single collectors and
one double collector.

Influence of peak 45 to 46. In Fig. 3 the mass spectrum of 5%
CO_2 within nitrogen is given for m/e between 44 and 46. The valley
between the peaks 45 and 46 amounts to 12% of the height of peak 46.
The contribution to the peak 46 by 45 may be approximately 0.3%. This
means absolute determinations of the abundance ratios are difficult.
But if the abundance of the components forming the peak 45 remain
constantly, changes of the $C^{16}O^{18}O/C^{16}O_2$ ratio can be analysed. We
have carried out investigations with this aspect in mind. Samples
of expired CO_2 of rabbits and humans were analysed on the ratio of
CO_2 of the mass peak 45 to 44. The analysis was performed with the
isotope ratio mass spectrometer (Micromass 602 C of Vacuum Generators)
after separating the CO_2 from the other components of expired air.
The measurements are given as δ-values relative to an internal
standard gas of CO_2. The results on 6 rabbits are given in Table 1.

Double collector

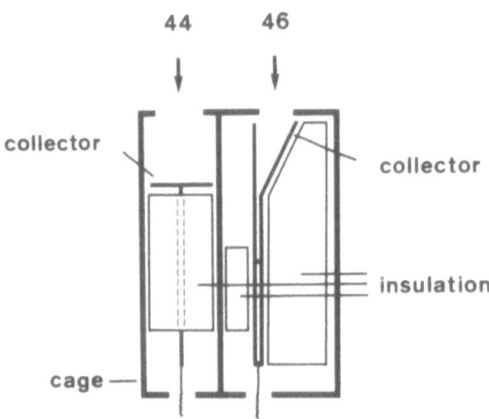

FIG. 2. Double collector for the simultaneous detection
        of m/e = 44 and m/e = 46.

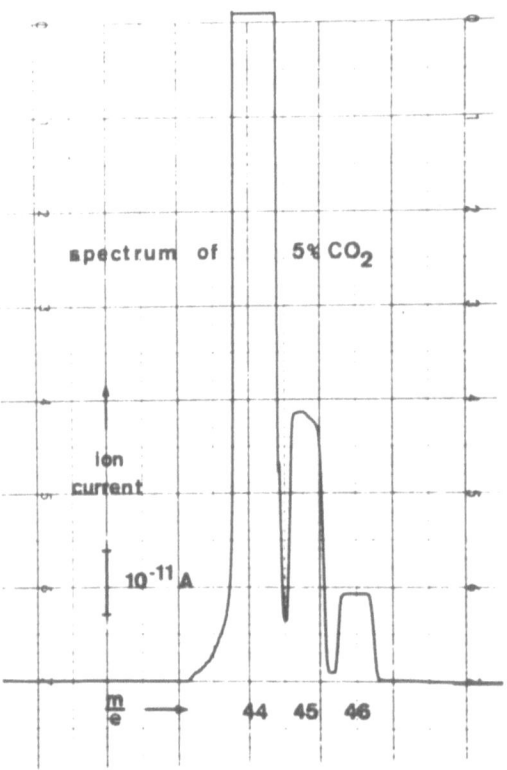

FIG. 3. Mass spectrum of
a gas mixture (5% CO_2,
95% N_2) of natural
isotopic composition in
the range 43<m/e<47.
During the detection of
the peak 44 the amplifier
is overdriven.

The mean value amounts to $-17.9\%_0$ with the standard deviation of 0.9. The differences between the individuals are minimal. Other authors have discovered a small dependence of the $^{13}C/^{12}C$ ratio by phylogenetic characteristics, type of diet and metabolic conditions (11, 12). The investigations on human subjects were performed at rest and at low levels of ergometer work; these results are given in Table 2.

TABLE 1.
^{13}C-content of expired CO_2 of rabbits. The δ-values are given relative to an internal standard gas of CO_2.

Rabbit	Ra 1	Ra 2	Ra 3	Ra 4	Ra 5	Ra 6
δ (‰)	-17.5	-16.3	-18.5	-18.1	-18.6	-18.2
mean:	-17.9					
SD:	0.9					

TABLE 2.
$^{13}C/^{12}C$ ratios of expired CO_2 of human subjects in comparison to the $^{13}C/^{12}C$ ratio of an internal standard gas of CO_2. The measurements have been performed at three levels of metabolic rate.

person	rest ‰	ergometer work 100 W ‰	200 W ‰
SB	- 18.3	- 18.9	- 17.6
SC	- 18.6	- 18.7	- 18.5
MÜ	- 18.3	- 18.8	- 18.4
mean	- 18.4	- 18.8	- 18.2
SD	0.2	0.1	0.5

The mean values amount to $-18.4\%_o$, $-18.8\%_o$ and $-18.2\%_o$ respectively. There are no remarkable changes in the different physiological situations. One can conclude from these investigations that the influence of peak 45 to peak 46 is negligible in experiments with the natural abundances of ^{13}C and ^{17}O at the given resolving power.

IMPROVEMENT OF SENSITIVITY

The sensitivity of a mass spectrometer can be improved by increasing the ion current or decreasing the noise of the analysing system.

Increase of the ion current. The M3 is equipped with an electron--bombardment ion source. The ion current produced by such a source depends on the gas pressure within the source and the current of the bombarding electrons. The gas pressure in a mass spectrometer is restricted by an upper limit: the mean free path of the molecules must be much larger than the length of the ion orbit. If it is too short the focussed beam is dispersed again by interactions with the molecules. In the M3 the mean free path of the molecules can be calculated to be in the order of the length of the ion orbits at a pressure of 10^{-2} Pa. The inlet system of the M3 consists of two stages (Fig. 1). First a bypass of a viscous leak connected to a vacuum pump reduces the atmospheric pressure to about 100 Pa. A second reduction to $5 \cdot 10^{-4}$ Pa is caused by a molecular leak. Between the viscous leak and the vacuum pump we inserted a dosage valve (Fig. 1) to increase the pressure in the mass spectrometer by a factor two.

The path of the ions is much shorter within the ion source than between source slit and collector slit. Therefore the pressure may be higher in the ion source than in the vacuum chamber. We used components of another mass spectrometer (M230 of Varian MAT) to make the ion source tighter against the vacuum chamber. From this modification an increase of the ion current resulted by a factor of about 5.

The maximum current of the bombarding electrons amounted to 225 µA. We adjusted the electron current to 550 µA and 1130 µA by increasing the current of the filament. Such an increase of the current necessarily decreases the lifetime of the filament. Operating with an electron current of 1130 µA and a pressure in the vacuum chamber of $5 \cdot 10^{-4}$ Pa the lifetime of the filament reached 40 hours. The same sensitivity was recorded using the half electron current but the double pressure within the chamber. Therefore the lifetime of the filament amounted in this case to about 120 hours. Most experiments were performed under the latter conditions. The ion current was increased by a factor of 20. It amounts to 10^{-11} A for the natural occuring $C^{16}O^{18}O$ molecules in alveolar air which are present with a partial pressure of approximately 21 Pa.

Decrease of noise. The valve preamplifiers of the M3 were replaced by the FET-amplifiers 42 K of Analog-Devices. The noise of the latter amounts to $5 \cdot 10^{-15}$ A peak to peak and 8 uV rms in the frequency range of 10 Hz. The principal arrangement to record the masses 44 and 46 is given in Fig. 4.

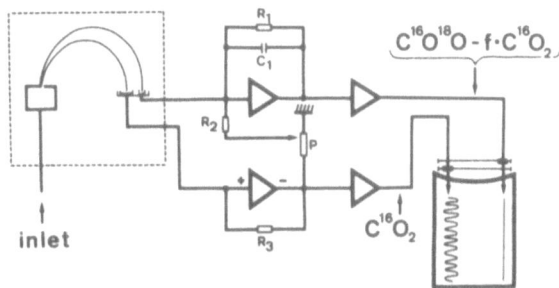

FIG. 4. Circuitry for the dynamic compensation of $C^{16}O^{18}O$ by a fraction of $C^{16}O_2$.
Preamplifiers: type 42 K of Analog Devices
$R_1 = R_2 = 10^{10}\Omega$, $R_3 = 6 \cdot 10^8 \Omega$, $P = 50$ kΩ,
$C_1 = 10$ pF.

The preamplifiers are working as feedback ammeters. The more sensitive channel is equipped with a feedback resistor of $10^{10}\Omega$. Therefore the current noise should become the limiting factor of the current detection.

After having improved the amplifier unit another source of noise raised its head at the level of $2 \cdot 10^{-14}$ A. The origin of this was found to be mechanical vibrations caused by rotary pumps and ventilators. To avoid this noise the whole analysing system was grounded on a separate base plate. Elastic tubes were used to connect the system with the rotary pumps. As a result the level of noise was decreased to $5 \cdot 10^{-15}$ A. An ion current of 10^{-11} A can be detected with a signal to noise ratio of 2000.

IMPROVEMENT OF LONG TIME STABILITY

Replacing the valve preamplifiers their poor long time stability was removed too, but there were some other effects which had caused a drifting output signal at a constant input.

Periodical output movements. Recording the isotopic ratio
$C^{16}O^{18}O/C^{16}O_2$ two independent periodical output movements had been
identified which were caused by changes in temperature of the inlet
system and the vacuum chamber respectively. Both these systems of the
M3 must be held at higher temperatures to avoid absorption of water
vapour at the walls. Temperature changes of the inlet system lead
to changes of the fractionation between $C^{16}O_2$ and $C^{16}O^{18}O$. This
effect was eliminated by stabilising the voltage of the heating
system.

The temperature of the vacuum chamber was controlled by an
electrical thermometer. On exceeding 393 degrees K, the heating
system was switched off. A remarkable overshoot in temperature
occurred because of the asymmetrical arrangement of the heating
elements relative to the chamber walls, to the magnet and to the
thermometer. Parallel to this effect shifts of the peak position
were discovered. These difficulties have been overcome by limiting
the heating current so that the point of actuation of the thermometer is not reached.

Stabilisation of the peak position. The plateau width of the peak
for mass 46 is smaller than 0.2% of the accelerating volatge. Hence
very small alternations in any parameter which influences the peak
position leads to movements of the output signal. If the peak shift
could be measured, the peak position could be controlled by the
accelerating voltage. A signal proportional to the peak shift can be
obtained by adjusting a collector on the flank of a peak. The
principle of the idea is given in Fig. 5. The first collector is
adjusted on the front flank of a peak. If the radius of the ion
orbit increases the signal measured by the first collector is increased too. This increase is used to decrease the accelerating
voltage. Two conditions must be fulfilled to guarantee a correct
run. First the gas component forming the used peak must be constant.
In most respiratory investigations the inspiratory nitrogen can be
taken as such a constant component. Secondly the production of ions
must be of high constancy in order to obtain a good stability in
peak height. This latter condition is not fulfilled with the
necessary accuracy. Therefore a signal being proportional to the
ion current was measured at the exit slit of the ion source. The
measured change of this signal was used to compensate the similar
change of the peak signal (Fig. 5).

COMPENSATION OF THE RESPIRATORY VARIATIONS OF $C^{16}O^{18}O$

The partial pressure of CO_2 changes its amplitude by a factor
of 180 between inspiratory and expiratory air. The changes in
expired CO_2 between two respirations may be 10%. It would be
desirable that the ratio $C^{16}O^{18}O$ to $C^{16}O_2$ be measureable with an
accuracy of up to 1‰. Hence the much larger respiratory oscillations

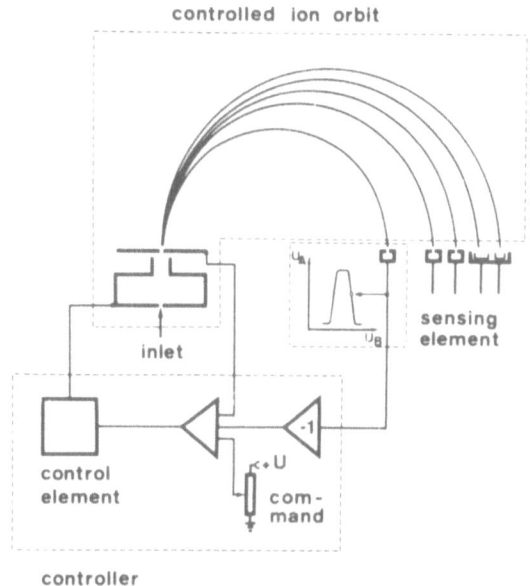

controlled ion orbit

sensing
element

inlet

control
element

com-
mand

controller

FIG. 5. Regulating circuit to control the ion orbits.
One collector is adjusted on a flank of a peak
symbolized by the diagram within the figure.
If the signal U_A, which is measured with this
collector, decreases the feedback of this change it
will cause an increase of the accelerating
voltage U_B. A signal proportional to the whole
ion current is measured at the ion source slit
to take into account alternations of the ion
production.

must be compensated. This could be done by dividing the signal of
$C^{16}O^{18}O$ or by subtracting it and a fraction of the signal of $C^{16}O_2$
by or from each other. The compensation by subtraction should be
preferred because the inspiratory partial pressure of CO_2 is almost
zero.

In first tests subtraction was performed by a difference amplifier
after preamplification of both signals. The time response of both
preamplifiers was tried to adjust to equal in the range of 0.1 sec
but without success. A more successful circuitry is given in Fig. 4.
This arrangement has been deviced by NIER et al. (13) to get a null
indication. But here the circuitry is considered with respect to
the dynamic compensation of a signal oscillating with a large
amplitude.

Both preamplifiers operate as ammeter. The feedback resistors R_1 and R_3 have the values 10^{10} and $6 \cdot 10^8$, the value of the capacitor C_1 amounts to 10 pF. Via the potentiometer P and the resistor R_2 a fraction of the inhibited signal of peak 44 is transferred to the input of the more sensitive preamplifier to compensate the signal of mass peak 46. This feedback of the peak-signal 44 to that of 46 is done with a very short time response in the range below a milli-second. This means the error in time response of the subtraction may be in the same order.

The preamplifier with the combination $R_1 \cdot C_1$ operates as low--pass filter for the compensated signal. The circuitry was tested by simulating the ion beams by currents produced with an external voltage and a resistor circuit. Changing the input currents by a transient the step-function response was within the noise background of the circuitry. Testing the whole mass spectrometer system by transients between gas mixtures of expiratory and inspiratory quality a drift effect results with a time response of approximately 20 seconds. This effect is supposed to be caused by the ion source due to the different gas mixtures.

In experiments on rabbits this effect had an influence of only 1.5 to 2 times the noise background because of its large response time in comparison to the frequency of respiration.

RESULTS

The resolving power of the M3 has been increased to 50 (5%--valley between two adjacent peaks of equal height) so that measurements of changes in the $^{18}O/^{16}O$ ratio of expired CO_2 can be performed. The ion current has been increased by a factor of 20. It has as a result an ion current of 10^{-11} A from $C^{16}O^{18}O$ molecules in alveolar air of natural isotopic abundances. The noise background has been decreased to $5 \cdot 10^{-15}$ A peak to peak. This means the signal to noise ratio amounts to 2000 for the record of $C^{16}O^{18}O$ in alveolar air. The dynamic sensitivity is reduced by a factor of 2. Continous determinations of the $^{18}O/^{16}O$ ratio in expired CO_2 can be performed with an accuracy up to 1‰ with a long time stability better than 1‰ per hour. This corresponds to a dynamic sensitivity of 0.2 ppm.

DISCUSSION

It has been suggested that in medical research on human subjects the use of radioactive isotopes should be restricted and replaced whenever possible by stable isotopes.

But a general application of stable isotopes in medicine depends above all on having simple methods of analysis. The continuous and immediate measurement in the respiratory air by a respiratory mass spectrometer can be considered to be such a method. Therefore a respiratory mass spectrometer has been adapted to the problem of isotopic determinations. This was done by a set of improvements most of which are known to us through the relevant literature. It is to be presumed that similar improvements could be considered in the construction of new respiratory mass spectrometers without having to increase their cost.

A useful sensitivity has been reached which allows the use of only small amounts of enriched isotope material in tracer experiments. Because of the possibility of analysing continuously and immediately, new methods could be developed in clinical medicine and physiology. First steps in this direction have already been taken (6).

Some problems are open for further investigation, for example the increase of the lifetime of the filament and the short time drift of the signals after changes in the gas mixture to be analysed.

REFERENCES

1) W.W. Shreeve, in "Proceedings Third International Conference on Stable Isotopes in Chemistry, Biology and Medicine", May 23-26, Argonne National Laboratory, Argonne, Illinois, 1978.
2) H. Helge, B. Gregg, E. Jäger, G. Borchert and S. Nigam, in "Proceedings Third International Conference on Stable Isotopes in Chemistry, Biology and Medicine", May 23-26, Argonne National Laboratory, Argonne, Illinois, 1978.
3) J.F. Schneider, D.L. Hachey, B.B. Schreider, A.N. Kotake, D.A. Schoeller and P.D. Klein, in "Proceedings Third International Conference on Stable Isotopes in Chemistry, Biology and Medicine", May 23-26, Argonne National Laboratory, Argonne, Illinois, 1978.
4) Y. Sasaki, H. Oh-hara, S. Takahashi, K. Someya and K. Sano, in "Proceedings Third International Conference on Stable Isotopes in Chemistry, Biology and Medicine", May 23-26, Argonne National Laboratory, Argonne, Illinois, 1978.
5) R. Park, J.B. Watkins, J. Perman and D.S. Schoeller, in "Proceedings Third International Conference on Stable Isotopes in Chemistry, Biology and Medicine", May 23-26, Argonne National Laboratory, Argonne, Illinois, 1978.
6) K.D. Schuster, K.P. Pflug, H. Förstel and J.P. Pichotka, in "Proceedings Third International Conference on Stable Isotopes in Chemistry, Biology and Medicine", May 23-26, Argonne National Laboratory, Argonne, Illinois, 1978.
7) K.P. Pflug, K.D. Schuster, H. Förstel and J.P. Pichotka, Biomed. Technick., Aachen 1977.

8) J.H. Beynon, in "Mass spectrometry and its application to organic chemistry", Elsevier, Amsterdam, 1960, p. 52.

9) J. Blears and A.K. Mettrick, in "Proceedings XIth International Congress Pure and Applied Chemistry", Vol. 1, 1947, p. 333.

10) C. Brunnëe and H. Voshage, "Massenspektrometrie", Verlag Karl Thiemig KG, München, 1964, p. 112.

11) J. Duchesne and A. van de Vorst, Acad. Sci., Ser. D, 1968, 266, 522.

12) B.S. Jacobsen, B.N. Smith and A.V. Jacobsen, Biochem. Biophys. Res. Commun., 1972, 45, 398.

13) A.O. Nier, E.P. Ney and M.G. Inghram, Rev. Sci. Instr., 1947, 18, 294.

THE DEVELOPMENT OF A FLEXIBLE MASS SPECTROMETER CATHETER

T.D. Johnson, G.M. Watkins, J. Holsinger, M.P. Roberts

and D.D. Thomas; University of South Florida, Veterans

Administration Hospital, Tampa, Florida, U.S.A.

INTRODUCTION

Measurement of blood oxygen and carbon dioxide tensions is routinely employed to assist clinicians in patient evaluation and treatment. This method necessitates either arterial or venous puncture or indwelling hollow catheter to obtain the sample. Ex vivo Clark Polarographic O_2 and the Severinghaus CO_2 electrodes are the actual measuring devices. Many attempts have been made at continuous analysis using the O_2 and CO_2 electrodes with specially designed flow through cuvettes or miniature catheter tip electrodes. Most of these methods suffer from lack of repeatability and/or loss or linearity.

The mass spectrometer is well established for accurately measuring respiratory gas concentrations. Recently it has been used to monitor blood and tissue gases in both research and clinical environments. The first use of a mass spectrometer for blood gas analysis was reported by Strang (1). The method he employed was equilibration of the dissolved gases in the blood with a gas bubble. The bubble was then analysed using the mass spectrometer to determine change in its gas concentrations. A simpler and more efficient technique was developed and tested in animals by Woldring et al. (2) and later in humans by Wald et al. (3). This procedure allowed for continuous in vivo measurement of any gas dissolved in blood or any other fluid, without drawing a sample. The system consisted of a gas permeable membrane mounted on the tip of a gas impermeable stainless steel cannula which was then inserted into the bloodstream. The gases that dissolved in the blood continuously diffused across the membrane into the cannula. Due to the inherent high sensitivity of the mass spectrometer, only small amounts of gas were actually withdrawn from the surrounding medium, on the order of 4×10^{-6} cc/sec for a catheter with a Silastic membrane.

Continuous in vivo monitoring of gases other then O_2 and CO_2 has been reported in the literature; Roberts et al. (4) monitored halothane in dogs, Dyken (5) reported using argon desaturation techniques to measure cerebral blood flow, and Wald et al. (3) employed nitrous oxide to study cerebral blood flow changes which occurred after cerebral trauma. Also, the mass spectrometer has been utilized to study cardiac output, renal and lingual blood flow, and biliary oxygen tension.

While the application of the mass spectrometer to blood gas analysis is well-documented, the use of catheter material other than stainless steel appears much less frequently in the literature. Woldring, (6) reported theoretically that the only sufficiently gas impermeable flexible catheter material was polyvinylidine chloride, or Saran. Wald et al. (3) published results using a polymeric catheter material but cautioned that variations in the background (that portion of the signal which is due to gas transmission across the catheter tubing) resulted in decreased measurement precision.

The catheter described herein was fabricated using Saran, (poly-vinylidine chloride) as the carrier tubing, and Silastic as the membrane material. This combination of materials resulted in a flexible mass spectrometer catheter which proved to be experimentally equal to the stainless steel Silastic catheter which has been traditionally employed for in vivo blood gas monitoring. Since the size of the catheter is only 1.4 mm in diameter, this catheter can easily be inserted in a peripheral artery through a standard size introducer. The smooth profile and flexibility minimize the possibility of puncturing the vessel wall while maintaining precision and accuracy necessary for clinical application.

The details of the theoretical analysis which preceeded the development will be presented in a subsequent publication.

DESIGN AND DEVELOPMENT

The first objective was to develop a flexible mass spectrometer catheter which allowed for accurate on line measurement of blood gases. For the measurement to be accurate, it was necessary for the gas in the blood to diffuse only through the sampling membrane located at the tip of the catheter. Gas diffusion through the wall of the flexible catheter tubing would have produced a signal which added to the membrane signal, thus acting as a parallel leak. Because the exposed surface area of the tubing exceeded the membrane area by several orders of magnitude it was mandatory to use tubing material that was virtually impermeable to gases and water vapor.

Theoretically, if the ratio of permeabilities per unit area of tubing and membrane is approximately 0.001, more than half of the recorded signal can be produced by gas leaking through the wall of the tubing when the surface area of the tubing exceeds the membrane area by a factor of 1000.

Stainless steel tubing is ideal, since it is, for all intents and purposes, impermeable to gases and water vapor. However, it is too stiff for safe, routine catheterization purposes unless its diameter is reduced to capillary dimensions. Such capillary-sized tubing would compromise the conductance and response time as to render the catheter useless for on line measurements.

Of the numerous polymers available today, polyvinylidine chloride, or more commonly known as Saran, was the only tubing sufficiently impermeable to gas and water vapor. Nylon tubing coated with polyurethane had been used by Key (7); however, no detailed data was available regarding the activaction energy or experimental permeability. Another polymer used by Wald et al. (8), designated X-129, had been marketed; however, the price was prohibitive, and other characteristics were inferior to those of Saran. Other plastics, polyethylene, polypropylene, polyvinyl, and rubber were much too permeable to gas and water vapor to be used as tubing.

A disadvantage inherent to any polymer used as a mass spectrometer catheter is the time necessary for the total outgassing of the catheter. Outgassing is the release of absorbed gas or organic material, i.e. solvents, benzene, toluene, xylene, acetone, etc., used in the extrusion process or used in fabrication of the catheter. Until the catheter is totally outgassed, erroneous values appear on the mass spectrometer. A finished 18" Saran/Silastic catheter with a Silastic membrane at the tip takes usually four to six hours to fully outgas. Once the catheters are fully outgassed there is no need to repeat the procedure. No significant reabsorption of gases appears to take place in the Saran when it is repeatedly subjected to air at atmospheric pressure and then attached to the mass spectrometer. Hence, an outgassed catheter can be stored in air, then attached to a mass spectrometer and the Saran catheter is ready for use in the same time as stainless steel catheters.

The most critical element of the system is the catheter membrane through which the blood gases diffuse for analysis in the mass spectrometer. It is necessary that the membrane be permeable to all or most gases soluble in blood. Also, the membrane should be relatively inert so that there is a linear relationship between concentration across the membrane and flow through the membrane. Furthermore, the membrane must be relatively insensitive to temperature changes, be physically rugged, and have sufficient rapid response time so that changes in blood gases can be on a real-time basis. The catheter must be nonthrombogenic for patient safety and accuracy of measurement.

The most commonly used membrane materials for mass spectrometer catheters are Silastic (silicone rubber) and Teflon (fluorinated – ethylene-polypropylene). The combination of Saran (polyvinylidine chloride) and Silastic (silicone rubber) theoretically would minimize the transmission of gas through the catheter tubing and maximize the transmission of gas through the membrane. Although the signal to background ratio per unit area was the major determining factor in choosing a suitable membrane material, Silastic and Saran appear to be quite acceptable as to strength, flexibility, water vapor transmission, etc. See Encyclopedia of Polymer Science and Technology (9) for a complete list of specifications. Silastic is a very rugged, resilient material which absorbs less than 1% of water (by volume) and lends itself to heparinization via a simple coating procedure. Silastic medical grade tubing is essentially non-reactive to body and tissue fluid, and also the surface is non-wetting which minimizes platelet destruction and contact hemolysis.

The term signal means all the gas that passes across the membrane at the tip of the catheter. Background means that amount of gas which passes across the Saran tubing. Since the background contributes to the amount of gas the mass spectrometer analyzes, part of the output of the mass spectrometer is due to the extraneous background. An estimate of this contribution in quantitative terms can be made. Table 1 details the permeabilities of Saran and Silastic.

The theoretical signal-to-background per unit area ratio is presented in Table 2.

TABLE 1
Comparison of permeabilities

Saran	0.01	0.05	0.29
Silastic	2000	6000	28000
in	$\dfrac{cm^3-mm}{cm^2 - sec - mmHg}$	x	10^{-11}

TABLE 2
Theoretical signal-to-background ratios

	N_2	O_2	CO_2
S/B	$2x10^5$	1.2×10^5	$9.6x10^4$

The area of the mebrane is:

$$A_{mem} = \pi DL \quad \text{where} \quad \begin{aligned} \pi &= 3.1415 \\ D &= \text{diameter: 0.127 cm.} \\ L &= \text{length: .2 cm.} \end{aligned}$$

Hence $A_{MEM} = 7.98 \times 10^{-2} \text{ cm}^2$.

The area of the catheter tubing according to the lengths is found in Table 3.

TABLE 3.
Area comparisons

Length cm	45.72	91.44	121.92
A_{tubing} cm^2	18.24	36.48	48.64
A_{tubing}/A_{mem}	229	457	610

As can be seen from the Tables 1, 2 and 3 the pemeability of the membrane is almost 10^5 times greater than that of the Saran tubing, yet the area of the tubing is only 610 times greater than the area of the membrane.

Gas flow through the membrane and the tubing can be calculated by using the relation:

$$q = \frac{KPA}{d}$$

where: q = flow in cm^3/sec
      K = permeability in $\dfrac{\text{cm}^3 - \text{mm}}{\text{cm}^2 - \text{sec-mmHg}}$

      P = Pressure in mmHg
      A = area in cm^2
      d = Thickness of the membrane in mm.

Given that:
    $d_{mem} = 0.0889$mm
    $d_{tubing} = 0.254$mm
    P = 760 mmHg

the simplifying assumption is that each catheter is subjected to 100%
of the gas at atmospheric pressure. The results are summarized in
Table 4 for dry gas.

TABLE 4
Theoretical gas flows

			N_2	O_2	CO_2
$\frac{cm^3}{sec}$	q_{mem}	25°C	1.36×10^{-5}	4.08×10^{-5}	1.90×10^{-4}
		37°C	2.38×10^{-5}	6.51×10^{-5}	2.73×10^{-4}
		40°C	4.32×10^{-5}	1.06×10^{-4}	4.02×10^{-4}
	q_{tubing}	25°C	3.65×10^{-8}	1.82×10^{-7}	1.05×10^{-6}
		37°C	1.07×10^{-7}	5.08×10^{-7}	2.32×10^{-6}
		40°C	3.38×10^{-7}	1.49×10^{-6}	5.36×10^{-6}
	q_{total}	25°C	1.36×10^{-5}	4.09×10^{-5}	1.91×10^{-4}
		37°C	2.39×10^{-5}	6.56×10^{-5}	2.75×10^{-4}
		40°C	4.35×10^{-5}	1.07×10^{-4}	4.07×10^{-4}
%	$\frac{q_{tubing} \times 100}{q_{total}}$	25°C	0.26	0.44	0.54
		37°C	0.44	0.77	0.84
		40°C	0.77	1.31	1.39

$\hat{}\hat{}$48" length.

Gas flow q varies as K varies, which is dependent upon
temperature according to:

$$\log K_2 = \log K_1 + \frac{1000}{4.0} Ep \left[\frac{1}{T_1} - \frac{1}{T_2} \right]$$

where: K = pemeability in $\frac{cm^3 - mm}{cm^2 - sec - mmHg}$

   T = temperature in °K

   Ep = activation energy in Kcal/mole

In theory the background will never exceed 1.4% of the total
signal. Experimental data indicated this was a conservative figure.
In addition when a fluid is saturated with any of the above gases,
gas flow will be approximately 10-20% lower.

The next developmental aspect of the catheter to consider was
the design of the sampling port. It must be omni-directional; that
is, whether the flow is coming toward or going against the catheter,

the sampling port should expose the same amount of area to gas diffusion. The size (length) of the active area should be small so that the sampling port would not protrude too far past the inflated ballon when the catheter is incorporated into the flow directed catheter. The development of the flow directed catheter is documented in another paper. Theoretical gas flows for various designs were calculated. The criteria for acceptance of a design employed here were subjective. If the catheter with the experimental tip attached repeated values for O_2, N_2, and CO_2 determined by a commercially available steel catheter at approximately the same emission current, then the exposed area was judged to be correct. If the values obtained from the experimental catheter did not change when the direction of flow was reversed, then the catheter was judged acceptable for further testing against the commercially available steel catheter.

The first attempt consisted of a piece of Saran tubing with the ends flared. The catheter was fitted with a Silastic membrane either by coating it with a liquid Silastic and letting the Silastic form a drop across the open end or with a sheet of .006" thick Silastic which was held in place with ligatures. Although the catheter made with the liquid Silastic provided a smooth profile, the permeability was low and membrane thickness uncontrollable. A high emission current had to be used indicating the low gas flow through the membrane and at the same time decreasing the efficiency of the ion source in the mass spectrometer. The other design increased the gas flow but was highly dependent on orientation in the flow.

The next catheter was Saran with slots on the 13 mm. terminal spaced 2 mm apart. This catheter fitted with a Silastic membrane, proved to be a successful design. In order to maintain structural strength the sampling area covered more than 13 mm of the tip. This long protrusion was not suitable for incorporation into the balloon catheter.

The final design consisted of a series of catheter all built around a central design. In this design the membrane was supported by a stainless steel spring fitted on the tip of the catheter. Fig. 1 shows the final design Saran catheter. Note the coiled steel membrane support at the tip (arrows). After a series of experiments, it was determined that a length of 2 mm provided the proper gas flow. The final fabrication procedure can be summarized as follows:

Fabrication. The steel spring consisted of 34 gauge stainless steel orthopaedic wire wound on a mandril of the appropriate dimensions. This steel spring was then fitted by friction into the Saran tubing. After expanding Silastic medical grade tubing, I.D. 0.5 mm. O.D. 1.0 mm, in xylene for approximately three minutes, it was dried of the excess xylene and then pushed over the steel coil on

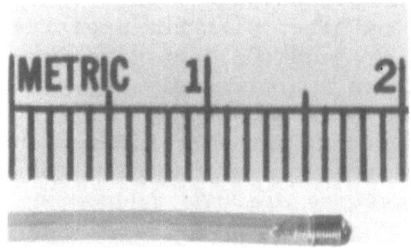

FIG. 1. Sensing tip of the Saran catheter.

to the Saran tubing. The Silastic tubing was pulled up as far as
necessary to insure a good fit, usually about 2-3 inches. The
catheter was air dried 1-2 hours to "shrink-fit" the Silastic onto
the Saran tubing. After thorough drying, the catheter tip was sealed.
Using a syringe attached to the opposite end of the catheter, the
membrane tip was placed in contact with the Silastic medical grade
adhesive, and a small bead of adhesive was drawn in to just close
the open end of the Silastic tubing. The catheter was set aside to
cure for 24 hours. When the Silastic was cured completely and the
Saran tubing cut to the desired length, heparinization of the membrane
was carried out. The procedure to heparinize the membrane was modification of that of Grode et al. (10) as modified by Brantigan et al.
(11). In this procedure, dodecyltrimethyl ammonium chloride, TDMAC,
was deposited in the Silastic by solution in a solvent, in this case
xylene. The membrane was soaked in a 10% solution of TDMAC and
xylene for 2.0 minutes, air dried for 20-60 minutes. The membrane
is placed in a solution of heparin (40,000 units/100 ml) and 50%
methanol and water for 5 minutes, then air dried for 1-2 hours. This
heparinization procedure proved successful in preventing clotting
of the membrane for up to seven hours in unheparinized animals.

Before the catheter was fitted into the mass spectrometer
connecting cannula, a stainless steel tube approximately 1" long
was inserted into the end so that the compression fittings did not
crush the Saran tubing. The next step was to outgas the catheter, by
letting the mass spectrometer pump on it for 4-8 hours until the
readings were stable for 30 minutes.

RESULTS AND DISCUSSION

In vitro procedure. Proper calibration of the mass spectrometer
is a critical step, since any error at this point will be carried
through the entire experiment. Calibration for the experiments
described below was performed under the exact same conditions as the
experiments themselves; that is, the same temperature, pressure, and
catheter. Also, a calibration check was performed at the end of each
experiment so that machine drift could be determined. In all cases,
drift never exceeded 1 mm Hg. The mass spectrometer used in this
investigation was a Medspect II, Model 7402 (Chemetron Medical
Products, Baltimore, Maryland).

The mass spectrometer and catheter were subjected to four
different conditions; (a) dry gas at 25°C, (b) flowing water at 23°C,
(c) flowing water at 37°C and (d) flowing water at 40°C. This covered
the most probable range of in vitro and in vivo use of the mass
spectrometer. According to Brantigan et al. (11) there is no
appreciable difference between blood, saline, or water calibrations.
The tonometer apparatus for calibration of the catheters which was
supplied with the Medspect II, utilized flowing gas saturated water
in a tube approximately 1 cm. in diameter into which the catheter
was placed.

Flow in the tonometer was considered to be laminar and governed
by the Hagen-Poiseville relation; that is, the velocity profile was
parabolic in shape. To avoid depletion, in an area of low flow, i.e.
near the walls, the catheter had to be placed in and remain in the
center of the flow (tube). Signal deterioration was demonstrated by
when the tip of catheter moved closer to the wall where the velocity
of blood approaches zero.

After correction for water vapor, the mass spectrometer and
catheter of choice were calibrated using a gas which had weighed
concentrations of oxygen, carbon dioxide, and nitrogen; guaranteed
accurate to .01% (Ohio Medical Gases, Cleveland, Ohio). Calibration
to the gases was done at values of 50 to 55 mmHg for both O_2 and CO_2
depending on the temperature and barometric pressure. Following
calibration, the mass spectrometer was subjected to the other four
gases in a random order. Then, the entire procedure, beginning with
calibration, was repeated at least four times for each catheter at
each of the four conditions of temperature. Figs. 2 through 5
depict graphically the quantitative response of the steel and the
Saran catheters. Statistical analysis was carried out using a
multivariate analysis of variance. There was no significant
difference between the catheters and the analyzed tank concentrations
for both O_2 and CO_2, (P<.05). The only significant interaction
effect, which was not directly attributable to temperature changes
and, therefore, vapor changes, was that between the different oxygen
concentrations and the steel and Saran catheters. The difference

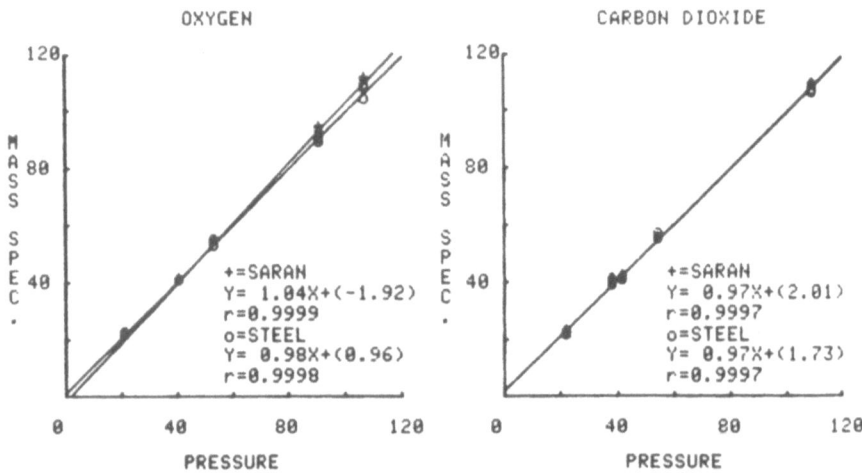

FIG. 2. Quantitative response of the Saran and steel
        catheters in dry gas at 25°C.

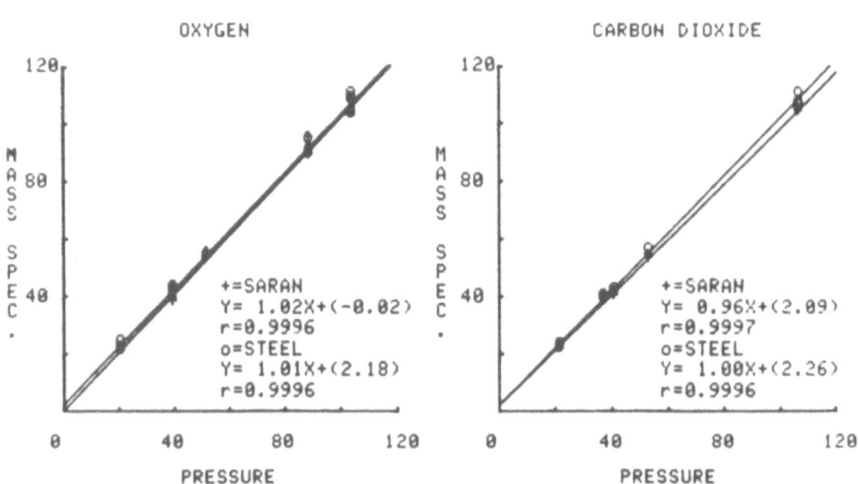

FIG. 3. Quantitative response of the Saran and steel
        catheters in flowing water at 23°C.

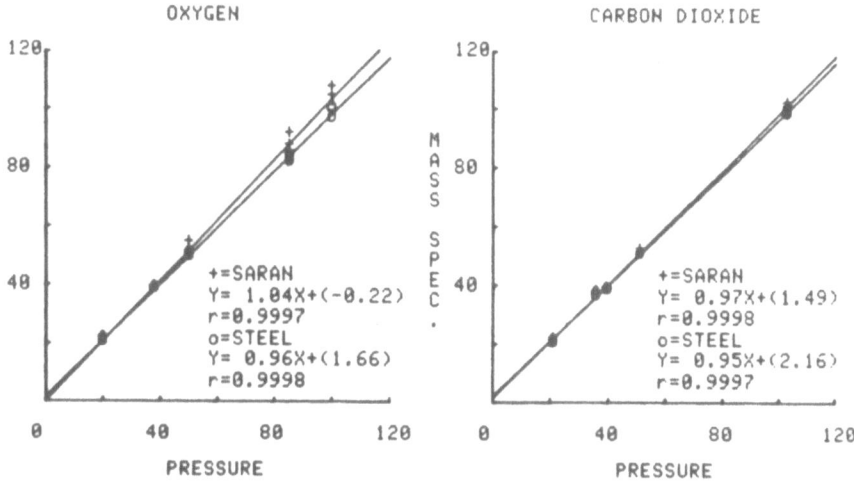

FIG. 4. Quantitative response of the Saran and steel
catheters in flowing water at 37°C.

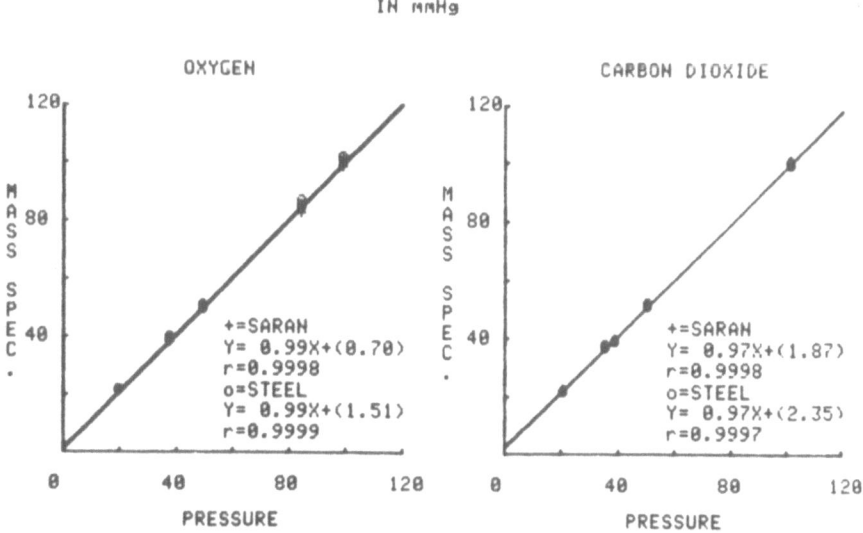

FIG. 5. Quantitative response of the Saran and steel
catheters in flowing water at 40°C.

between the two catheters could be divided into two groups, one consisting of oxygen below 40 mmHg and the other consiting of oxygen
above 89 mmHg. In the lower group, the difference between the Saran
steel catheter values for oxygen was approximately -1 mmHg; for the
second group, the difference was approximately 1.6 mmHg.

An important characteristic of the steel and the Saran catheters
is their lag and response times. Lag time is usually defined as that
length of time between the start of a step change in partial pressure
and the time the mass spectrometer first "sees" that change. The
response time is defined as that length of time necessary for the
signal to rise or decay to 63% of its final value. Table 5 summarizes
the response time, t_1 and the lag time, t_2 for the Saran and steel
catheters.

TABLE 5
The empirical response time date

	lag t_2 time (sec)	Response time t_1			Total time $t_2 + t_1$		
Catheter length		CO_2	N_2	O_2	CO_2	N_2	O_2
18" Chemetron steel	11.75	30.4	31.0	31.6	42.5	42.75	43.35
18" Saran	6.1	31.5	33.5	34.2	37.6	39.6	40.3
36" Saran	5.9	50.5	51.2	52.4	56.4	57.1	58.3
48" Saran	8.75	57.6	59.7	60.4	66.35	68.5	69.15

The transient responses are reproduced in Figs. 6 through 9.
These were obtained by subjecting the catheters of various lengths
to step changes in gas concentrations. These changes were accomplished
by having the catheter in flowing dry gas, adjusting the gains of
the recorder to give relative concentrations equal to one, then
rapidly pulling the catheter out of the flowing gas into an
atmosphere of 100% Helium; the change takes less than a second to
implement. Therefore, decays are shown here rather than rises; the
arrows to the left of the origin indicate the point where the step
change was made; the origin indicates that point where the mass
spectrometer first began to react.

In Table 5, one can see that the 18" Saran catheter reacted
more rapidly, in terms of total time, than the 18" steel catheter.
This better conductance was due to the larger ID of the Saran. The
36" and 48" Saran catheters have total times of approximately 58
and 69 sec. These sizes are of suitable lengths for heart catheters.

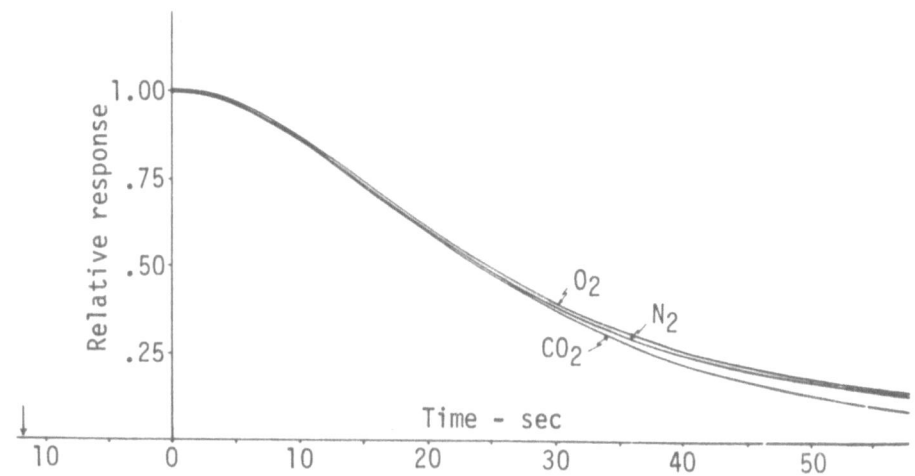

FIG. 6. Transient response of the 18" Chemetron steel
catheter.

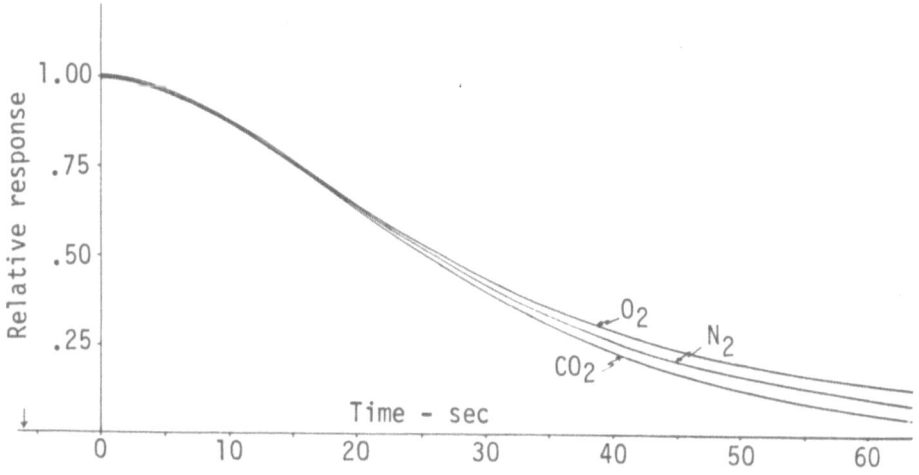

FIG. 7. Transient response of the 18" Saran catheter.

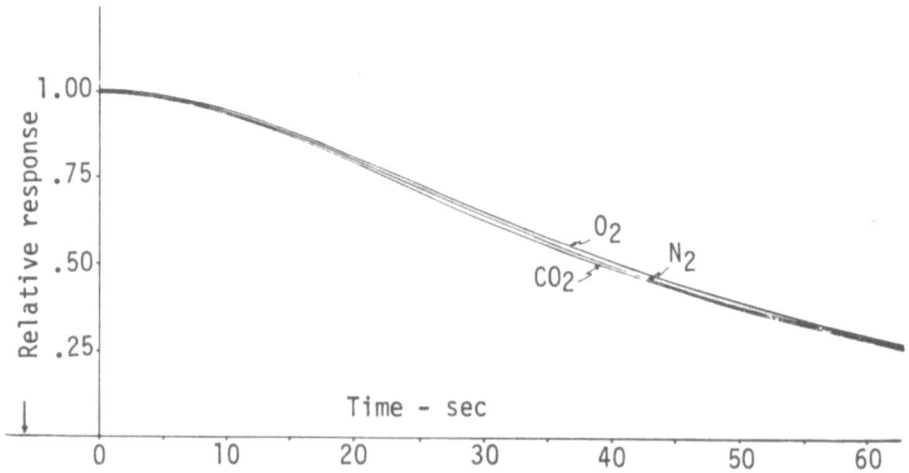

FIG. 8. Transient response of the 36" Saran catheter.

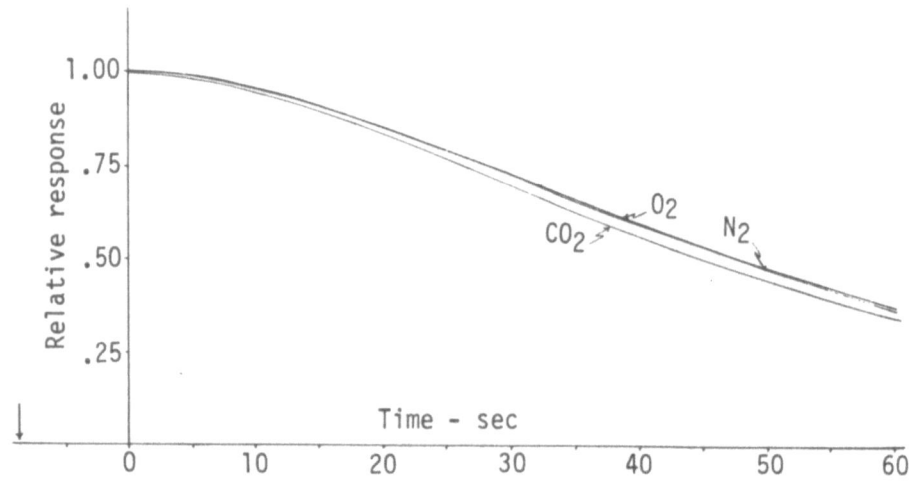

FIG. 9. Transient response of the 48" Saran catheter.

In an earlier section, the theoretical signal to background
ratios were presented. Due to unavoidable leaks in the system, the
experimental signal-background ratio was much smaller. Table 6 is
a summary of experimental signal-to-background test conducted
according to the procedure the Chemetron Corp. recommends for their
steel catheters.

TABLE 6
Empirical signal-to-background ratios

	18" Saran	18" SRI steel	Acceptable Range (20)
Signal/machine background	43.38	44.81	30.40+
Signal/total background	36.85		

According to the standard calibration procedure one should insert a
catheter into one inlet and block the other, allow the machine to
equilibrate, then sample a known gas from the catheter inlet (air
+ CO_2), increase the emission until the O_2 channel reads 600, set
the second (blocked) inlet to the same emission, and then switch
inlets. Background due to the machine is read off of inlet 2 and
the signal off of inlet 1. The ratio of the two should fall between
30 and 40+.

The signal to total background was obtained in basically the
same manner. In this case, instead of comparing the signal to a
blocked inlet, a connecting cannula and a Saran catheter sealed at
the end (no membrane) was inserted into the second inlet. Therefore,
a comparison of the total background was obtained. Table 6
demonstrates that the Saran catheter still fell in the acceptable
region.

On the basis of the above results, it was concluded that: (a)
the mass spectrometer is an accurate means to measure partial
pressures, and (b) that a flexible, Saran/Silastic catheter is
comparable to a rigid steel/Silastic catheter in terms of signal to
background ratios, response and lag time, variability and linearity.

In vivo procedure. The Saran/Silastic catheters were inserted
into peripheral arteries of twenty-five dogs via a 14 gauge Medicut
(TM). The catheter was threaded up the artery until it had entered
the aorta. With the gain and emission adjustments on the mass
spectrometer set to normal values, the position of the catheter tip

was optimized through the following procedure. The catheter was advanced a few centimeters at a time until the mass spectrometer reading decreased in value. The catheter was pulled back to the position which registered the highest reading. This simple technique minimized the possibility that the catheter tip was in an area of low flow which would result in unreliable, inaccurate data.

After the optimizing procedure, a sample was drawn from the site of placement and analysis of oxygen and carbon dioxide performed on an Instrumentation Laboratories Model 213 Blood gas analyzer (Instrumentation Laboratories, Lexington, Mass.). The gains on the mass spectrometer were adjusted to obtain agreement between the two instruments. This calibration procedure was repeated at least three times to reduce the possibility of a mis-calibration.

The animals were then placed on breathing gases containing various partial pressures of oxygen, carbon dioxide and nitrogen. After an equilibration period of thirty minutes, samples were drawn in triplicate and compared to the mass spectrometer readings. Figs. 10 and 11 graphically depict the results.

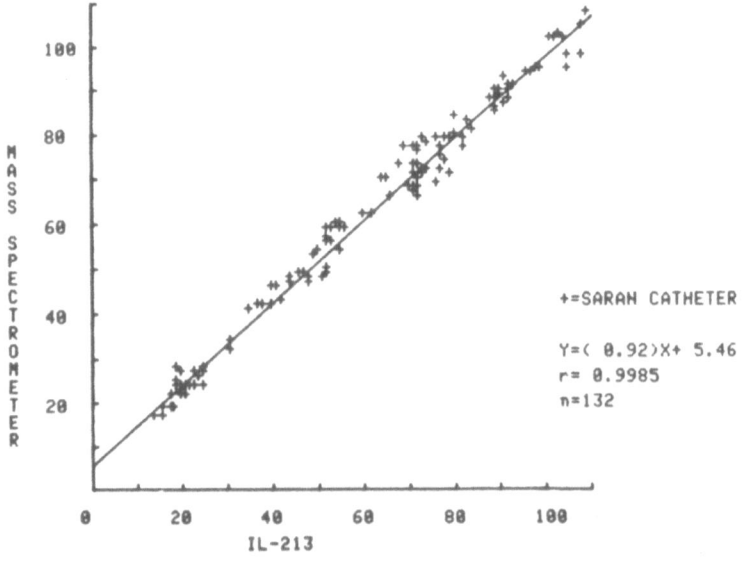

FIG. 10. In vivo partial pressure of oxygen as measured with a Saran catheter.

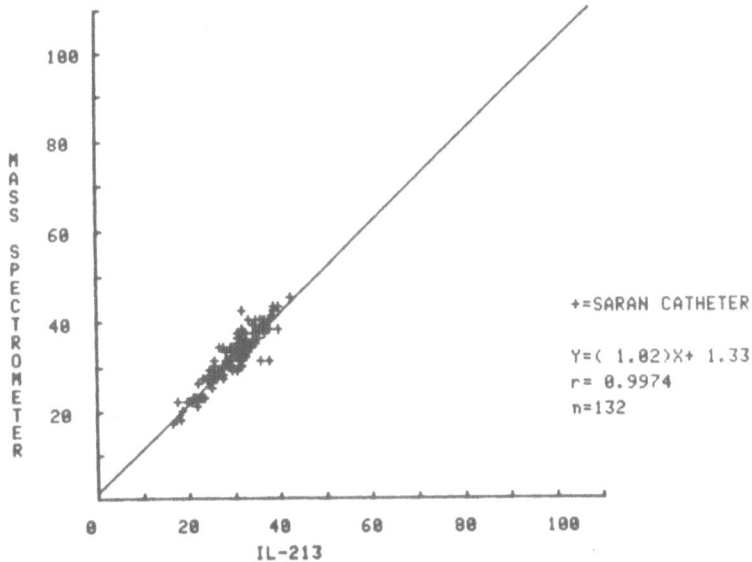

FIG. 11. In vivo partial pressure of carbon dioxide as
measured with a Saran catheter.

The intercept for the oxygen regression line is slightly above zero.
In view of the in vitro test results, the authors believe offset is
caused by one or both of two factors; the orientation of the catheter
in the flow and the comparison device itself, the blood gas analyzer.
The minor changes are not caused by the use of Saran as carrier
tubing. The Saran/Silastic catheter is an acceptable tool for
research use at this time.

CONCLUSIONS

Polyvinylidine chloride, Saran, tubing is an excellent gas
barrier and a suitable material for use in fabrication of a flexible
mass spectrometer catheter. The combination of Saran tubing and
Silastic as a membrane material results in a catheter that is
durable yet flexible, rapid responding and, in a flowing medium,
non-depleting. The Saran/Silastic catheter is quantitatively
comparable to the steel/Silastic catheter and does not present the
hazard of possible vessel damage. In vivo the catheter performed
to acceptable standards clinically and experimentally. The total
spectrum of interaction of Saran and oxygen in its environment is

an area yet to be explored. It is likely to be a minor problem
since the error induced by the interaction is of little significance.

REFERENCES

1) L.B. Strang, J. Appl. Physiol., 1961, 16, 562.
2) S. Woldring, G. Owens and D. Woolford, Science, 1966, 153, 885.
3) A. Wald, W.K. Hass and J. Ranschoff, JAAMI, 1971, 5, 325.
4) M. Roberts, E.T. Colton, G. Owens, D.D. Thomas and G.M. Watkins,
   Med. and Biol. Eng., 1975, 13, 535.
5) M.L. Dyken, Stroke, 1972, 3, 279.
6) S. Woldring, JAAMI, 1970, 4, no. 2.
7) A. Key, Med. and Biol. Eng., 1975, 13, 583.
8) A. Wald, W.K. Hass, F.P. Siew and D.H. Wood, Med. and Biol.
   Eng., 1970, 8, 111.
9) "Encyclopedia of Polymer Science and Technology: Plastics,
   Resins, Rubbers, Fibers", Vol. 13, pp. 560-576, pp. 630-290,
   Vol. 2, p. 324, Wiley, New York, 1965 and 1970.
10) G.A. Grode, S.J. Anderson and R.D. Falb, Trans. Am. Soc.
    Artificial Internal Organs, 1969, 15, 1.
11) J.W. Brantigan, V.L. Gott, M.L. Vestal, G.J. Fergusson and W.H.
    Johnston, J. Appl. Physiol., 1970, 28, 375.

Contributor Index